Optoelectronics: Principles and Practices

Optoelectronics: Principles and Practices

Editor: Zander Luther

MURPHY & MOORE
www.murphy-moorepublishing.com

Published by Murphy & Moore Publishing
1 Rockefeller Plaza,
New York City, NY 10020, USA
www.murphy-moorepublishing.com

Optoelectronics: Principles and Practices
Edited by Zander Luther

International Standard Book Number: 978-1-63987-414-9 (Hardback)

Cataloging-in-Publication Data

Optoelectronics : principles and practices / edited by Zander Luther.
p. cm.
Includes bibliographical references and index.
ISBN 978-1-63987-414-9
1. Optoelectronics. 2. Electronics. 3. Photonics. 4. Optoelectronic devices. 5. Electrooptical devices.
6. Electron optics. 7. Electrooptics. I. Luther, Zander.
TA1750 .O683 2022
621.381 045--dc23

Contents

Preface

I am honored to present to you this unique book which encompasses the most up-to-date data in the field. I was extremely pleased to get this opportunity of editing the work of experts from across the globe. I have also written papers in this field and researched the various aspects revolving around the progress of the discipline. I have tried to unify my knowledge along with that of stalwarts from every corner of the world, to produce a text which not only benefits the readers but also facilitates the growth of the field.

Optoelectronics, a subfield of photonics, is the study and application of electronic equipment and systems that produce, detect and control light. In addition to visible light, it also includes the study and use of gamma rays, X-rays, ultraviolet rays and infrared rays. Optoelectronic devices can be optical-to-electrical or electrical-to-optical transducers, or devices that use such instruments in their working. Optoelectronics is centered on the quantum mechanical effects of light on electronic matter and involves the usage of concepts such as the photoelectric effect, photovoltaic effect, photoconductivity, stimulated emission, radiative recombination, etc. Optocoupler and optical fiber communications are some of the important applications of optoelectronics. This book traces the progress of this field and highlights some of its key concepts and applications. It is compiled in such a manner, that it will provide in-depth knowledge about the principles and practices of optoelectronics. Through this book, we attempt to further enlighten the readers about the new concepts in this field.

Finally, I would like to thank all the contributing authors for their valuable time and contributions. This book would not have been possible without their efforts. I would also like to thank my friends and family for their constant support.

Editor

1

A New FPN Cancellation Circuit for Time-Domain CMOS Image Sensors

Fernando de S. Campos, José Alfredo C. Ulson,
José Eduardo C. Castanho,
Bruno Albuquerque de Castro and Rudolf Riehl

Abstract

A fixed-pattern noise correction technique for time-domain CMOS imagers with high dynamic range is presented in this chapter. Analytical derivations are presented showing how the circuit variations affect the time measured. The error in the time measured can be reduced by using lower reference voltages achieving values smaller than 4%. The fixed-pattern noise correction technique proposed is based on a new readout method for time-domain imagers employing two reference voltages for the discharge time measurement. This new technique is non-sensitive to circuit parameter variations that contribute to fixed-pattern noise such as hold voltages of transistors. A simple electronic circuit is proposed to implement the technique. Circuit and simulation results are presented to demonstrate the feasibility of the proposed technique.

Keywords: CMOS image sensors, Active Pixel Sensor, fixed-pattern noise, photo-diode

1. Introduction

CMOS image sensors have been an important source of research and industrial development in the area of image sensors due to the present high level of integration and low power consumption compared to charge-coupled devices (CCDs). The technology advances of CMOS image sensor results in image sensors with high dynamic range (>=100dB). Higher dynamic range become other important advantage of CMOS image sensors over CCDs. However, most of CMOS imager architecture with high dynamic range neglect the fixed-pattern noise (FPN) and do not present any kind of cancelling circuit or impact on such architectures. Therefore, although conventional linear CMOS sensors incorporate FPN cancellation techniques as correlated-double sampling (CDS), in general, it is still missing in the literature of FPN cancellation techniques for high dynamic range CMOS imagers.

Several techniques were proposed to achieve high dynamic range using CMOS image sensors [1-6]. In [2], by integrating a comparator and a dynamic flip-flop in each pixel, a multi-sampling technique was proposed. In [3], a self-reset technique using one comparator per pixel was described. In general, these techniques are effective as they achieved a dynamic range >100dB, but, none of them show the impact of FPN or present FPN cancellation techniques.

Recently, high dynamic range CMOS imagers using analog-to-digital converters have been proposed and they can incorporate CDS techniques for FPN cancellation although they do not present any FPN impact analysis [7-12]. However, these approaches achieve dynamic ranges of only 70–80dB, which is too low for some applications that typically demand dynamic range higher than 100dB.

In this chapter we present a FPN impact analysis and a technique of FPN cancellation for time-domain CMOS imagers with high dynamic range.

2. Principle of operation of time-domain CMOS imagers

Figure 1a shows the typical time-domain CMOS pixel. It comprises a photodiode, a reset transistor and a comparator. The pixel operation has two stages. The first stage is the reset period. During the reset period the reset transistor operating as switch is closed and the photodiode voltage (V_{pd}) is charged to an initial voltage V_{reset} close to V_{dd}. At the beginning of the second stage, the integration period, the reset transistor is turned off and the photodiode at high impedance starts to discharge. The discharge speed during discharge is according to the photocurrent intensity (see Figure 1b). The discharge time is smaller for higher photocurrent values.

The time-domain CMOS sensor operation consists of measuring the discharge time of the photodiode during the integration period by comparison of photodiode voltage with a reference voltage (see Figure 1b). This comparison is made by using a comparator as seen in Figure 1a. The discharge time is defined as the instant in which the photodiode voltage drops below (ou to) the reference voltage and the output comparator goes high. According to [2], the discharge time as function of the photocurrent for a reference voltage V_{ref} is given by:

$$t_d = \frac{C_{pd}}{\left(I_{dark} + I_{ph}\right)}\left(V_{reset} - V_{ref}\right) \tag{1}$$

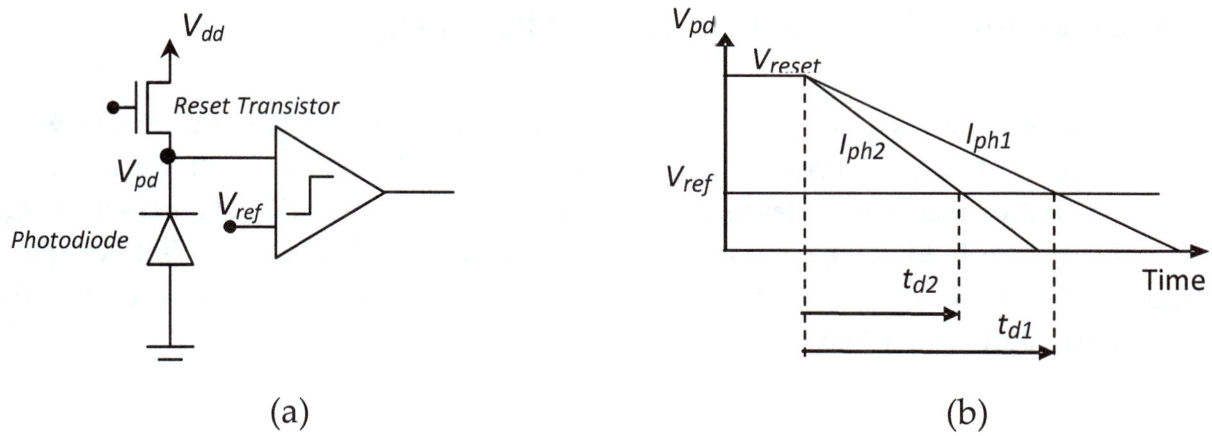

Figure 1. (a) Time-domain CMOS pixel; (b) main waveforms.

where C_{pd} is the photodiode's capacitance, I_{dark} is the dark current, I_{ph} is the photocurrent, V_{reset} is the initial photodiode's voltage and V_{ref} is the reference voltage.

Figure 2 shows the discharge time for different values of reference voltage assuming C_{pd}=30fF and I_{dark}=100fA. As one can see, the discharge time is smaller for higher reference voltages. For photocurrent smaller than 100fA, the curve saturates due to the presence of the dark current.

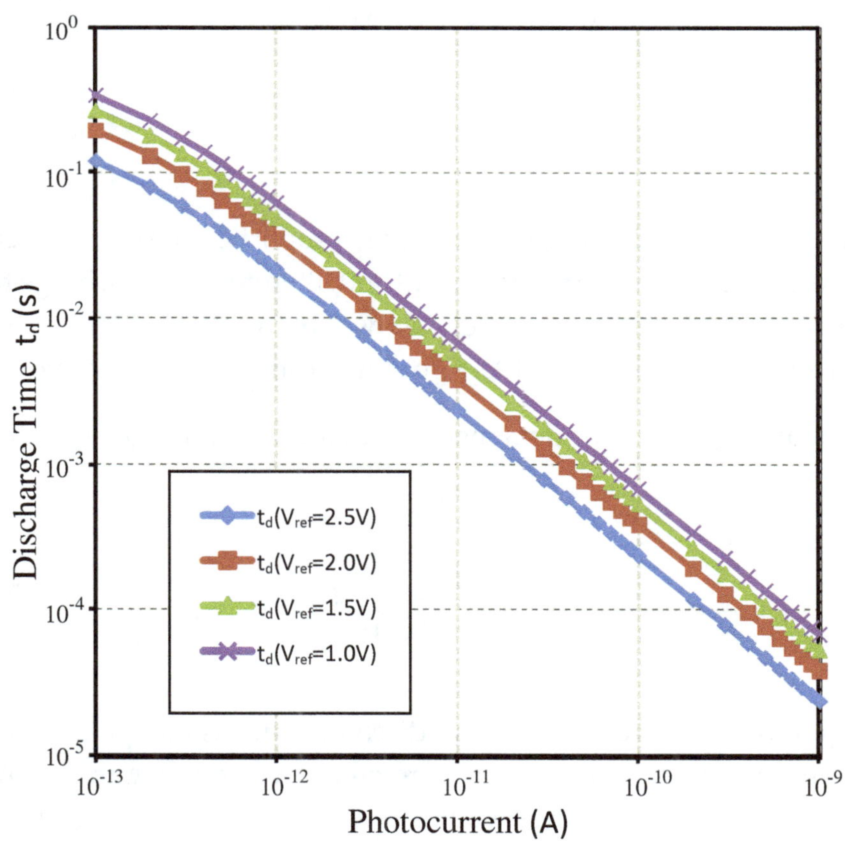

Figure 2. Discharge time of a time-domain CMOS sensor at different values of reference values.

3. Fixed-pattern noise in time-domain CMOS imagers

The fixed-pattern noise is defined as the variation of pixel's photoresponse in a matrix when a uniform light intensity focuses the matrix. Time-domain CMOS imagers have two main sources of variation that contributes to fixed pattern noise; (1) the variation of the reset voltage (ΔV_{reset}) due to threshold voltage variation of the reset transistor and (2) the offset voltage of voltage comparator (ΔV_{offset}). The threshold voltage variation causes a variation on the discharge time that can be written as

$$t_d + \Delta t_d = \frac{C_{pd}}{(I_{dark} + I_{ph})}\left(V_{reset} \pm \Delta V_{th} - V_{ref} \pm \Delta V_{offset}\right) \tag{2}$$

where Δt_d is the error reflected in the discharge time measured, ΔV_{th} is the variation introduced by the threshold voltage variation of reset transistor and ΔV_{offset} is the voltage variation introduced by the offset voltage of the voltage comparator. It is possible to demonstrate from the previous equation that the error reflected in the discharge time (Δt_d) can be written as

$$\Delta t_d = \frac{C_{pd}}{\left(I_{dark} + I_{ph}\right)}\left(\Delta V\right) \tag{3}$$

where, $\Delta V = \Delta V_{th} + \Delta V_{offset}$.

According equation (3), the discharge time error is proportional to voltage error introduced (ΔV), and inversely proportional to photocurrent and. therefore the discharge time error varies according light intensities changes. Figure 3 shows the discharge time error as function of photocurrent assuming C_{pd}=30pF, I_{dark}=100fA for three different values of voltage errors.

Manipulating equation (3), it can be demonstrated that the percentage of relative error is given by

$$\frac{\Delta t_d}{t_d}\% = \frac{\Delta V}{\left(V_{reset} - V_{ref}\right)}100\% \tag{4}$$

According to equation (4) the relative error of discharge time is independent of light intensity (photocurrent) and inversely proportional to reference voltage. Figure 4 shows the relative discharge time error for three different values of voltage error. As one can see, the relative discharge time error is smaller for small reference voltage values. It is smaller than 4% at V_{ref}=0.5V.

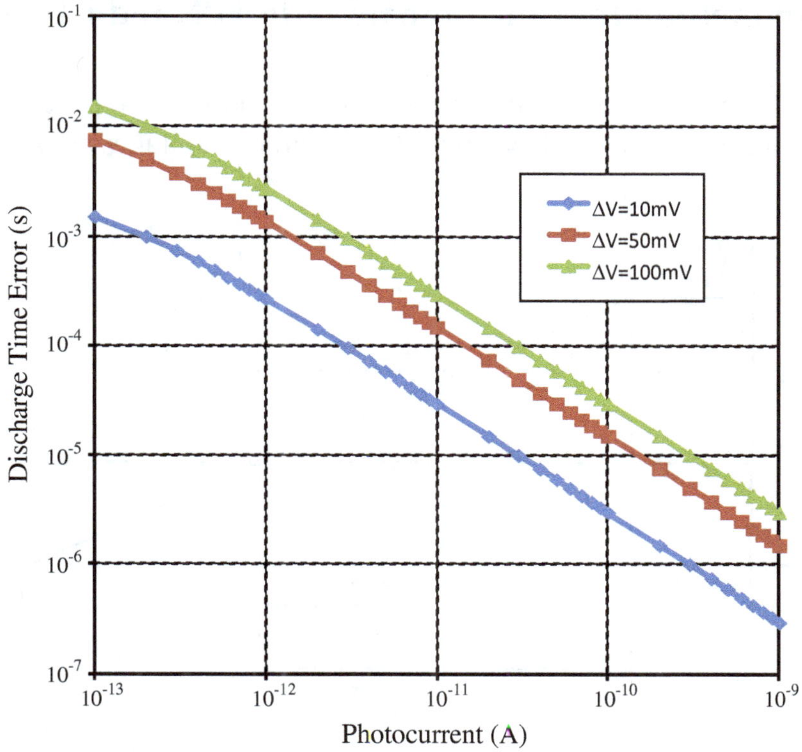

Figure 3. Discharge time error.

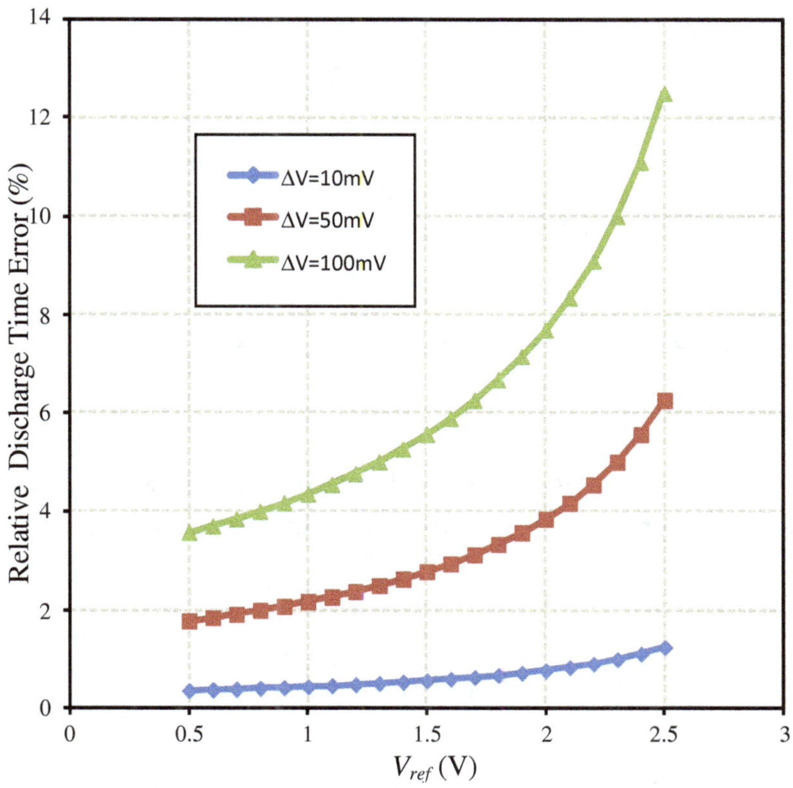

Figure 4. Relative discharge time error.

4. New readout method for time-domain with FPN reduction

In order to cancel the voltage error introduced (see equation 2), we propose to measure the discharge time between the comparisons with two reference voltages as shown in Figure 5.

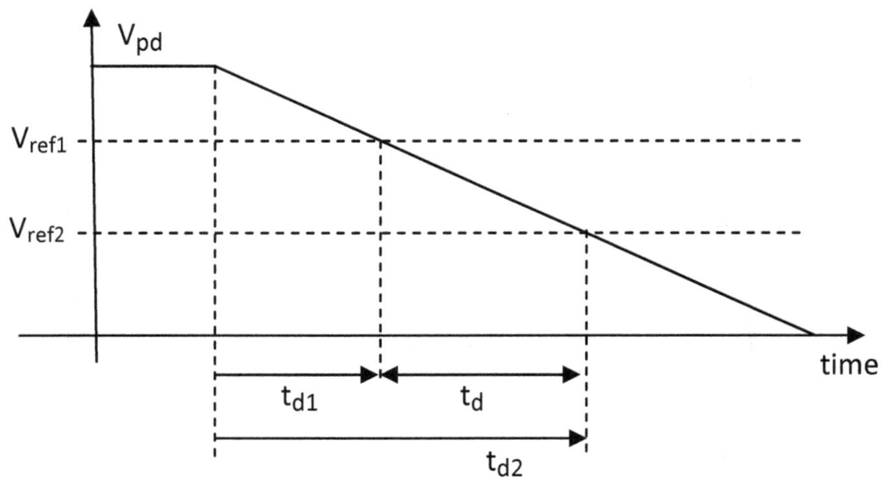

Figure 5. New readout method main waveforms.

The instant of comparison are two, t_{d1} and t_{d2}. For V_{ref1} the comparison time is given by

$$t_{d1} = \frac{C_{pd}}{\left(I_{dark} + I_{ph}\right)}\left(V_{reset} \pm \Delta V_{th} - V_{ref1} \pm \Delta V_{offset}\right) \tag{5}$$

For V_{ref2} the comparison time is given by

$$t_{d2} = \frac{C_{pd}}{\left(I_{dark} + I_{ph}\right)}\left(V_{reset} \pm \Delta V_{th} - V_{ref2} \pm \Delta V_{offset}\right) \tag{6}$$

The discharge time is now given as the interval time between t_{d1} and t_{d2}. Subtrating equations (5) of (6), it is possible to demonstrate the discharge time given by

$$t_d = t_{d2} - t_{d1} = \frac{C_{pd}}{\left(I_{dark} + I_{ph}\right)}\left(V_{ref1} - V_{ref2}\right) \tag{7}$$

According to equation (7) the new discharge time is independent of the source of fixed-pattern noise because it is independent of reset voltage and the offset voltage of voltage comparator. However, the same comparator must be used in order to achieve the offset voltage cancelling.

Figure 6 shows the transfer function graph of t_d for $V_{ref}1$-V_{ref2}=2V, C_{pd}=30fF and I_{dark}=100fA. The characteristic of time versus photocurrent is the same of conventional time-domain CMOS imager as can be seen in Figure 3.

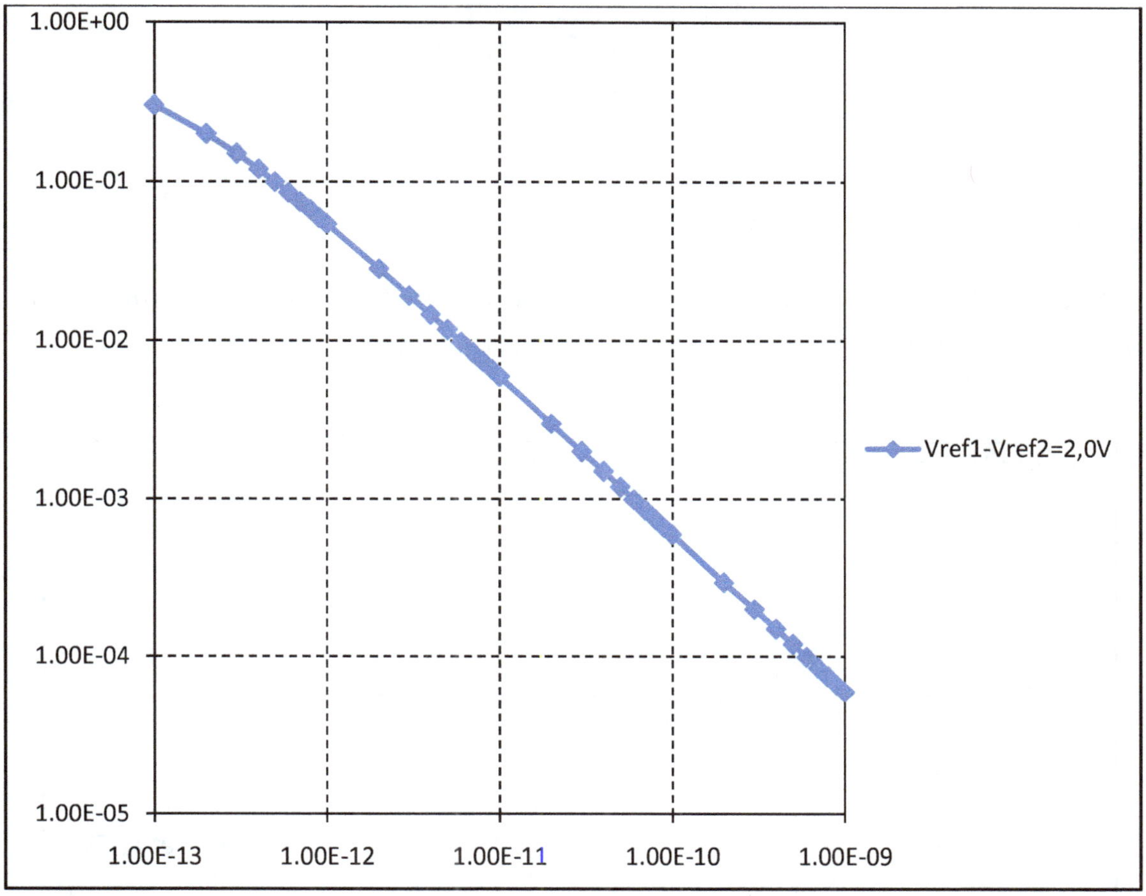

Figure 6. Discharge time (s) versus photocurrent (A).

5. Simulation results

5.1. FPN time-domain characteristic without FPN cancellation

Figure 7 shows the equivalent circuit of the CMOS sensor image used for simulations. The simulated pixel is comprised of a reset transistor and capacitance and a source current as equivalent circuit of photodiode. We assume that the photocurrent is constant and independent of voltage as usual in the literature. For simulations we used a design kit of a 0.35μm 3.3V CMOS technology.

In order to achieve the transfer characteristic time versus photocurrent simulated, we use a constant reference voltage of 1.5V, reset voltage of 3.3V and a typical capacitance of 30fF. By varying the photocurrent different comparison times for reference voltage of 1.5V were found

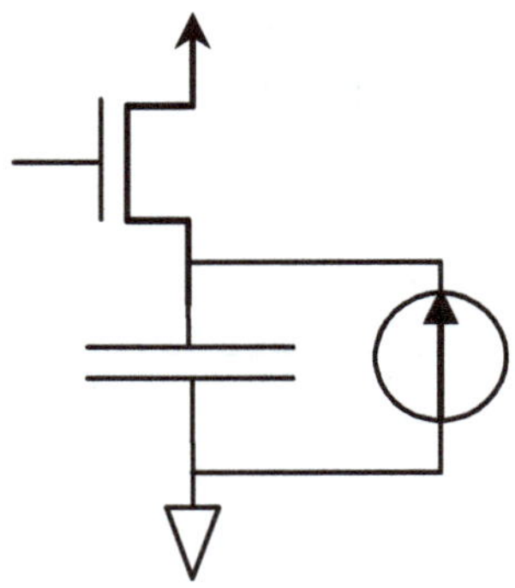

Figure 7. Discharge time (s) versus photocurrent (A).

and the obtained transfer characteristic is shown in Figure 8. Our analysis indicates that the simulation results are in agreement with equation (1).

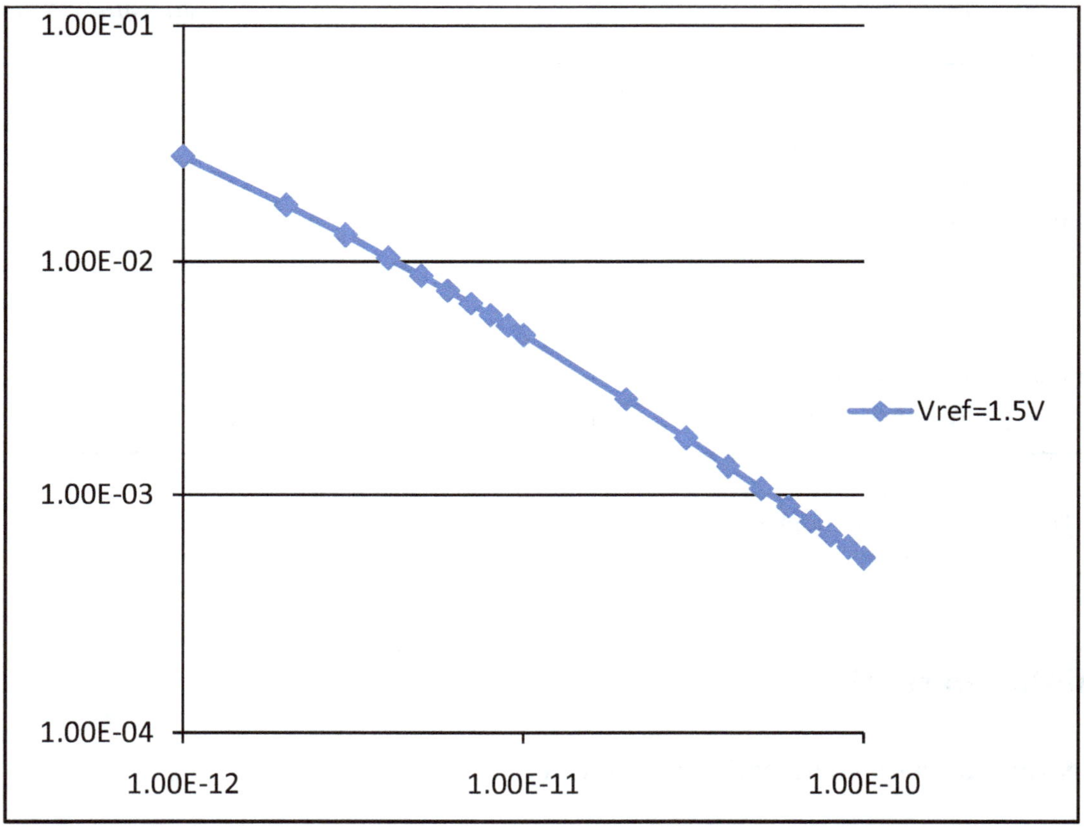

Figure 8. Discharge time (s) versus photocurrent (A).

For simulating the error voltage introduced by circuits, we choose to change the reset voltage (initial voltage) from 3.3V to 3.2V. Therefore, we simulate a error voltage of 100mV caused by the variation of the threshold voltage of the reset transistor and/or by the offset voltage of the comparator. Using a reference voltage of 1.5V and changing the initial voltage in the simulations we found a comparison time error from 6% to 10% as shown in Figure 9.

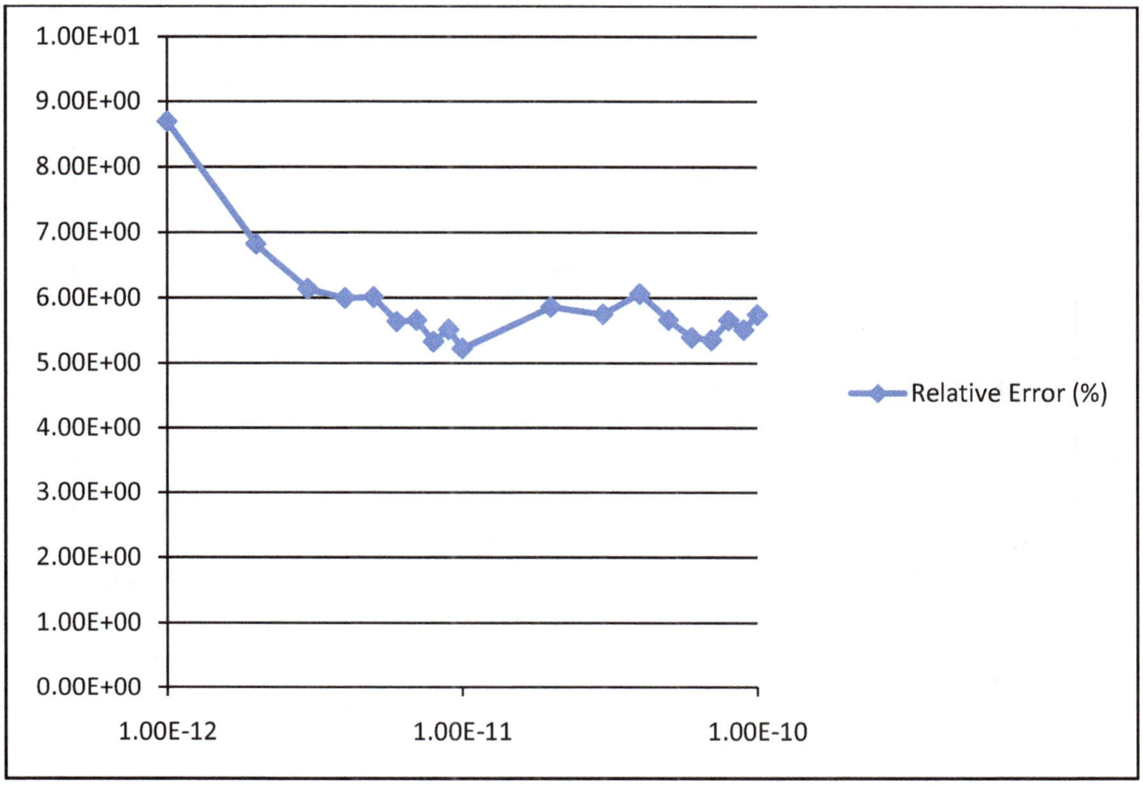

Figure 9. Relative error (%) versus photocurrent (a) for 100mV of variation in reset voltage.

This result indicates that the FPN in time domain for an entirely pixel matrix is less than 10%. The error is calculated as

$$error(\%) = 100 * \frac{t_d(3.3V) - t_d(3.2V)}{t_d(3.3V)} \tag{8}$$

5.2. FPN time-domain characteristic with FPN cancellation

Two references voltages are used for FPN cancellation. We use V_{ref1}=2.5V and V_{ref2}=2.0V as reference voltages. For this case, the transfer characteristic discharge time versus photocurrent is shown in Figure 10. The discharge time t_d was measured as the difference between the discharge time to V1 and the discharge time to V2 ($t_d = t_d(2.5V) - t_d(2.0V)$). Our analysis indicates that the transfer characteristic indicated in Figure 10 is in agreement with equation (7).

A simulation with 100mV of error introduced again was performed and the error was calculated as equation (8). For this case, the relative error found is shown in Figure 11. This simulation result indicates that the relative error is less than 2% percent. Therefore, the simulation result shows that this technique can reduce the FPN in time-domain imagers.

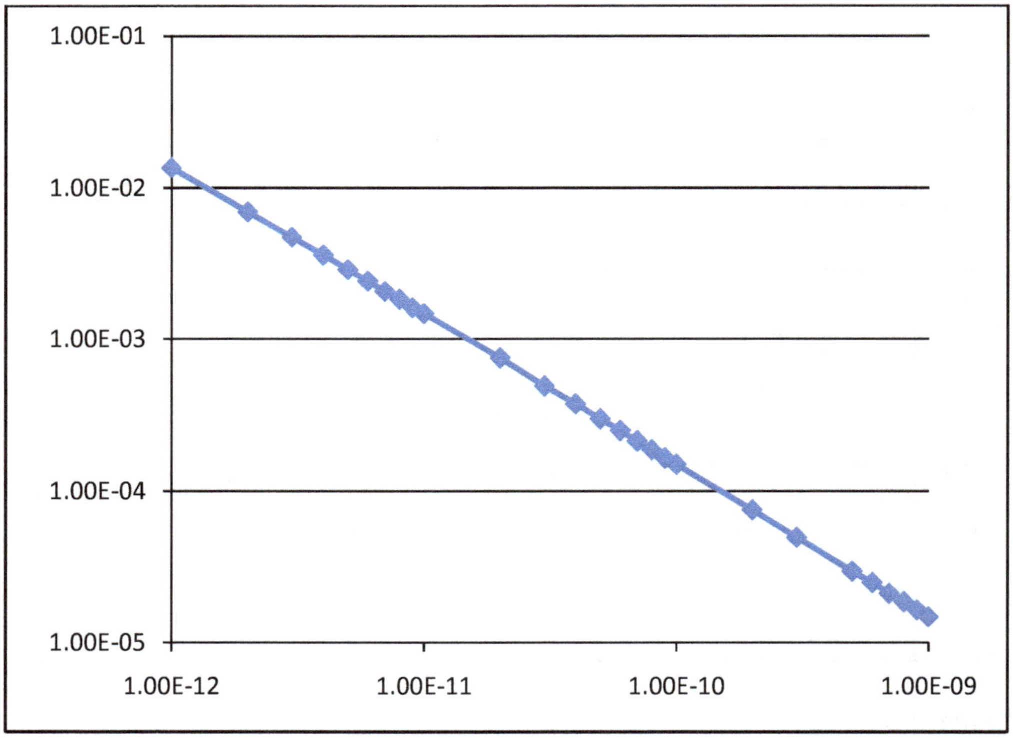

Figure 10. Discharge time (s) versus photocurrent (A) for Vref1=2.5V and Vref2=2.0V.

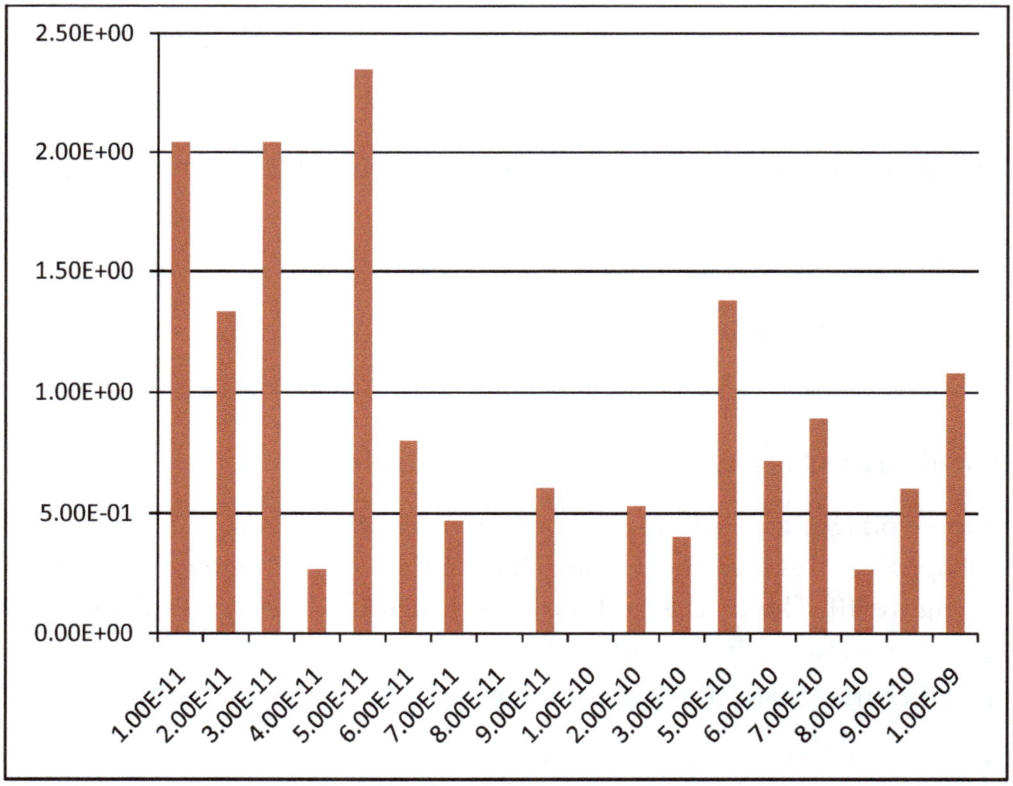

Figure 11. Relative error (%) versus photocurrent (A).

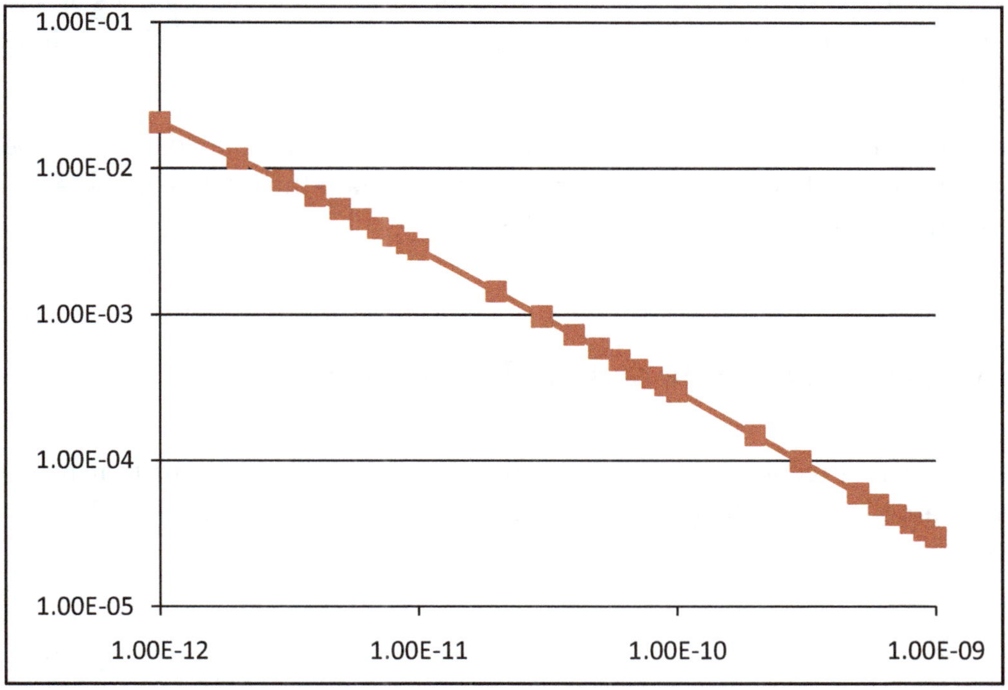

Figure 12. Discharge time (s) versus photocurrent (A) for Vref1=2.5V and Vref2=1.5V.

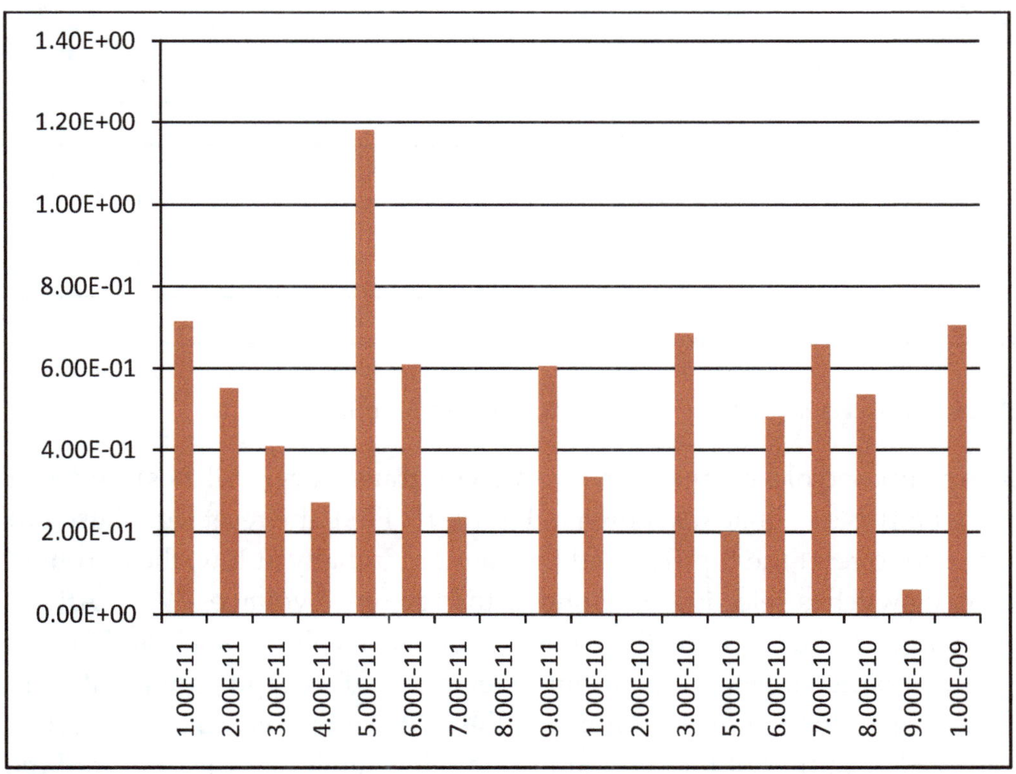

Figure 13. Relative error (%) versus photocurrent (A).

In order to confirm the results, we decide to simulate using reference voltages of V1=2.5V and V2=1.5V. As one can see in Figure 12, for this case the transfer characteristic is in agreement with equation (7) and the transfer characteristic in Figure 10.

For this case, an error of 100mV was introduced in the circuit simulated as stated before. The relative error found in this case is shown in Figure 13. Again the relative error is less than 2% showing that the technique can reduce the FPN.

6. Electronic circuit for time domain FPN cancellation

The challenge in this technique implementation is to develop a circuit that uses the same comparator (same offset voltage) to compare two reference voltages. For this purpose we propose the circuit shown in Figure 14.

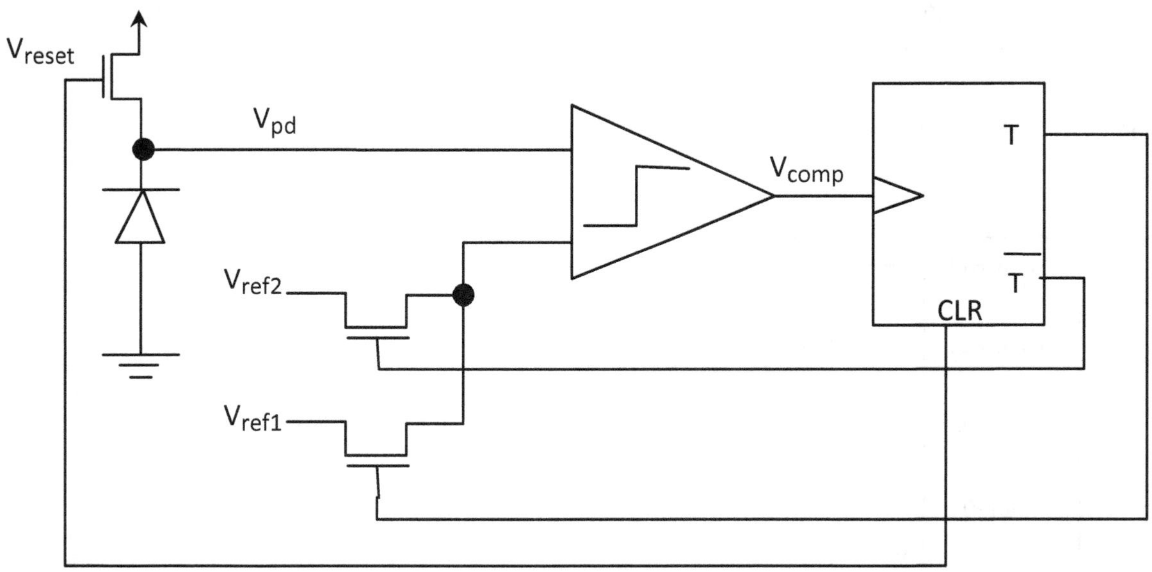

Figure 14. Electronic circuit of new FPN cancelling for time-domain imagers.

The circuit is composed of a comparator, two selector transistors (switches), and a T Flip-Flop. The main waveforms of signal are shown in Figure 15. First, the reset signal used to reset the pixel is also used to reset the flip-flop and the counter. Initially at low, the output of flip-flop control the two switches selecting only one of the reference voltages (V_{ref1}). After reset, and during the beginning of integration time, the V_{ref1} reference voltage is selected for comparison. At instant of comparison, when the output pixel is V_{ref1}, the output comparator triggers the flip-flop that changes to high. At this moment, the switches are biased disconnecting V_{ref1} and connecting V_{ref2}. The output of the comparator goes low again but the output of the flip-flop remains high. When the output pixel voltage reaches V_{ref2} the comparator output pulses again trigger the flip-flop output to low. At this moment the counter stops and the reference voltage V_{ref1} returns to be connected to the comparator input.

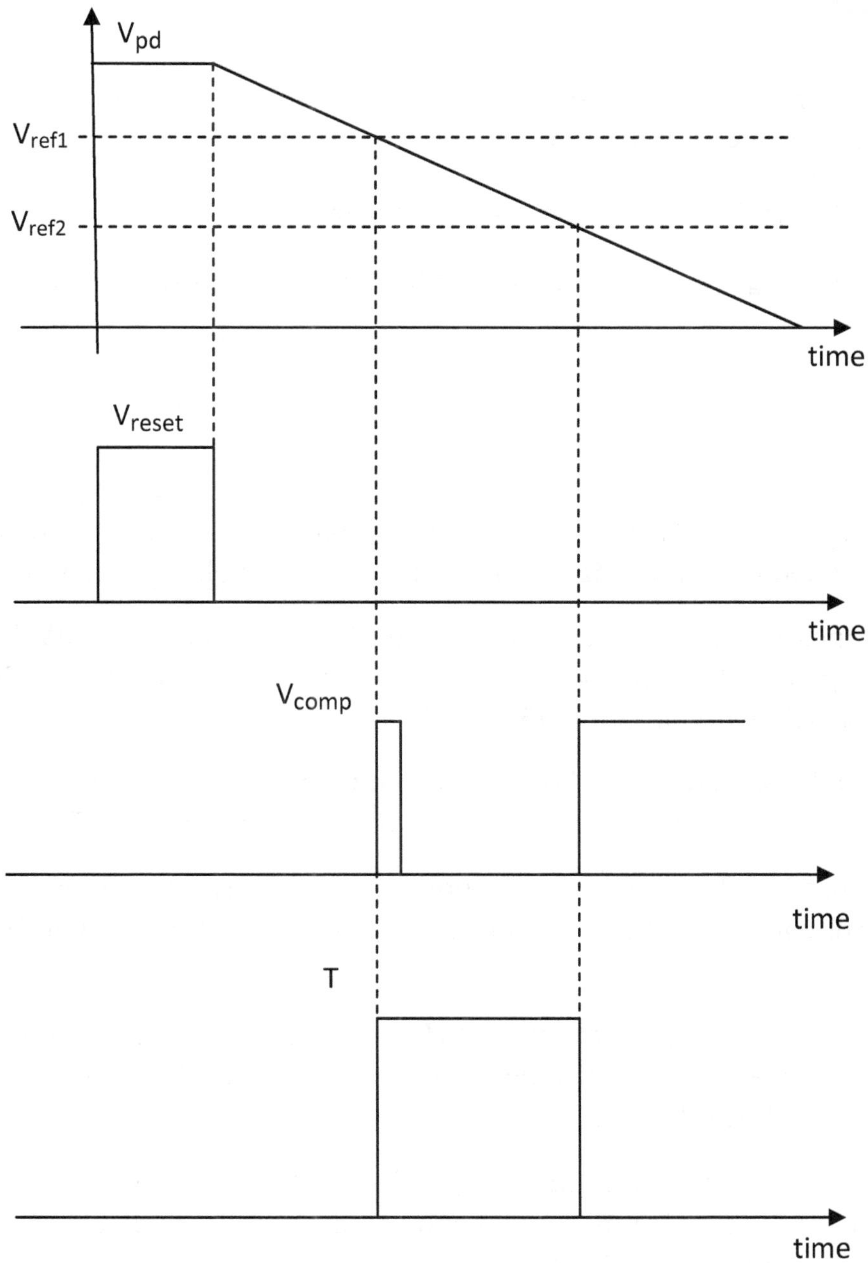

Figure 15. Electronic circuit main waveforms.

7. Conclusions

High dynamic range imager CMOS image sensors are one of the most important applications of CMOS image sensors. Most of the high dynamic range CMOS image sensors proposed in the literature neglect the fixed pattern noise, which is one of the major drawbacks of CMOS image sensors compared to CCDs. In this work, a fixed-patter noise reduction technique is presented for high dynamic range CMOS image sensors operating in time domain. According to simulation results, the new technique can reduce the fixed patter noise down to 2%.

Author details

Fernando de S. Campos*, José Alfredo C. Ulson, José Eduardo C. Castanho,
Bruno Albuquerque de Castro and Rudolf Riehl

*Address all correspondence to: fcampos@feb.unesp.br

Univ. Estadual Paulista "Júlio de Mesquita Filho" (UNESP) – Bauru campus, Brazil

References

[1] A. Belenky, A. Fish, A. Spivak, and O. Yadid-Pecht, "A snapshot CMOS image sensor with extended dynamic range," *IEEE Sensors Journal*, vol. 9, pp. 103-111, 2009.

[2] F.S. Campos, O. Marinov, N. Faramarzpour, F. Saffih, M.J. Deen, J.W. Swart, "A multisampling time-domain CMOS imager with synchronous readout circuit," *Analog Int. Cir Signal Proc*, vol. 57, pp. 151-159, 2008.

[3] D. Park, J. Rhee, and Y. Joo, "A wide dynamic-range CMOS image sensor using self-reset technique," *IEEE El Dev Lett*, vol. 28, no. 10, pp. 890-892, 2007.

[4] L.G. McIlrath, "A low-power low-noise ultrawide-dynamic range CMOS imager with pixel-parallel A/D conversion," *IEEE J. Solid-State Circuits*, vol. 36, pp. 846-853, 2001.

[5] D. Yang, A. El Gamal, B. Fowler, and H. Tian, "A 640x512 CMOS image sensor with ultra wide dynamic range floating point pixel level ADC," *IEEE J. Solid-State Circuits*, vol. 34, pp. 1821-1834, 1999.

[6] S. Decker, R.D. McGrath, K. Brehmer, and G. Sodini, "A 256x256 CMOS imaging array with wide dynamic range pixels and column parallel digital output," *IEEE J. Sol-St Cir*, vol. 33, pp. 2081-2091, 1998.

[7] M.-W. Seo, T. Sawamoto, T. Akahori, Z. Liu, T. Iida, T. Takasawa, T. Kosugi, T. Watanabe, K. Isobe, and S. Kawahito, "A low-noise high-dynamic-range 17b 1.3-Megapixel 30-fps CMOS image sensor with column-parallel two-stage folding-integration/cyclic ADC," *IEEE Trans. Electron Devices*, vol. 59, no. 12, Dec. 2012.

[8] Z. Ignjatovic, D. Maricic, M.F. Bocko, "Low power, high dynamic range CMOS image sensor employing pixel-level oversampling ΣΔ analog-to-digital conversion, " *IEEE Sensors Journal*, vol. 12, no. 4, pp. 737-746, April 2012.

[9] M.W. Seo, et al., "A low-noise high intrascene dynamic range CMOS image sensor with a 13 to 19b variable-resolution column-parallel folding-integration/cyclic ADC", *IEEE J Sol-St Cir*, vol. 47, no. 1, pp. 272-283, Jan 2012.

[10] Y.C. Chae, J.M. Cheon, S.H. Lim, M.H. Kwon, K.S. Yoo, W.K. Jung, D.H. Lee, S.H. Ham, and G.H. Han, "A 2.1 M pixels, 120frames/s CMOS image sensor with column-parallel ΔΣ ADC architecture," *IEEE J. Solid-State Circuits*, vol. 46, no.1, pp. 236-247, Jan. 2011.

[11] J. H. Park, S. Ayoama, T. Watanabe, K. Isobe, and S. Kawahito, "A high-speed low noise CMOS image sensor with 13-b column-parallel single-ended cyclic ADCs," *IEEE Trans. Electron Devices*, vol. 56, no. 11, pp. 2414-2422, Nov. 2009.

[12] M. Furuta, Y. Nishikawa, and S. Kawahito, "A high-speed, high-sensitivity digital CMOS image sensor with a global shutter and 12-bit column-parallel cyclic A/D converter," *IEEE J. Solid-State Circuits*, vol. 42, no. 4, pp. 766-774, Apr. 2007.

2

Excitonic Crystal and Perfect Semiconductors for Optoelectronics

Sergei L. Pyshkin

Abstract

This chapter demonstrates the growth of perfect and contamination-free gallium phosphide (GaP) crystals and discusses the influence of crystallization conditions on their quality and properties. The long-term ordered and therefore close to ideal crystals replicates the behavior of the best nanoparticles exhibiting pronounced quantum confinement effect. These perfect crystals are useful for application in top-quality optoelectronic devices as well as they are a new object for the development of fundamentals of solid state physics.

Since samples of gallium phosphide doped by nitrogen (GaP:N) were originally prepared by the author in the 1960s, followed by the introduction of the excitonic crystal concept in the 1970s, the best methods of bulk, film and nanoparticle crystal growth have been elaborated. The results of semi centennial evolution of GaP:N properties are compiled here. Novel and useful properties of perfect GaP including its stimulated emission, very bright and broadband luminescence at room temperature were observed. These results provide a new approach to selection and preparation of perfect materials for optoelectronics and a unique opportunity to realize a new form of solid-state host — the excitonic crystal as high intensity light source with expected low threshold for the generation of non-linear optical phenomena.

Using the example of GaP here is proposed as a cheap, resource-saving and impactful way to develop optoelectronics through a special transformation of an ordinary semiconductor into the base material for various device structures.

Keywords: GaP, long-term ordering, excitonic crystal, perfect semiconductors for optoelectronics

1. Introduction

Single crystals of semiconductors grown under laboratory conditions naturally contain a varied assortment of defects such as displaced host and impurity atoms, vacancies, dislocations, and impurity clusters. These defects result from the relatively rapid growth conditions and inevitably lead to the deterioration of mechanical, electric, and optical properties of the material, and therefore to degradation in the performance of the associated devices. Note, the deterioration of optical properties of any luminescent material for application in optoelectronics may appear in the complete absence of light emission in the spectral region, where the perfect material gives an excellent luminescence, in too narrow emissive band, in a very weak light emission, in impossibility to control its shape and brightness, as well as in fast degradation of a device prepared on the base of this material. In order to partly overcome the noted preceding difficulties, industry uses expensive but only palliative decisions such as limitless extension of the list of materials for the device making or small improvement of technologies for growth and preparation of electronic materials. Huge material, time and mental resources already have spent and will be spent further in our efforts to support or improve achieved parameters and reliability of electronic devices. Therefore, finding of alternative drastic methods for device making is one of the main priorities of electronic industry development. This paper describes the experience of the author in this field.

The pure and doped GaP crystals discussed herein were prepared about 50 years ago [1]. Throughout the intervening decades they have been periodically re-evaluated in order to investigate the marked changes over time in their electro- and photoluminescence, photoconductivity, behavior of bound excitons characteristic for doped GaP, nonlinear optics, and other phenomena. Accordingly, it was of interest also to monitor the change in crystal quality over the course of several decades while the investigated crystals are held under ambient conditions.

Over time, as it is confirmed by the author during 50 years of the relevant experiments, that driving forces such as diffusion along concentration gradients, strain relaxation associated with clustering, and minimization of the free energy associated with properly directed chemical bonds between host atoms result in an ordered redistribution of impurities and host atoms in a crystal. In the particular case of GaP and some other chemical compounds, having in their compositions highly volatile components, any attempt to accelerate these processes through annealing at increased temperatures cannot be successful because high-temperature processing results in thermal decomposition (in GaP — due to P desorption) instead of improved crystal quality. Therefore, successful thermal processing of these compounds can only take place at temperatures below the sublimation temperatures of their volatile constituents, requiring a longer annealing time. For instance, evaluated in the framework of the Ising model, the characteristic time of the substitution reaction during N diffusion along P sites in GaP:N crystals at room temperature constitutes 15–20 years [2]. Hence, the observations of luminescence and some other phenomena in the crystals made in the 1960s–1970s and in the 1980s–1990s were then compared with the results obtained in 2005–2014 under similar experimental conditions.

The long-term ordering of doped GaP and other semiconductors has been observed as an important accompanying process, which can only be studied using the same unique set of samples and the interest to observe them over decade time scales. More specifically, the optical and mechanical properties of single crystalline GaP, and some other semiconductors also grown in the 1960s, have been analyzed. Comparison of the properties of the same crystals has been performed in the 1960s, 1970s, 1980s, and 1990s [1, 3-17] along with those of newly made GaP nanocrystals [18-20] and freshly prepared bulk single crystals [21-24]. Jointly with the references [25-27], this review provides a generalization of the results on long-term observation of luminescence, absorption, Raman light scattering, and microhardness of the bulk single crystals in comparison with the same properties of the top quality GaP nanocrystals. It is shown that the combination of these characterization techniques elucidates the evolution of these crystals over the course of many years, the ordered state brought about by prolonged room-temperature thermal annealing, and the interesting optical properties that accompany such ordering. It is demonstrated that long-term natural stimuli that improve the perfection of crystals prevail over other processes and can lead to novel heterogeneous device systems and new semiconductor devices with high temporal stability.

Additionally, it is worth noting, that semiconductor nanoparticles for optoelectronic applications also were synthesized mainly to avoid limitations inherent to freshly grown bulk semiconductors with a wide range of different defects. For instance, different defects of high concentration in freshly prepared GaP single crystals completely suppress any luminescence at room temperature due to the negligible free path for non-equilibrium electron-hole pairs between the defects and their non-radiative recombination, while the quantum theory predicts their free movement in the field of an ideal crystal lattice. However, the long-term ordered and therefore close to ideal crystals even at 300K demonstrate bright luminescence and stimulated emission equivalent to the best nanoparticles. These perfect crystals, due to their unique mechanical and optical properties, are useful for application in high-quality optoelectronic devices as well as they are a new object for the development of fundamentals of solid state physics, nanotechnology, and crystal growth.

Also noted is the application of GaP/polymers nanocomposites in device structures for accumulation, conversion and transport of light energy that has only recently received attention while bulk and thin GaP films have been successfully commercialized for many years. Therefore, for completeness, during the recent years, since 2005, the author and colleagues continued their efforts on the preparation of GaP nanoparticles in order to improve their quality and to apply their composites with appropriate polymers for advanced light emissive structures [18-20, 28-32].

In preparing this review, the author did not intent to evaluate the works on GaP of the other authors, but their works are cited and used here when it is necessary for explanation and interpretation of new phenomena observed during long-term ordering of impurities and host atoms in the crystal lattice. Elaborating optimal methods of preparation of GaP bulk crystals, nanoparticles and their light emissive composites with compatible polymers, we use our own experience and literature data [33-42]. The main goal of these 50 year efforts and this review are the observation and description of very interesting results of the long-term evolution of GaP properties and the relevant idea to propose for many years ahead an alternative and resource-saving way for the development of electronics, as well as to propose and justify the excitonic crystal [26, 27] as a new optical media for the future optoelectronic devices used in optical data processing, storage, and transmission as well as for the generation of non-linear optical effects at rather modest thresholds for nonlinearities. Interesting and very useful for application results of long-term evolution of GaP properties as well as the unique collection of tested and stored for years pure and doped perfect GaP crystals are demonstrated and proposed to academic researchers, engineers, and managers of electronic industry for intensification of collaboration in patent activity, reorganization of the material and device making processes, reduction in price, improvement of parameters, and reliability of devices.

Perfect GaP single crystals, excitonic crystal on the base of the GaP crystal doped by nitrogen (GaP:N), as well as understanding of properties and available application of these new objects coming these days into science and industry are the result of intense many years work of headed by the author groups of top specialists on crystal growth, investigation of their properties, and application in optoelectronics in Russia, Moldova, the USA, and Italy. This activity was stimulated and followed by the natural processes and phenomena elapsed with time in the crystals.

According to the chosen plan of presentation, this review is divided into the next sections:

1. Introduction
2. Properties of GaP
3. Growth Technology for Perfect GaP Bulk and Nano-Crystals
4. Optical Properties of Perfect, Long-term Ordered GaP:N Crystals
5. Comparison of Properties of GaP Nanocrystals and GaP Perfect Bulk Single Crystals
6. Excitonic Crystal and Its Importance in Optoelectronics
7. Already Discovered and Possible Nonlinear Optical Phenomena in GaP
8. Conclusions
9. Acknowledgments
10. References

2. Properties of GaP

GaP crystallizes in zinc blende structure, where Ga and P atoms create two interpenetrating face-centered lattices spaced ¼ of the (111) cube diagonal apart.

Brillouin zone of GaP and other III-V compounds represent the truncated octahedron (Figure 1) having the next high symmetry points:

$\Gamma - \overline{K} = (000)$ — center of Brillouin zone

Δ — along the (100) axis inside the zone

X — (100) Brillouin zone edge

Σ — along the (110) axes inside the zone

K — (110) Brillouin zone edge

Λ — along the (111) axes inside the zone

L — (111) Brillouin zone edge

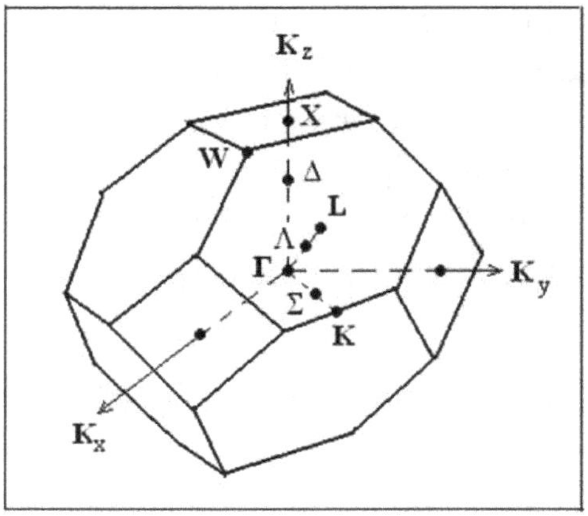

Figure 1. Brillouin zone of gallium phosphide.

A concrete band structure for each III-V representative in the limits of common for them Brillouin zone depends on the type of symmetry of the wave functions of valence electrons of the atoms, creating the compound. The most reliable data on GaP band structure were obtained from the experiments on light absorption and reflection as well as using the spectral distribution of photoconductivity in the region of intrinsic absorption.

According to W. Paul's empirical rule [36], the energy gaps equally depend on the pressure for the relevant electron states. Using this rule and experiments on dependence of electron transitions on pressure, the authors of Ref. [37] have proposed the band structure of GaP presented in Figure 2.

Experimental data confirm the details of the GaP band structure. So, the absolute minimum of the conductance band (the X-point) presented in Figure 1 lays at the edge of the Brillouin zone in the (100) direction, while the valence band maximum (the Γ-point) position is the center of the zone. The absolute minimum value of the forbidden gap (Figure 2) corresponds to the indirect optical transition $\Gamma_{15}{}^{v} \rightarrow X_1{}^{c}$; this value depends on the temperature changing between 2.354 eV at 4.2K (liquid helium) and 2.328 at 77K (liquid nitrogen) until 2.248 eV at 300K (room temperature) [37]. Minimum gap for direct optical transition $\Gamma_{15}{}^{v} \rightarrow \Gamma_1{}^{c}$ at 300K is equal to 2.78eV [37]. The valence band, taking into account its spin-orbit splitting, consists of two confluent bands and another one shifted downward. The other details of GaP band structure are shown in Figure 2.

According to the quantum selection rules for optical transitions, the lattice phonons do not participate in the direct transitions, while in the indirect transition lattice phonons participate, the type and energy of which are determined in Ref. [38] together with the low- and high-frequency dielectric constants, 10.182 and 8.457, respectively. The data on GaP phonon spectrum are widely used at interpretation of its light emissive and absorption spectra. In indirect optical transitions [38, 40] participate transversal and longitudinal acoustic and optic phonons with energies 12.8 (TA), 31.3 (LA), 46.5 (LO), and 50.0 meV (TO). Note, at low temperatures, when the thermal energy, kT, is less than the respective energies of the free and bound exciton creation (10 and 21 meV, respectively, for free and N bound excitons [4, 7]), the indirect optical transitions occur mainly through the excitonic states.

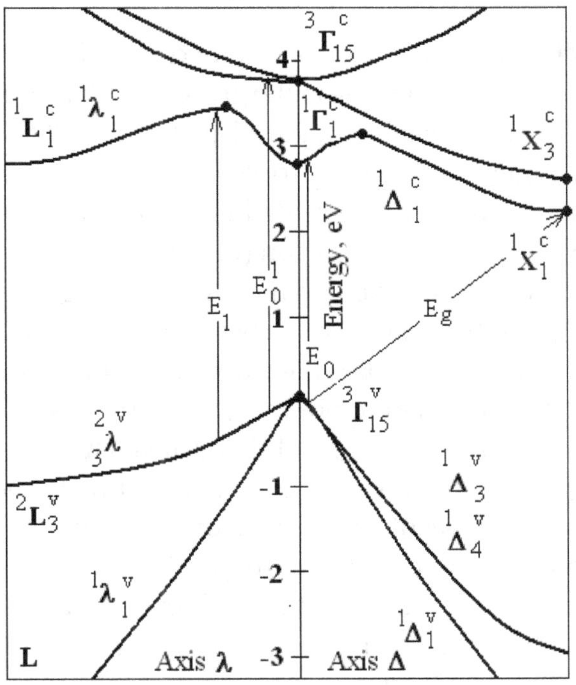

Figure 2. Band structure of gallium phosphide.

3. Growth Technology for Perfect GaP Bulk and Nano-Crystals

Single crystals of gallium phosphide, in principle, can be obtained in several ways [1, 3, 7]. The method for obtaining gallium phosphide from solution-melt, chosen by us, has several significant advantages:

1. A significant temperature reduction of the process and the presence of large amounts of solvent dramatically reduce crystal pollution by material of the container.
2. The light sources creating on their basis have high efficiency.
3. Due to specifics of the method, gallium phosphide at the appropriate level of the experiment can be obtained in the form of the relatively large lamellar crystals of a definite crystallographic orientation. Note, lamellar crystals are the most convenient and economical material in the manufacture of many semiconductor devices.

Consider the peculiarities of growth of lamellar crystals of gallium phosphide with the set-up properties, formed under uniform cooling of the P solution in Ga. The influence of the following factors has been investigated: 1) quantity and chemical nature of impurities; 2) geometric shape of the container and conditions of heat rejection; 3) accuracy of the temperature control; 4) cooling rate and the law of temperature change in time.

The aim of this work was to elucidate the mechanism of crystallization of lamellar GaP crystals and experimental confirmation stated earlier assumptions about the nature of this process.

I. The solution-melt method of GaP growth is described in detail in Refs.[1, 3, 7]. So we focus here only on the characteristics for this study associated of the crystal growth.

Synthesis, alloying, and crystallization were combined in a single cycle. Experiments on the establishment of the influence of impurities on the growth of GaP crystals was performed using Ga000 and phosphorus B5, industrially prepared in former USSR, the highest purity produced at that time (1960). When the growth process was thoroughly investigated, the growth of perfect crystals has been conducted in a sealed and drained simple container from optical quartz that was flushed with spectral pure argon (Figure 3a). Container 1 was introduced into the programmable furnace with rod 2 fixed on the vibrator to agitate the mixture of Ga, P, and any chosen dopant, to facilitate chemical reaction between components, and to avoid a possible explosion of the container in dangerous temperature points of the process of mass crystallization (Figure 3a). However, for investigation of the growth process some other containers were used. For instance, the container having a good thermal isolation of the walls and a Cu metal rod on its bottom was used to create the center of crystallization for needle-like crystals (Figure 3b). The container, comprising the set of necessary growth components 4, thermoisolating layer of Al_2O_3 6, Ga etalon 7, and thermocouples in the critical places 5, 8, has been used for the investigation of thermal processes during the crystal growth (Figure 3c). At last, the upside-down rotatory container with the thin quartz net 9 was used to see and investigate the crystal forms at different growth regimes (Figure 3d).

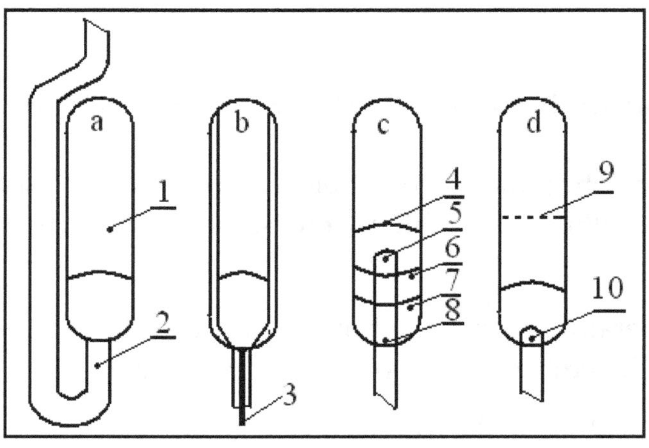

Figure 3. The containers for growth of perfect GaP crystals and investigation of the growth process. **a, b, c, d.** Types of containers. 1. Silica ampule. 2. Silica inlet rod, fixing the ampule to the vibrator. 3. Cu metal rod. 4. Container for investigation of thermal processes. Points 5, 8, 10 — locations of thermocouples. 6. Thermoisolating Al_2O_3 layer. 7. Ga etalon. 9. Silica net.

The impact of the temperature control accuracy was investigated at the facility, which allows to reproduce with controlled accuracy the necessary law of the temperature change in the growth container [1, 3]. Permissible fluctuation of temperature can be set within ± 0.5 ± 10° C interval. In order to study the form and quality of growing crystals, the crystallization process could be interrupted at any temperature below the point of liquidus. This interruption was achieved with the help of a special designed growth container (Figure 3d) and its programmable heating furnace, which allow to stop the change of temperature of the solution-melt, to identify and examine the crystals grown to this time (see details in [1, 3]).

Let us now describe and explain the peculiarities of temperature regime during the growth of lamellar perfect GaP single crystals from approximately 10 cm^3 of 5 at.% P – Ga melt solution (Figure 4). The smoothly heated slightly over 400°C solution must be kept for around 2 hours at this temperature for safe, non-dangerous explosion transformation of red phosphorus into its white modification. Then the heating process may be smoothly prolonged until a temperature of approximately 1200°C that increases the liquidus point for 5% P solution in Ga (1123°C) and creates good conditions for Ga-P reactions. After 1–3 hrs. soaking interval start to gradually cool the solution with a velocity of around 30°C/hrs. until 600–700°C and shutoff of the furnace. For perfect quality of the growing crystals during this 12–15 hrs. cooling of the solution, it is extremely important to support very fine, ±0.5°C temperature control possible with the specially elaborated heating and temperature control installation [3].

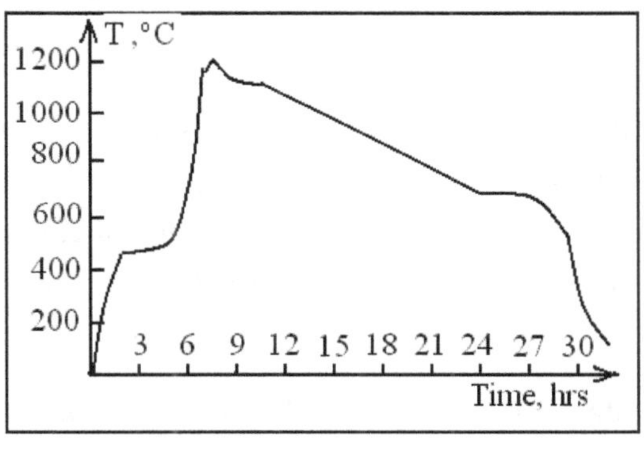

Figure 4. Programmable temperature changes in growth container.

With the introduction into the Ga solvent of various impurities, the choice of which was determined by practically important properties of the obtained crystals, at the same time the effects of contaminants on the nature of solidification have been investigated. The following impurities were introduced one by one in the solvent: Cu, Zn, Cd, In, Si, Ge, Sn, S, Se, Te, Ni, Cr, Fe, Co, Sm, La, Gd, and Sm_2O_3.

Typical features of the doped GaP crystals (size, color, polarity, lamellar structure, twinning, the dislocation structure, and others) were compared with similarly grown but undoped crystals.

It was found that the properties of the crystals depend on the chemical nature as well as on the quantity of the selected impurities. Their small admixtures did not significantly change the form of the crystals, while significant change of the crystallization environment composition created changes in forms of growth. For instance, the introduction into the solution of large quantities of Sn, the element of another valence compared with Ga, led to the emergence of the crystals having octahedral isometric forms. Isometric forms with simultaneous reduction of the sizes of crystals occurred with the introduction of solvent refractory impurities — samarium oxide (Sm_2O_3), solid units of which, apparently, were the centers of crystallization. Introduction into the solution of significant quantities of In, the chemical analogue of Ga, only slightly influenced on the morphology of the crystals, changing the lattice parameter, indicating the formation of 0.01 at. % In-GaP solid solution.

It was established that an increase in concentration of Zn and Gd in the Ga-P solution leads to the increase in the concentration of carriers (holes), reducing the size of crystals and their chemical resistance with a simultaneous increase in microhardness on the plane (111). At the introduction of S (1 at. % and more) the forms of the crystals varied from lamellar to volumetric. By increasing the concentration of tellurium in solution from 0,007 to 0.02 at. % the lattice constant varied from 5.4511 to 5.4524Å. At small quantities of impurities Zn and Te in GaP, until about 10^{19} cm^{-3}, a direct proportional relationship between the amount of impurities added to the solution and the concentration of charge carriers in the crystals is observed. Doping of the Ga-P solution by small quantities of rare earths and elements of the Fe group did not impact significantly on the forms of crystals, however, led to a significant change in the electrical and luminescent properties of GaP crystals. It is interesting to note that the density of etching pits on the plane (111) of doped crystals sharply increased in comparison with undoped ones.

III. The following factors influenced on the morphology of GaP crystals grown from the Ga-P solution of a certain concentration in conditions of the mass crystallization: the degree of supersaturation, which depends on the temperature of the solution and its first time derivative, the conditions of the heat crystallization removal and the accuracy of the temperature control in the solution.

Changing conditions of crystallization, mainly the rate of cooling of the melt, significantly affects the morphology of crystals. So, increasing the cooling rate reduced the size of the crystals, but the dislocation density was not changed and was primarily the determined concentration and chemical nature of impurities. The number of crystals having isometric forms increased with the decreasing cooling rate at simultaneous improvement of quality of the crystal surface. Uneven removal of heat from walls of the container has led to the formation of up to 30 mm crystal needles, elongated in the direction of the heat removal.

Significant impact on the crystal quality provides more accurate temperature control. The crystals obtained with the accuracy of temperature control ±0.5°C had a perfect plane (111), low dislocation density, and dimensions in 2–3 larger than the crystals obtained at the same conditions, but with the accuracy ±5°C or according to the regimes described in the literature. Crystals reached 25 mm in length and had the dislocation density ~ $10^3 cm^{-2}$, which is significantly less than that of the crystals obtained with low accuracy of the temperature control.

IV. In order to clarify the mechanism of crystallization of lamellar GaP crystals, differential record of temperature change between the standard of pure Ga and 5 at.% Ga-P solution near the point of liquidus as well as interruption of the crystallization process at different temperatures below the point of liquidus were conducted in special containers presented in Figure 3c and 3d (please see details described in Refs.[1, 3]). The record of temperature was conducted in the locations of the Ga standard and the Ga-P solution, interruption of the process was carried out by turning the container with quartz net from the top down. The pattern of temperature change in the solution-melt is presented in Figure 5.

Let us now present an overview of the growth of lamellar GaP crystals basing on the findings of these experiments. Total quantity of the solution-melt in the experiments was 10–12 cm^3, 1 cm^3 of this solution (5 at.% of P) at a temperature above the line of the liquidus contained $3x10^{21}$ and $6x10^{22}$ of P and Ga atoms, respectively. The first crystals, having the form of dendritic needles with the length of the order of 5 mm and thickness of 0.15 mm, were registered at a temperature of 1107°C at approximately 16°C supercooling. The crystals, found on the net inside the growth container [1] at lower temperatures of the process interrupting, represent thin plates in the form of rhombs, triangles, or hexagons. The twinning on the transverse cross-sections was observed at the study of microsections and cleaved facets. It turned out that the lamellar GaP crystals contain the plane of twinning, parallel to the planes (111).

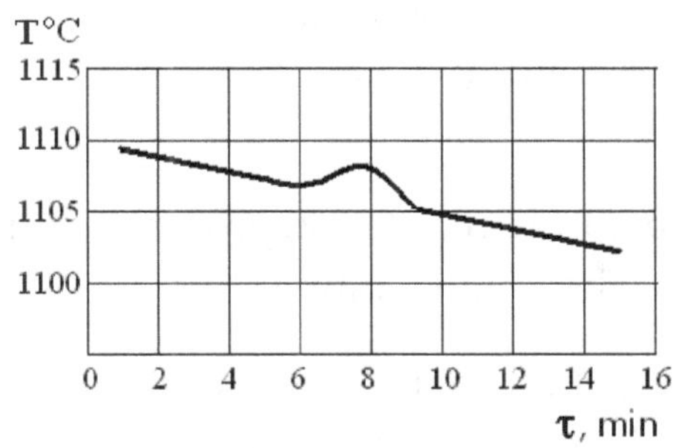

Figure 5. Change of temperature in the container when cooling with 30 deg/hour the 5 at.% solution of P in Ga.

The growth process of lamellar crystals can be formally divided into two components: the tangential growth in the plane (111) and layer-by-layer growth in the direction normal to the plane (111). Because the ratio of Ga and P atoms differed significantly from the stoichiometric, it is natural to assume that the process determining the rate of growth is diffusion. According to our estimates, the coefficient of diffusion of phosphorus atoms in Ga at T~1400K is equal $D_p \sim 3x10^{-3}\ cm^2sec^{-1}$.

Upon cooling the system to 1107°C degrees, the supercooling of the solution was approximately $1.6x10^{20}\ cm^{-3}$. Accepting the growth of needle-shaped crystals ends during recalescence time about 2 min (Figure 3), we get the diffusion length $L = (D_p \cdot \tau)^{1/2} = 0.6$ cm, and, further, the dimensions of the crystal in diffusion approximation: needle length 6 mm and thickness 0.56 mm; this value is in a good agreement with the experimental data. With further lowering of the temperature, the crystals acquire lamellar form with mainly developed planes (111) having the greatest reticular density and composed of the same atoms. Note that the removal of heat of crystallization is facilitated by the lamellar form of the crystals.

Thus, accepting the above presented two-stage model of lamellar crystal growth, let us consider the ratio of the contributions of the tangential growth in the plane (111) and layer-by-layer growth in the direction normal to the plane (111). The growth in the tangential direction is limited by diffusion and by the size of the previous layer, that is, it is determined by the initial conditions and the shape and size of the plates, formed during the cooling of the solution-melt 10–20 degrees below the point of liquidus. Since the formation of the initial crystals occurs in a very short time, the growth in the tangential direction is difficult to control, as it is in the case of the dendritic crystal growth. The growth in the direction normal to the plane (111) is sensitive to the fluctuations of the degree of supersaturation, resulting from the thermal motion of atoms and temperature fluctuations. In general, uniformity and perfection of the surface of lamellar GaP crystals the better, the less the time ratio needed for formation a new layer on a flat nucleation center to the time for which will be created the next nucleation center. The latter, obviously, in an extreme extent depends on fluctuations of the degree of supersaturation in different points of the growing plane and at different points in time. With the deterioration of precision temperature control, the number of fluctuations increases and bad crystals grow in the conditions of "entanglement" of the above mentioned characteristic times. Small fluctuations of supersaturation also are probably the cause of twinning planes. Indeed, assuming that the probability of fixation of the twins on the plane (111) is $W = \exp(-n\varepsilon/kT)$, where n is the number of atoms in a flat nucleus and ε is the energy of packing defect per atom, and knowing the ratio of the number twin planes to normal equals $10^{-5.5}$, we obtain $W = \exp(-n\varepsilon/kt) = 10^{-5.5}$, $k = 1.4 \times 10^{-16}$ erg/degree, T = 1400K, and $n\varepsilon = 1.6$ eV or n = 10, $\varepsilon = 0.16$ eV/atom that under the order of the values is reasonable.

Increase of the accuracy of temperature regulation promotes some reduction in the number of twins and this fact also confirms the notable influence of fluctuations on the twinning.

Thus, deteriorations of quality or shape defects of GaP plates during their growth from Ga – P melt solution are observed in the next cases: 1) at a considerable increase of concentration of impurities in the GaP solution and dependently on the chemical nature of the chosen impurity; 2) at the change of the crystallization conditions, namely, at the sufficient deterioration of the cooling velocity control or in the case of nonuniform heat removal from the container for the crystal growth.

Processes of creation of the dendritic needles and their transformation to platelet crystals take place in a small temperature interval and for a short time, therefore they are hardly controlled, while crystal growth is easily controlled in the direction normal to the plate (111), because it is very sensitive to the external factors such as velocity of the solution cooling and accuracy of the temperature control.

All the above-stated means that only very perfect growth equipment and top quality growth experience will give an opportunity to grow good and high service ability GaP crystals. However, in spite of all precautions and high experience of the specialists, growing the crystals, freshly prepared GaP crystals cannot be successfully applied in electronics due to their low operational data, such as absence of high photosensitivity, bright and broadband luminescence, satisfactory transparency, and mechanical characteristics. Further, it will be shown how to considerably increase the quality and service ability of the freshly prepared GaP crystals.

4. Optical Properties of Perfect, Long-term Ordered GaP:N Crystals

It is necessary to note that the very important for optoelectronics long-term ordering and considerable improvement of the semiconductor crystal lattice and accompanying phenomena have been discovered and observed over decade time scales only with the help of the same unique collection of samples. This collection of long-term ordered perfect GaP single crystals gives opportunities to find deep fundamental analogies in properties of the perfect single crystals and nanoparticles as well as to predict and to realize in nanoparticles and perfect bulk crystals new interesting properties and applications.

Investigating the samples grown by the author in the 1960s, it was noted [1, 7], gallium phosphide (GaP) crystals for the first time clear improved their optical and mechanical properties only after 10–12 years since their preparation. It was unusual and interesting in the situation when all non-living matter around us usually deteriorates in time; therefore the author decided to investigate, to understand, and to use in the future this phenomenon. Shortly, the investigation process can be described as following.

Grown about 50 years ago pure and doped GaP crystals throughout the intervening decades have been periodically re-evaluated in order to see and investigate the change over time in their electro- and photoluminescence, photoconductivity, behavior of bound excitons, giving interesting and bright luminescence, nonlinear optical, and other phenomena. Accordingly, it is of interest also to monitor the change in crystal quality over the course of several decades while the crystal is held under ambient conditions.

Figure 6 demonstrates Raman spectra in GaP and GaP:N in 1989–1993 (a) and in 2006 (b, c). Concerning the general differences in the Raman spectra produced by the ordered and disordered forms of the crystals, note in accordance with Refs. [10, 17, 21-24] that the spectrum of the most ordered heavily N doped crystal (spectra 2 in Figure 6b) shows a considerably more intense narrow LO line than the less ordered pure or doped crystals (spectra 3, 4 in Figure 6b). Whereas the distribution and the environment of the P or N anions at a particular site in the unit cell in the ordered crystal are uniform, the great variability in this environment from site to site exists in a disordered or less ordered crystal.

Note that the theory of Raman light scattering in GaP predicts the LO phonon decay into two longitudinal acoustic phonons LA. LA phonons with a frequency LO/2 [41], and 2-phonon processes of 2TO and TO+LO emission also can be observed in perfect crystals [42]. This observation of a multi-phonon process and a decay of LO phonon, having a low intensity, confirm the high quality of the host lattice, uniform impurity distribution, and as a consequence, low noise background in the Raman scattering.

Figure 7 provides the Raman spectra obtained in 2005–2006 from pure and heavily doped GaP at 300K (a) and 80K (b). It can be seen only in the GaP:Bi crystal at 300K that a very narrow LO phonon is slightly shifted (approximately 2 cm^{-1}), while the TO and LO peaks maintain their spectral positions at 80K for the pure and doped GaP crystals independently of the type of impurity and its concentration. The LO phonon line is narrower in the doped crystal than in the undoped (pure) one and also is more intense than the TO phonon line. Note that these results are obtained despite the fact that the maximum concentrations of N, Sm, and Bi in these

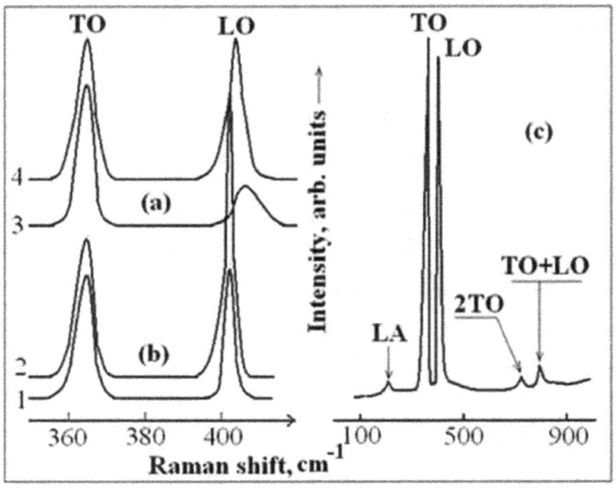

Figure 6. Raman spectra in GaP and GaP:N in 1989–1993 (a) and in 2006 (b, c). 1, 4—pure GaP, 2, 3—GaP heavy doped by N. Spectra 3, 4 are taken from Ref. [11] and obtained by one of the authors (SLP) in 1991. A new phenomenon observed in the Raman spectra that has developed in the crystals over 40 years are the peaks denoted by us here as LA, 2TO, and TO+LO (Figure 6c).

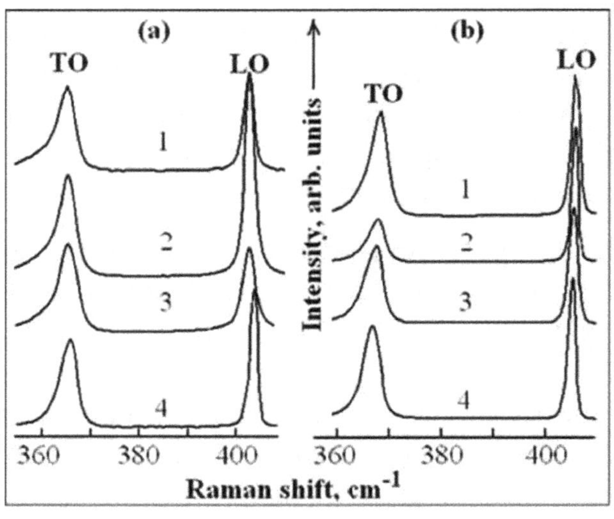

Figure 7. Spectra of Raman light scattering of pure and doped GaP at 300K (a) and 80K (b). 1. Pure. 2-4: GaP doped by N, Sm, and Bi, respectively

crystals are close to their limit of solubility (~ $10^{19}cm^{-3}$), the masses of impurity atom are very different, (the atomic weight 14, 150, and 208 for N, Sm, and Bi, respectively), and the N and Bi impurities substitute lattice points whereas Sm occupies interstitials.

The position and the line-width of the TO line in the ordered crystals do not depend on type and concentration of impurity nor on temperature. The temperature-independent TO line-width in the aged crystals implies that the impurities do not perturb the order of the lattice, which is possible only for a very uniform environment.

It is known that the broadening and frequency shift of the phonon lines are due to anharmonicity of the lattice vibrations and disorder in the crystal. Note that the TO line-shapes in the crystals again measured after 17–19 years are more close to Lorentzian than in [41] and LO lines are narrower in the crystals with big impurity concentrations. This most likely means that the anharmonicity of the lattice vibrations in the presence of ordered impurities is less than that in the pure unordered crystals. An increase in temperature only slightly distorts the TO line-shape in these crystals implying that the anharmonicity also is weak. Considering these results, it is assumed that the anharmonicity is larger in perfect undoped crystals than in doped crystals with periodically disposed impurities. Thus, it has been observed that in long-term ordered GaP crystals different impurities presented there at high concentrations do not distort crystal lattice, lead to narrowing and increase of longitudinal optical mode LO, and do not change the symmetrical form and positions of the TO line.

After 40 years since their fabrication the impurities in these doped GaP crystals create a superlattice with a period that depends on their concentration. Absence of luminescence from bound pair excitons (which need for their creation considerably less NN spacing than it is possible at the N concentration in our GaP:N crystals) as well as the spectral position of the narrow zero phonon line A shifting along the luminescent spectrum dependently on the N impurity concentration in exact correlation with the relevant theoretical ratio (see [40] and Figure 8a) only clearly show that in the 25–40 years aged GaP:N crystals the impurity atoms are located with the equal r_{NN} spacing. Thus, we get as a minimum the evidence of anti-clustering of nitrogen atoms as the crystals age. However, taking into account that the N impurity atoms substitute the host P atoms in very perfect diamond-like cubic crystal lattice (it is really perfect according to our data on the position of extremely narrow phonon replica in heavily doped aged GaP:N crystals), we may assert from very simple geometrical consideration that in the case of the equal spacing measured with high accuracy, the N impurity system also forms a cubic crystal superlattice that means not simply anti-clustering, but the ordering of impurities as well as the perfection of the host lattice modified by impurities.

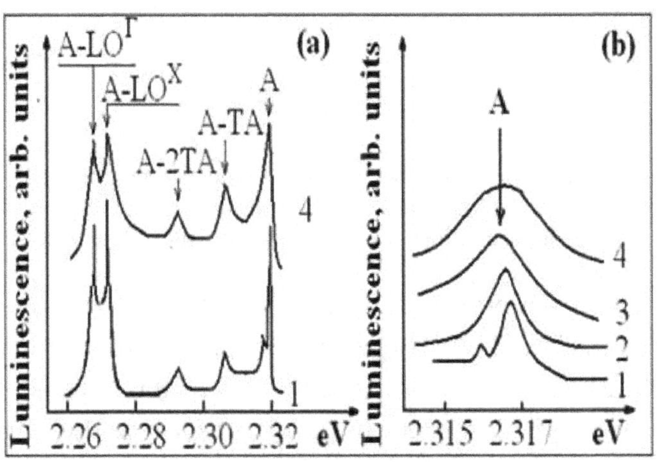

Figure 8. Evolution of the GaP:N luminescence with time and nitrogen concentration at a temperature of 15K. **a.** Zero-phonon line of the bound exciton A and its transversal acoustic (TA) and longitudinal optic (LO) phonon replica in as-prepared (4) and 25-year long-term ordered (1) crystals. **b.** Zero-phonon line A as a function of nitrogen (N) concentration. 1-3: 25-year-old crystals. 4: as-prepared. Curves 1 through 4 represent samples with nitrogen concentrations of 10^{17}, 10^{18}, 10^{19}, and 10^{18} cm^{-3}, respectively.

Figure 8 provides a comparison of the evolution in luminescence spectra at low temperatures (80K and below) from GaP:N over a period of 25 years (original measurement in about 1963). Over this period, zero-phonon line A of single N impurity-bound excitons and their phonon replica are narrower in their line-widths when compared to the freshly prepared single crystals (Figure 8a). Further, as expected, zero phonon line and phonon replica in samples aged at room temperature for 25 years shift spectral position depending upon concentration of N impurities (Figure 8b, spectra 1-3) according to Ref. [40], while the same freshly prepared crystals exhibited broader luminescence line-widths with increasing nitrogen content (Figure 8b, spectrum 4). These, along with other half-centennial findings, including modifications of luminescence kinetics, spontaneous Raman scattering, X-ray diffraction, absorption spectra, micro-hardness, and density of dislocations, which are reported elsewhere [1, 3-24], strongly suggest that close-to-ideal GaP:N crystals are formed over time due to the equally spaced disposition of N impurities from their chaotic distribution in the same freshly prepared crystals.

As first noted in Ref. [23], these results suggest a new type of crystal lattice in which the host atoms occupy their proper (equilibrium) positions in the crystal, while the N impurities, periodically substituted into the lattice, portion it into short chains of equal length. According to the data obtained from Raman light scattering [10, 14], host atoms of this new lattice develop harmonic vibrations, and a high degree of lattice perfection leads to an abrupt decrease in the non-radiative recombination and an increase of efficiency and spectral range of luminescence.

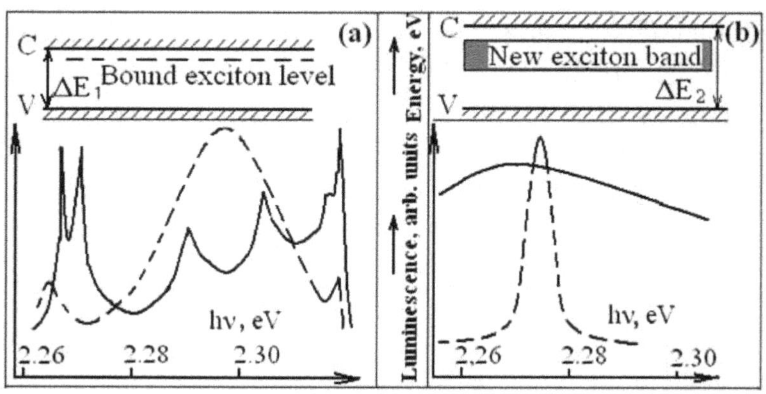

Figure 9. Luminescent spectra and schematic representation of the forbidden gaps (ΔE_1, ΔE_2) in the nitrogen-doped GaP aged for (**a**) 25 years and (**b**) 40 years. The dotted lines correspond to highly optically excited crystals. C and V represent the positions of the bottom of the conductance and the top of valence bands, respectively.

Additionally, stimulated emission of light in these temporally ordered crystals (Figure 9b) is observed. As also shown in Ref. [23], the GaP:N crystals aged for at least 40 years possess no discrete impurity level for N-bound excitons in the forbidden gap. They also demonstrated a uniform luminescence from a broad excitonic band instead of the narrow zero-phonon line and its phonon replica as observed from the less-aged 25-year-old crystals.

Thus, long-term ordered GaP:N crystals demonstrate uniform bright luminescence from a broad excitonic band instead of the narrow zero-phonon line and its phonon replica in disordered and partly ordered (25-year-old) crystals. This is due to the fact that ordered crystals have no discrete impurity level in the forbidden gap. To the best of our knowledge, such transformation of a discrete level within the forbidden gap into an excitonic band (Figure 9 a, b) is observed for the first time. In this case, the impurity atoms regularly occupy the host lattice sites and affect the band structure of the crystals, which is now a dilute solid solution of GaP-GaN with regular disposition of N atoms rather than GaP doped by occasionally located N atoms. Note that the increase of luminescence excitation in case of partly ordered GaP:N (Figure 9a, dotted line) leads to a broad luminescence band as a result of bound exciton interaction [9], while in the case of perfectly ordered crystals (Figure 9b) one can see an abrupt narrowing of the luminescence band, probably due to stimulated emission in defect-free crystals. Earlier, in freshly prepared crystals, we observed a clear stimulated emission from a GaP:N resonator at 80K [5], as well as the so-called superluminescence from GaP single crystals having natural faceting. Presently, our ordered crystals have a bright luminescence at room temperature that implies their perfection and very low light losses. In our studies [14, 23], we demonstrate that the stimulated emission in GaP also developed even at room temperature by direct electron–hole recombination of an electron at the bottom of the conduction band with a hole at the top of the valence band and the LO phonon absorption.

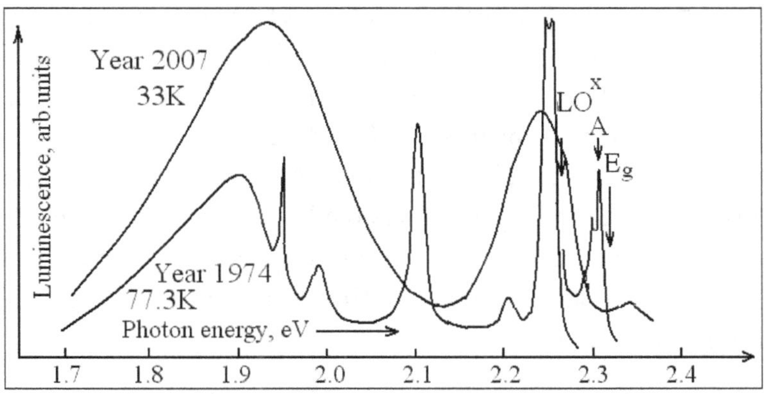

Figure 10. Evolution of GaP:N:Sm luminescence between 1973 and 2007. The crystals are grown in 1965.

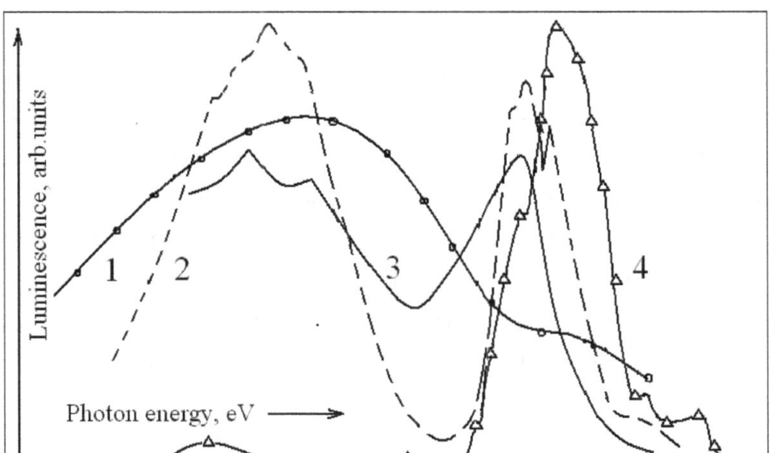

Figure 11. Luminescence of long-term ordered GaP:N:Sm. 1, 2—low; 3, 4—room temperatures. 1, 3—low; 2, 4—high excitation.

Investigating earlier luminescence of GaP doped by Sm and Ge, we have demonstrated bright luminescence in infrared region and opportunities to change in the wide wavelength diapason the position of luminescent maximum. For the first time, the GaP:N:Sm crystals were investigated in the year 1974 when their luminescence could be observed only at 80K and below [6]. The relevant spectrum, obtained in the year 1974, is shown in Fig. 10. Then it consisted of two parts: 1) the green part reflecting recombination of N bound excitons through the emission of the narrow zero-phonon A line and its phonon replica, arising in the emission of LO and other lattice phonons, and 2) the red and yellow parts reflecting the absorption of the bound exciton irradiation by Sm centers and its characteristic radiative recombination. In the year 2007 (Fig. 10, [13]), the spectrum of luminescence also consisted of these two parts, but instead of narrow excitonic lines we see a broad green excitonic band and also the broad red and yellow band without any fine structure.

Efficiency of Sm excitation at the N bound exciton recombination highly depends on a degree of uniformity of the N-Sm mixture. The long-term ordering gives a uniform mixture of Sm and N impurities and according to R. L. Bell [43], it provides up to 100% quantum efficiency of radiation from GaP:N:Sm system and bright luminescence at 300K. The bright emission bands at 300K, shown in Figure 11, curves 3, 4, arise as a result of formation during the period 1963–2010 of the uniform mixture of N recombination and Sm activation centers. The luminescence spectra of GaP:N:Sm at low temperatures (35K and below, Figure 11, spectra 1, 2) have the same maxima in green and red and yellow regions where the ratio of their intensities depends on the concentrations of N and Sm impurities as well as on the level of excitation. For instance, at a chosen level of excitation and an N/Sm ratio of concentration one can realize pure red or green emission of high efficiency or any combination of these colors that is an important property of a tunable light source.

Thus, doping GaP simultaneously with Sm and N, choosing necessary temperature and intensity of excitation we can get light emission from green to red. However, further expansion of the emissive spectrum in GaP can be achieved using some of the following other dopants or specific methods of crystal preparation.

Let us now discuss a possibility to converse visible emission into infrared region. Obviously, it is possible, creating in GaP an additional to the basic green another channel for light emissive electron-hole recombination in infrared region. For instance, it is known that the donor-acceptor pair Ge^A-O^D on the base of oxygen (O^D) and germanium (Ge^A) emits the band with the maximum at 1.38 eV [44]. It was shown in Ref. [7] and in some other works fulfilled under the author's supervision, the existence of this band and its intensity, compared with the basic green emission, depends on the concentration of Ge-O pairs and intensity of excitation, obtained with the help of a Q-switched Nd glass laser supplied with the frequency doubler generating 2.34 eV photons.

Increasing the intensity of excitation, one can decrease the contribution into luminescence of light emissive recombination through the Ge-O pairs when the concentration of these pairs will be equal and less than the concentration of electron-hole pairs generated by the laser excitation source. In these conditions will be also clear detected the growing with excitation contribution into the whole light emission from the other recombination channels depressed

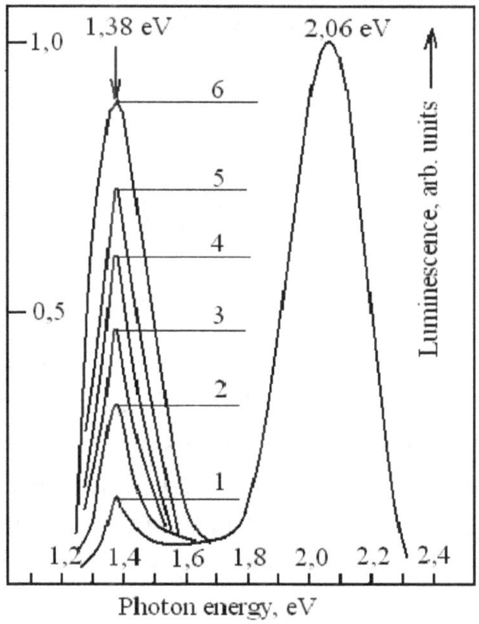

Figure 12. Luminescent spectra of GaP doped by Ge with a concentration of approximately $10^{18}cm^{-3}$ at 77.3K and different excitation intensity I_{ex} x 10^{-24} $cm^{-2}sec^{-1}$: 16(1); 10(2); 4,5(3); 2,3(4); 1,2(5); and 0,46(6).

by the recombination through Ge-O pairs. Thus, the switching of light emissive recombination channels, important for application in optoelectronic devices, will be observed and demonstrated in the luminescence of GaP doped by Ge. Figure 12 demonstrates depending on excitation intensity the switching of luminescence at 77.3K from infrared with maximum at 1.38 eV to green-yellow region with maximum at 2.06 eV [7].

Moving to infrared region with the help of GaP doped by Ge, we have reached 1.2 eV in the emissive spectrum that 1 eV less than the forbidden gap of GaP crystals. Taking into account that our goal is to propose a cheap, resource-saving, and impactful way for the development of optoelectronics with the help of the special transformation of an ordinary semiconductor into the base material for various light emissive devices structures with broad and bright light emissive spectra, let us consider opportunities to expand light emission as much as possible to the ultraviolet (UV) region, using only GaP bulk and nanocrystals instead of a lot of the other semiconductors with their complicated, labor-consuming, and expensive technologies of preparation of materials with necessary reliable radiative properties.

As we have shown previously [8-17], different defects of high concentration in freshly prepared GaP single crystals completely suppress any luminescence at room temperature due to negligible small free path for non-equilibrium electrons and holes between the defects with their non-radiative recombination, while the quantum theory predicts free movement of electrons and holes in the field of an ideal crystal lattice. It was also shown that the long-term ordered and therefore close-to-ideal crystals demonstrate bright luminescence and stimulated emission repeating behavior of the best nanoparticles with pronounced quantum confinement effects. Therefore, there are only two ways of preparing a material with bright luminescence from IR to UV regions: to prepare very perfect, defectless single crystals or nanocrystals with the dimensions less than electron-hole free path in this material of standard quality. Correctness of the chosen way can be confirmed by the following comparison of optical properties of the best GaP nanocrystals and GaP perfect bulk single crystals.

5. Comparison of Optical Properties of GaP Nanocrystals and GaP Perfect Bulk Single Crystals

Jointly with the references [3, 7, 9-17, 21-27, 32] here we present a generalization of the results on long-term observation of luminescence, absorption, and Raman light scattering in bulk semiconductors in comparison with some properties of the best to the moment GaP nanocrystals. The combination of these characterization techniques elucidates the evolution of these crystals over the course of many years, the ordered state brought about by prolonged room-temperature thermal annealing, and the interesting optical properties that accompany such ordering. We demonstrate that long-term natural stimuli improve the perfection of our crystals, which can lead to novel heterogeneous systems and new semiconductor devices with high temporal stability. Raman light scattering confirms high quality of the long-term ordered crystals.

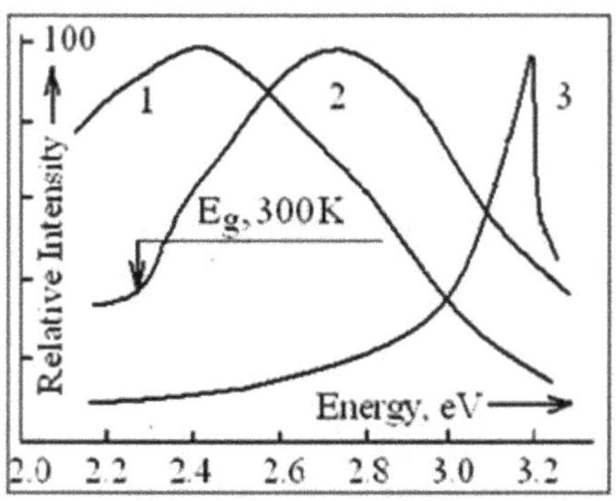

Figure 13. Luminescence of perfect bulk GaP single crystals (1) in comparison with the luminescence of GaP nanoparticles and GaP/polymers nanocomposites (2, 3). Nanoparticles were prepared from white P by mild aqueous or colloidal synthesis at decreased temperature, stored as the dry powder (spectrum 2) or suspension in a liquid (spectrum 3). Details: [15, 17].

We further improved upon the preparation of GaP nanocrystals using the known methods of hydrothermal and colloidal synthesis [30-32] by taking into account that the success of our activity depends on the optimal choice of the types of chemical reactions, necessary chemicals and their purity, conditions of the synthesis (control accuracy, temperature, pressure, duration, etc.), methods and quality of purification of the nanocrystals, and storage conditions for nanoparticles used in the further operations of fabrication of the GaP nanocomposites. The best quality GaP nanoparticles have been prepared by hydrothermal or colloidal synthesis from white phosphorus at decreased temperature (125°C) and intense ultrasonication; it was established that the maximum shift of their luminescence to ultraviolet and the best quality in general have the nanocomposites obtained from the nanoparticles of the same dimensions stored as a suspension in a suitable liquid.

Figure 13 compares the luminescence spectra of our long-term (up to 50 years) ordered GaP single crystals (spectrum 1) to that from high quality GaP nanoparticles and their GaP nanoparticles/polymers nanocomposites [15, 17]. Nanocrystals of the different dimensions, stored as dry powder, demonstrate rather broad luminescent band with maximum at 2.8 eV (Figure 13, spectrum 2), while the nanocrystals of about 10 nm sizes, thoroughly separated and distributed in a suspension, that prevent their coagulation, mechanical, and optical interaction, exhibit bright narrow-band luminescence with a maximum at 3.2 eV, approximately 1 eV above the position of the absorption edge in GaP at 300K (Figure 13, spectrum 3). The thoroughly washed, ultrasonicated and dried nanopowders as well as their specially prepared suspensions have been used for fabrication of blue light emissive GaP nanocomposites on the base of some optically and mechanically compatible with GaP polymers [15, 17, 30-32]. According to our measurements, the matrix polymers PGMA-co-POEGMA or BPVE used in this work provide no contribution to the spectra of luminescence of the based on these matrixes, so, the nanocomposite spectra coincide with those obtained from the relevant GaP powders or suspensions. We note that in the GaP/BPVE nanocomposite, the position of the luminescent maximum can be changed between 2.5 and 3.2 eV and the brightness is 20–30 more than in the PGMA and PGMA-co-POEGMA matrixes.

Long-term ordering leads to the creation of perfect bulk GaP crystals with considerably expanded and bright emissive band, practically the same as in the perfect GaP nanoparticles. We explain the broadening of the luminescence band and the shift of its maximum to low photon energies in luminescence of the nanocomposite based on the GaP powder by presence in the powder of the nanoparticles with the different dimensions between 10 and 100 nm. Meanwhile, the nanocomposites on the base of the suspensions containing only approximately 10 nm nanoparticles exhibit bright luminescence with a maximum at 3.2 eV due to a high transparency of 10 nm nanoparticles for these high energy emitted photons and pronounced quantum confinement effect.

In accord with our data [15, 31] the shift due to the quantum confinement effects is about a few tenths of eV and, obviously, it is impossible to explain only through this effect the dramatic 1 eV expansion of the region of luminescence at 300K to the high-energy side of the spectrum. In order to explain this interesting phenomenon, we postulate that the nanocrystals, much like the ideal long-term ordered bulk GaP single crystals, where this effect is not so strong, exhibit the huge increase in blue-shifted luminescence due to: (a) negligibly small influence of defects and non-radiative recombination of electron-hole pairs and very high efficiency of their radiative annihilation, (b) high perfection of nanocrystal lattice, and (c) high transparency of nanocrystals due to their small dimensions for the light emitted from high points of the GaP Brillouin zones, for instance, in the direct transitions Γ_1^c–Γ_{15}^v between the conductive and valence bands with the photon energy at 300K equal to 2.8 eV [37], and (d) high efficiency of this so-called "hot" luminescence that means direct radiative recombination of electrons without their preliminary thermalization into the nearest particular point of the conductance band. Taking into account the high light absorption coefficient equal to approximately 10^5 cm^{-1} for photons with the energy in the vicinity of maximum at 3.2 eV [38], we can explain the difference in the spectrum 1 from perfect bulk GaP crystal and the spectrum 3 from its 10 nm perfect nanoparticles. Really, a big GaP single crystal, even very perfect one, in principle, cannot emit many photons in UV region, because the overwhelming majority of those photons will be immediately absorbed in the crystal; only tiny defectless 10 nm GaP spheres, transparent for this UV region, distributed in the transparent suspension or a polymer film will easily emit this UV light.

Note, our first attempts to prepare GaP nanoparticles [18, 28] yielded room temperature luminescence with the maximum shifted only to 2.4 eV in comparison with the achieved now new maximum at 3.2 eV and it confirms our significant progress in preparation of GaP nanoparticles and GaP/polymers nanocomposites. The perfect quality of the nanoparticles prepared by improved technologies is confirmed by all the used methods of characterization, while investigation of Raman light scattering evolution during 25 years (since 1989) clear confirms considerable improvement of GaP single crystal quality and the existence of new interesting phenomena characterizing only very perfect crystals.

On the base of these improved technologies for the preparation of GaP nanoparticles and GaP/polymer nanocomposites, we can change within the broad limits the main parameters of luminescence and create a framework for novel light emissive device structures using dramatic 1 eV expansion of GaP luminescence to UV region. Besides that, using all the noted in the presented review opportunities, including specially doped GaP and the necessary level of luminescence excitation, we can change the position of maximum and bandwidth of luminescence in wide, from infrared 1.2 to UV 3.2 eV, limits.

6. Excitonic Crystal and Its Importance in Optoelectronics

The role and application of bound excitons in nanoscience and technology are discussed in this chapter. Bound excitons are well studied in semiconductors, especially in gallium phosphide doped by nitrogen (GaP:N) [3, 4, 7, 38, 40, 45]. Doping of GaP with N leads to isoelectronic substitution of the host P atoms by N in its crystal lattice and to the creation of the electron trap with a giant capture cross-section. Therefore, any non-equilibrium electron in the vicinity of the trap will be captured by N atom, attracting a non-equilibrium hole by Coulomb interaction and creating the bound exciton — short-lived nanoparticle with the dimension of the order of 10 nm (it is the Bohr diameter of bound exciton in GaP:N). Note, that none of nanotechnology methods are used in the creation or selection of dimensions of these nanoparticles — only natural forces of electron–hole interaction and electron capture by the traps are necessary for the creation of these nanoparticles. As the result we get something like neutral short-lived atom analogue — a particle consisting of heavy negatively charged nucleus (N atom with captured electron) and a hole. So-called zero vibrations do not destroy the possible solid phase of bound excitons having these heavy nuclei that give an opportunity to reach in GaP:N a new crystal state — the short-lived excitonic crystal appeared at the necessary level of excitation, N impurity ordering and concentration, energy of photons and temperature.

Taking into account the above-mentioned preceding results, a model for the GaP:N long-term ordered crystal and its behavior at the relevant level of optical excitation for 40-year-old ordered N-doped GaP (Figure 14) can be suggested. At the relevant concentrations of N, the anion sub-lattice can be represented as a row of anions where N substitutes for P atoms with the period equal to the Bohr diameter of the bound exciton in GaP (approximately 10 nm) (Figure 14a). At some level of optical excitation, all the N sites will be filled by excitons, thereby creating an excitonic crystal (Figure 14b), which is a new phenomenon in solid-state physics and a very interesting medium for application in optoelectronics and nonlinear optics [4, 15, 17, 25, 26].

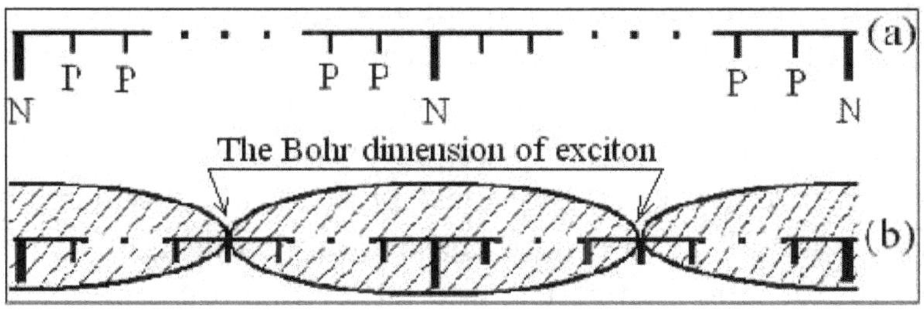

Figure 14. The models of 40 years ordered GaP doped by N. **a.** The new type of crystal lattice with periodic substitution of N atoms for the host P atoms. **b.** The excitonic crystal on the base of this lattice. The substitution period is equal to the Bohr diameter of exciton (~ 100Å) and optical excitation is enough for complete saturation of the N sublattice (please see details in Refs. [15, 17]).

Thus, using bound excitons as short-lived analogues of atoms and sticking to some specific rules, including the necessity to build in the GaP:N single crystal the excitonic superlattice with the identity period equal to the bound exciton Bohr dimension, we get a unique opportunity to create a new solid state media — consisting from short-lived nanoparticles excitonic crystal, obviously, with very useful and interesting properties for the application in optoelectronics, nanoscience, and technology. The following will discuss methods of preparation and possible application in optoelectronics of perfect GaP crystals, based on perfect GaP excitonic crystals and nanocrystals.

7. Already Discovered and Possible Nonlinear Optical Phenomena in GaP

Thus, confirmed by this semicentennial study, the impurities in doped long-term ordered GaP create a sublattice with a period that depends on their concentration. By periodically substituting for host atoms or occupying interstitial sites in the host lattice, the impurities become an intrinsic component of the modified crystal lattice and participate in the formation of a "new" phonon spectrum.

The ordered crystals with the host lattice modified by impurities could be very useful in various optoelectronic applications. Noted here are only a few potential applications in light emissive device structures. The properties of these structures will be very stable and independent on time. Uniform distribution of the recombination (N) and activator centers at the optimum concentration will yield the maximum efficiency for light emission. Further investigations of the quasi-crystalline state of excitons (or bi-excitons) bound to an impurity superlattice with a period equal to the Bohr dimension will be very interesting and useful because they should greatly strengthen nonlinear optical effects at low excitation intensities. This excitonic phase of high density of photons as an original accumulator of light also provides new opportunities for storage, transmission, and conversion of light.

With significant recent progress in semiconductor thin film deposition and growth techniques in some specific cases of device optoelectronic structure preparation, there obviously will be no longer a need to wait during years for such ordering to occur. For instance, superlattice from GaP/GaP:N with the period of the order of the bound excitons (their Bohr radius is equal to 5 nm) can be prepared by molecular beam epitaxy (MBE) or by MBE in combination with the laser-assisted epitaxy (LAE) [46-48]. Further, the preparation of a two- or three-dimensional arrangement of N impurities in a GaP film is difficult but also possible with the help of ion lithography. Of course, nowadays this technique is a frontier of our technological possibilities, but within the nearest future some very important progress likely will be obtained also in this direction. In this case, we will get a unique opportunity to design 3-dimensional impurity superlattices with configuration, symmetry, and lattice that are optimal for application in a concrete device structure or for efficient realization of specific linear or nonlinear optical phenomena. In any case, independently on the method of creation, the impurity modified crystal lattices, the excitonic (as well as bi-excitonic) phase with translational symmetry are very interesting objects, the properties and possible application of which are now under our investigation.

Nonlinear optics, starting its epoch-making development from the Nobel Prize Laureate N. Bloembergen transaction [49], with the appearance of lasers and highly supported personally by one of their inventors, the Nobel Prize Laureate A. Prokhorov, who helped the author to found in 1985 the Laser Research Laboratory in the Academy of Sciences of Moldova, since 1960s has taken its noteworthy place in investigations of GaP [5, 7, 50-52]. Here especially important for fundamentals and application in optoelectronic device structures were investigations of direct and indirect (with participation of the lattice phonons) many-quantum absorption between high symmetry points of GaP Brillouin zone (Figure 1), following the increase of photoconductivity and UV photon emission. So, exciting these high symmetry points with the help of infrared photon of a Q-switched laser, we get photoconductivity with participation of different energy bands and UV photon emission that is important equally for the investigation of band structure and for application in light frequency convertors. Addition of new opportunities due to elaboration of defectless perfect GaP bulk single crystals, its top-quality nanoparticles and multi-layered structures, discovery of the new nonlinear optic medium – excitonic crystal and its very interesting nonlinear optical phenomena [25-27] will surge of interest to this crystal, giving a new prospective industrial method of perfect crystal preparation, as well as opportunities for efficient realization in optoelectronics and electronics in general of remarkable properties of semiconductors due to a big commercial advantage from their fabrication.

8. Conclusions

Since the time of original preparation of gallium phosphide doped by nitrogen crystals (GaP:N) by the author in the 1960s, followed by the introduction of the excitonic crystal concept in the 1970s, the best methods of bulk, film, and nanoparticle crystal growth were elaborated. The results of semi-centennial evolution of GaP:N properties are compiled here and in the references to this paper. Novel and useful properties of GaP including an expected similarity in behavior between nanoparticles and perfect bulk crystals, as well as very bright and broadband luminescence at room temperature, are observed. These results provide a new approach to the selection and preparation of perfect materials for optoelectronics [25] and a unique opportunity to realize a new form of solid-state host — the excitonic crystal [26, 27]. In spite of the fact that the time necessary for natural long-term ordering (years) does not lead to optimism, the collected experience and results confirm expedience of the efforts directed to the formation in GaP of the N impurity superlattice having the identity period equal to the bound exciton dimension. As noted in Ref. [25], the process of preparation of top quality material for industrial electronics can be organized if the freshly grown crystals will be kept for years in a special storage and only old crystals (like to the old wine, if to use close to Moldova and other winemaking countries analogy) with the necessary properties will be annually retrieved for device fabrication while new portions of fresh crystals will be placed for long-term ordering.

Except natural aging of the relevant crystals for years, preparation of the N superlattice for excitonic crystal can be also realized by known methods of growth of multi-layer films, in particular by molecular beam and laser-assisted epitaxy [46-48].

The excitonic crystal, created by the long-term ordering or by the noted above methods of growth of multi-layer films, as well as the bulk top quality GaP crystals with the unique optic properties, obtained by the long-term ordering process of freshly prepared crystals, will be used in the new generation of optoelectronic devices, sometimes instead of nanoparticles and a lot of other materials. In particular, keeping in mind the low energy of the bound exciton creation, one can expect a low threshold for the generation of non-linear optical effects in the excitonic crystal and a good opportunity to create new and very efficient optoelectronic devices.

Note that semiconductor nanoparticles were introduced into materials science and engineering mainly in order to avoid limitations inherent to freshly grown semiconductors with a lot of different defects. However, it was shown [15] that this reason becomes unessential if, when justified, perfect long-term ordered semiconductor crystals are applied in electronics. Independently on their dimensions they demonstrate very interesting for application properties. Therefore, using the long-term ordered, perfect GaP crystals or similar on behavior and properties material in the electronic industry instead of the elaboration of very expensive and labor-consuming technologies for diverse materials and their nanoparticles with limited for application spectral region and other parameters, we get a big commercial advantage from their fabrication and application (for details please see the paper [25]).

All the obtained results presented here and included in summary reviews [15, 17, 25-27] may sufficiently change the approach to the selection of materials necessary for electronics, to make cheaper and simpler technology for the preparation of the selected materials and device structures based on them. This study of long-term convergence of bulk- and nanocrystal properties brings a novel perspective to improving the quality of semiconductor crystals. The unique collection of pure and doped crystals of semiconductors grown in the 1960s provides an opportunity to observe the long-term evolution of properties of these key electronic materials. During this half-centennial systematic investigation we have established the main trends of the evolution of their optoelectronic and mechanical properties. It was shown that these stimuli to improve the quality of the crystal lattice are the consequence of thermodynamic driving forces and prevail over tendencies that would favor disorder and destruction. Our long-term ordered and therefore close to ideal crystals repeat the behavior of the best nano-particles with pronounced quantum confinement effect.

For the first time, to the best of our knowledge, we have observed a new type of the crystal lattice where the host atoms occupy their proper (equilibrium) positions in the crystal field, while the impurities, once periodically inserted into the lattice, divide it in the short chains of equal length, where the host atoms develop harmonic vibrations. This periodic substitution of a host atom by an impurity allows the impurity to participate in the formation of the crystal's energy bands. In GaP it leads to the change in the value of the forbidden energy gap, to the appearance of a crystalline excitonic phase, and to the broad excitonic energy bands instead of the energy levels of bound excitons. The high perfection of this new lattice sharply decreases non-radiative electron-hole recombination, increases efficiency and the spectral range of luminescence, and promotes the stimulated emission of light due to its amplification inside the well-arranged, defect-free crystal. The development of techniques for the growth of thin films and bulk crystals with ordered distribution of impurities and the proper localization of host atoms inside the lattice are our high priority.

Semiconductor nanoparticles were introduced into materials science and engineering mainly to avoid limitations inherent to freshly grown semiconductors with a lot of different defects. Here and in other publications we show that this reason becomes unessential if we will apply in electronics, when it is justified, perfect long-term ordered semiconductors, which demonstrate independently on their dimensions very interesting for application properties. Especially important for application in the new generation of light emissive devices are the predicted and investigated by us crystalline state of bound excitons in GaP:N, the discovered in the framework of the STCU Project 4610 [16, 32] dramatic expansion of luminescence region in GaP perfect bulk single crystals as well as in the best prepared GaP nanocrystals and based on them composites with transparent polymers.

Using the long-term ordered GaP or similar on behavior and properties material in the electronic industry instead of elaboration of very expensive and labor-consuming technologies for diverse materials with their limited for application spectral region and other parameters, we get a big commercial advantage from their fabrication and application. So, the results of this long-term evolution of the important properties of our unique collection of semiconductor single crystals promise a novel approach to the development of a new generation of optoelectronic devices.

Besides the long-term ordering, the combined methods of laser-assisted and molecular beam epitaxies [46-48] will be applied to fabrication of device structures with artificial periodicity; together with classic methods of the perfect crystal growth, they can be employed to realize impurity ordering that would yield new types of nanostructures and enhanced optoelectronic device performance.

For the first time we demonstrate that well-aged GaP bulk crystals as well as high quality GaP nanoparticles have no essential difference in their luminescence behavior, brightness, or spectral position of the emitted light. The long-term ordered and therefore close to ideal crystals repeat the behavior of the best nanoparticles with pronounced quantum confinement effect. These perfect crystals are useful for application in top-quality optoelectronic devices and are a new object for the development of fundamentals of solid state physics.

Of course, waiting for improvement of the crystal quality for tens of years can be justified only in exceptional cases, but we propose to turn this perennial procedure of long-term ordering into the necessary one for the preparation of the top quality material for industrial electronics, which due to its unique properties will be used in electronic devices instead of a lot of various materials.

Acknowledgements

I am glad to note that the broad discussion and dissemination of our joint results stimulate the development of our further collaboration with reliable partners from the USA, Russia, Italy, Romania, France, and other countries. I express my cordial gratitude to my teachers, the late world-known scientists: Professor Nina A. Goryunova, Nobel Prize Laureate Alexander M. Prokhorov, Professors Rem V. Khokhlov, and Sergei I. Radautsan. The author is very grateful to the US Department of State, the Institute of International Exchange, Washington, DC, the US Air Force Office for Scientific Research, the US Office of Naval Research Global, Civilian R&D Foundation, Arlington, VA, Science & Technology Center in Ukraine, to his colleagues and co-authors from Clemson University, SC, University of Central Florida, FL, Istituto di elettronica dello stato solido, CNR, Rome, Italy, Universita degli studi, Cagliari, Italy, Lomonosov Moscow State University, Ioffe Physico-Technical Institute and State Polytechnical University, St. Petersburg, Prokhorov General Physics Institute of the Russian Academy of Sciences, Moscow, Russia, the Institute of Applied Physics of the Academy of Sciences of Moldova for support and attention to this protracted (1963–present time) research.

Author details

Sergei L. Pyshkin[1,2]

Address all correspondence to: spyshki@clemson.edu

1 Academy of Sciences of Moldova, Republic of Moldova

2 Clemson University, USA

References

[1] Goryunova NA, Pyshkin SL, Borshchevskii AS, et al. Influence of impurities and crystallisation conditions on growth of platelet GaP crystals. Symposium on Crystal Growth at the 7th Int Crystallography Congress (Moscow, July 1966). J Growth Crystals 1969; 8 68-72, ed. N.N. Sheftal, New York.

[2] Pyshkin SL, Radautsan SI, Zenchenko VP. Processes of long-lasting ordering in crystals with a partly inverse spinel structure. J Soviet Phys – Doklady 1990; 35(4) 301-4.

[3] Pyshkin SL. Preparation and properties of gallium phosphide. Ph.D. thesis 1967, State University of Moldova, Kishinev.

[4] Pyshkin SL, Zifudin L. Excitons in highly optically excited gallium phosphide. J Luminescence 1974; 9 302-8.

[5] Pyshkin SL (presented by Nobel Prize Laureate A.M. Prokhorov). Stimulated emission in gallium phosphide. J Soviet Physics – Doklady 1975; 19 845-6.

[6] Pyshkin S. Luminescence of GaP:N:Sm crystals. J Soviet Phys Semicond 1975; 8: 912-3.

[7] Pyshkin SL. Photoconductivity and luminescence of highly optically excited semiconductors. Dr.Sc. thesis 1978, Lomonosov Moscow State University.

[8] Pyshkin S, Anedda A, Congiu F, Mura A. Luminescence of the GaP:N ordered system. J Pure Appl Optics 1993; 2 499-502.

[9] Pyshkin S, Anedda A. Time-Dependent Behaviour of Antistructural Defects and Impurities in Cd-In-S and GaP. Institute of Physics Conference Series 1998, Ternary and Multinary Compounds; 152, Section E 785-9.

[10] Pyshkin S, Ballato J, Chumanov G. Raman light scattering from long-term ordered GaP single crystals. J Optics 2007, J Opt A Pure and Appl Optics; 9 33-6.

[11] Pyshkin SL, Ballato J, Bass M, Turri G. Luminescence of Long-Term Ordered Pure and Doped Gallium Phosphide. TMS Annual Meeting, invited talk at the Symposium: Advances in Semiconductor, Electro Optic and Radio Frequency Materials, March 9–13, New Orleans, LA. J Electron Mater 2008; 37(4) 388-95.

[12] Pyshkin SL, Ballato J, Bass M, Chumanov G. Time-dependent evolution of crystal lattice, defects and impurities in CdIn2S4 and GaP. J Phys Status Sol 2009; C(6) 1112-5.

[13] Pyshkin SL, Ballato J, Bass M, Turri G. Evolution of luminescence from doped gallium phosphide over 40 Years. J Electron Mater 2009; 38(5) 640-6.

[14] Pyshkin S, Ballato J. Evolution of optical and mechanical properties of semiconductors over 40 Years. J Electron Mater 2010; 39(6) 635-41.

[15] Pyshkin SL, Ballato J. Long-term convergence of bulk- and nano-crystal properties. In: Predeep P (ed.) Optoelectronics – Materials and Techniques 2011. Rijeka: InTech; 459-76.

[16] Pyshkin S, Ballato J. Dramatic expansion of luminescence region in GaP/polymer nanocomposites: Supplemental Proceedings, vol.1: "Materials Processing and Interfaces": 353-9, TMS Annual Meetings, Orlando, FL, USA, March 11-15, 2012.

[17] Pyshkin SL, Ballato J. Advanced light emissive device structures. In: Pyshkin SL and Ballato J (ed.) Optoelectronics – Advanced Materials and Devices 2013. Rijeka: InTech; 1-24.

[18] Pyshkin S, Ballato J, Chumanov G, DiMaio J, Saha AK. Preparation and Characterization of Nanocrystalline GaP: Technical Proceedings of the 2006 NSTI Nanotech Conference; 3 194-7.

[19] Pyshkin SL, Ballato J, Chumanov G. Optical Characterization of Long -Term Ordered and Nanocrystalline GaP, (invited), 3rd Int. Conf. on Materials Science and Condensed Matter Physics (Kishinev, Moldova, Oct 2006), Abstract Book, p172; Mold. J. of the Physical Sciences 2007; 6(2) 148-159.

[20] Pyshkin SL, Ballato J, Belevschii S, Rusu E, Racu A, Van DerVeer D. Synthesis and Characterization of GaP Nanoparticles for Light Emissive Devices: The 2011 NSTI-Nanotech Conference Proceedings: www.nsti.org, ISBN 978-1 4398-7142-3 1, 327-30, Boston, MA, June 13-16.

[21] Pyshkin SL, Ballato J, Bass M, Chumanov G, Turri G. Time-dependent evolution of crystal lattice, defects and impurities in $CdIn_2S_4$ and GaP. Phys Stat Sol 2009; C(6) 1112-5.

[22] Pyshkin S, Zhitaru R, Ballato J, Chumanov G, Bass M. Structural Characterization of Long-Term Ordered Semiconductors: The 2009 MS&T Conference Proceedings, International Symposium "Fundamentals & Characterization" 698-709, Pittsburgh, October 24-29.

[23] Pyshkin S, Ballato J, Bass M, Chumanov G, Turri G. Properties of the long-term ordered semiconductors: The 2009 TMS Annual Meeting and Exhibition Suppl. Proceedings; 3 477-84, San Francisco, February 15–19.

[24] Pyshkin S, Ballato J, Chumanov G, Bass M, Turri G, Zhitaru R, et al. Optical and Mechanical Properties of Long-Term Ordered Semiconductors. Moldavian J Phys Sci 2009; 8(3-4) 287-95. The 4th Int Conference on Materials Science and Condensed Matter Physics, Kishinev, Sept 23-26, 2008.

[25] Pyshkin SL. Gallium Phosphide - New Prospect for Optoelectronics. J Adv Optoelectron Mater 2013 (ISSN Online: 2327-7661, ISSN Print: 2327-767X); 1(4) 59-66.

[26] Pyshkin SL. Excitonic crystal and nanotechnology. Symbiosis J Nanosci Technol: Open Access 2014; 1(2), http://symbiosisonlinepublishing.com/nanoscience-technology08.pdf.

[27] Pyshkin SL. Excitonic Crystal, Nanotechnology and New Prospect for Optoelectronics, J Open Optics 2015; 9 26-37.

[28] Pyshkin S, Ballato J. Advanced Light Emissive Composite Materials for Integrated Optics: The 2005 MS&T Conference Proceedings, Symposium: The Physics and Materials Challenges for Integrated Optics - A Step in the Future for Photonic Devices 3-13, Pittsburgh, PA.

[29] Pyshkin S, Ballato J, Luzinov I, Zdyrko B. Fabrication and Characterization of the GaP/Polymer Nanocomposites for Advanced Light Emissive Device Structures: The 2010 NSTI-Nanotech Conference Proceedings; 1 772-5. www.nsti.org, ISBN 978-1-4398-3401-5, Anaheim, CA, Jun 21-24, 2010.

[30] Pyshkin S, Ballato J, Chumanov G, Tsyntsaru N, Rusu E. Preparation and Characterization of Nanocrystalline GaP for Advanced Light Emissive Device Structures: The 2010 NSTI-Nanotech Conference Proceedings; 1 522-5, www.nsti.org, ISBN 978-1-4398-3401-5.

[31] Pyshkin SL, Ballato J, Luzinov I, Zdyrko B. Fabrication and characterization of GaP/polymer nanocomposites for advanced light emissive device structures. J Nanoparticle Res 2011; 13 5565-70, ISSN: 1388-0764, doi: 10.1007/s11051-011-0547-0.

[32] Pyshkin Sergei L (Project Manager). Advanced Light Emissive Device Structures. STCU (www.stcu.int) Project 4610, 2009-2012.

[33] Shanmin Gao, et al. Aqueous synthesis of III-V semiconductor GaP and InP exhibiting pronounced quantum confinement. Chem Commun 2002 3064-5.

[34] Zhengang Liu, Yujun Bai, Deliang Cui, Qilong Wang. Preliminary Insight into the formation process of InP and GaP nanocrystals. Solid State Sci 2003; 5 1037-40.

[35] Zhao-Chun Zhang, Bao-Ping Wang. Diamond and oxidized disordered graphite on the surface of gallium phosphide nanoparticles. Syst Charact 2009; 26 53–7.

[36] Paul W. Band structure of the intermetallic semiconductors from pressure experiments. J Appl Phys 1961; 32 2082-94.

[37] Zallen R, Paul W. Band structure of gallium phosphide from optical experiments at high pressure. Phys Rev 1964; 134 A1628-41.

[38] Dean PJ, Thomas DG. Intrinsic absorption-edge spectrum of gallium phosphide. Phys Rev 1966; 150(2) 690-702.

[39] Kleinman DA, Spitzer WG. Infrared lattice absorption of GaP. Phys Rev 1960; 118 (1) 110-7.

[40] Allen WJ. Energy levels of nitrogen-nitrogen pairs in gallium phosphide. J Phys C 1968, 1 1136-8.

[41] Bairamov B, Kitaev Y, Negoduiko V, Khashkhozhev Z. Light scattering by optical phonons and non-linear optical properties of GaP. J Soviet Phys Solid State 1975; 16 1323-7.

[42] Pyshkin S, Radautsan S, Zenchenko V. Raman spectra of Cd-In-S with different cation-sublattice ordering. J Soviet Phys Doklady 1990; 35(12) 1064-67.

[43] Bell RL. A proposal for a DC pumped rare-earth laser. J Appl Phys 1963; 34 1563-4.

[44] Gershenzon M, et al. Evidence for radiative recombination between deep donor-acceptor pairs in GaP at room temperature. J Appl Phys 1966; 37(2) 483-6.

[45] Thomas DG, Hopfield JJ, Frosch CJ. Isoelectronic traps due to nitrogen in gallium phosphide. Phys Rev Lett 1965; 15 857-9.

[46] Pyshkin SL, et al. Preparation and structural properties of some III-V semiconductor films grown on (100) oriented Si substrates. J Appl Surf Sci 1992; 56-58 39-43.

[47] Budyanu VA, Chechuy SN, Pyshkin SL, et al. Investigation of III-V/Si heterojunction grown by laser deposition. Phys Stat Sol (a) 1985; 91 737-44.

[48] Budyanu VA, Chechuy SN, Pyshkin SL, et al. Laser vacuum epitaxy of III-V semiconductors on silicon. Rev Roumaine de Physique 1987; 32(1-2) 216-9.

[49] Bloembergen N. Nonlinear Optics. Edition: 4th. Paperback – June, 1996.

[50] Ashkinadze BM, Pyshkin SL, Radautsan SI. Some Non-linear Optical Effects in GaP: The IX Int. Conf. on the Physics of Semiconductors Proceedings; 2 1189-93, Moscow, July 1968.

[51] Pyshkin SL, et al. Multi-quantum absorption in GaP. Opto-electronics 1970; 2 245-9.

[52] Ashkinadze BM, Kretsu IP, Pyshkin SL, et al. Luminescence of Gallium phosphide near an indirect transition in the multiphoton excitation case. Soviet Phys Semicond 1969; 2(10) 1261-3.

3

Physical Characteristics, Sensors and Applications of 2D/3D-Integrated CMOS Photodiodes

Oscal T.-C. Chen, Yi-Yang Lee and Robin R.-B. Sheen

Abstract

Two-dimensional photodiodes are reversely biased at a reasonable voltage whereas 3D photodiodes are likely operated at the Geiger mode. How to design integrated 2D and 3D photodiodes is investigated in terms of quantum efficiency, dark current, crosstalk, response time and so on. Beyond photodiodes, a charge supply mechanism provides a proper charge for a high dynamic range of 2D sensing, and a feedback pull-down mechanism expedites the response time of 3D sensing for time-of-flight applications. Particularly, rapid parallel reading at a 3D mode is developed by a bus-sharing mechanism. Using the TSMC 0.35μm 2P4M technology, a 2D/3D-integrated image sensor including P-diffusion_N-well_P-substrate photodiodes, pixel circuits, correlated double sampling circuits, sense amplifiers, a multi-channel time-to-digital converter, column/row decoders, bus-sharing connections/decoders, readout circuits and so on was implemented with a die size of 12mm×12mm. The proposed 2D/3D-integrated image sensor can perceive a 352×288-pixel 2D image and an 88×72-pixel 3D image with a dynamic range up to 100dB and a depth resolution of around 4cm, respectively. Therefore, our image sensor can effectively capture gray-level and depth information of a scene at the same location without additional alignment and post-processing. Finally, the currently available 2D and 3D image sensors are discussed and presented.

Keywords: CMOS photodiode, active pixel, Geiger mode, time of flight, image sensor

1. Introduction

Nowadays, 3-dimensional (3D) images and videos are widely used in various applications, like games, robotics, cinema and so on. How to effectively capture 3D information becomes a critical issue. In general, stereo images from two slightly different viewpoints are used to establish 3D perception of which mechanism is based on binocular parallax. For example, a 3D camera of Panasonic LUMIX® DMC-3D1K takes two pictures simultaneously and then displays these two pictures to the left and right eyes of a human for 3D perception [1]. Since the parallax of two viewpoints for capturing is fixed, 3D perception may not be easily adapted to different viewpoints.

Another approach also adopts two cameras in which one captures a gray-level image, and the other seizes object depths where an active light source, like infrared, is utilized. Based on the characteristics of light reflected from objects, a 3D image can be built via post-processing computation. For instance, Kinects I and II from Microsoft Xbox use the structured light technique and Time Of Flight (TOF) technique to acquire object depths, respectively [2], [3]. Particularly, the TOF technique is used to estimate positions of objects in space according to the travel time of light emitted from a light source, reaching an object, reflected from an object and arriving at a sensor [4], as depicted in Fig. 1. Based on the different travel time captured by photodiodes, a depth map of a scene can be fairly attained. The information from the luminance (2D) and depth (3D) cameras is used to reconstruct complete 3D pictures which are observed from multiple viewpoints. Since these two cameras may be located at different positions, the capture-point difference need be compensated to yield a correct 3D picture.

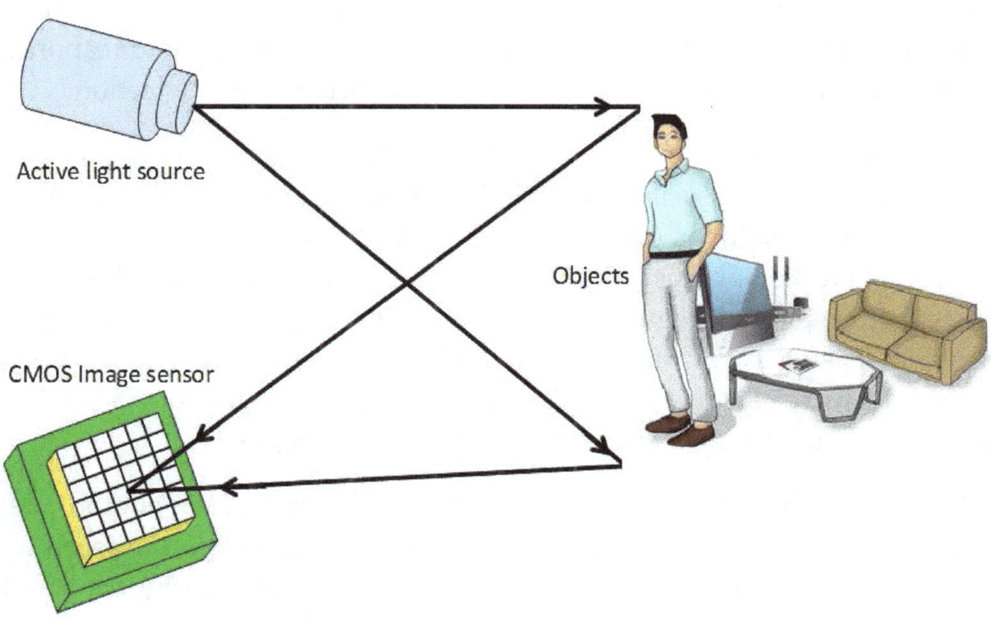

Figure 1. Concept of TOF.

In this chapter, physical characteristics of CMOS photodiodes are illustrated first. How to design high-efficiency photodiodes for 2D and 3D sensing is addressed. Second, a 2D/3D-integrated active pixel is presented where the same photodiodes are shared by 2D and 3D capturing. Additionally, the readout circuits used for 2D and 3D image sensors are integrated. Particularly, two feedback mechanisms are adopted to delay saturation of 2D luminance sensing, and accelerate response time of 3D depth sensing. Third, in order to reach a compact pixel array, high readout speed and the minimized coupling effect between transmission lines and photodiodes, the mechanism associated with bus sharing is developed to effectively accomplish parallel reading. Fourth, based on the TSMC 0.35μm 2P4M CMOS technology, the proposed 2D/3D-integrated image sensor was implemented with a die size of 12mm×12mm where 2D and 3D modes have the resolutions of 352×288 and 88×72 pixels, respectively, under a fill factor of 44%. Finally, the proposed image sensor, and the currently available 2D and 3D image sensors are illustrated, compared and concluded.

2. Physical characteristics of CMOS photodiodes for 3D sensing

The *p-n* junction of a diode under a reversely biased voltage is employed to perceive photos. Such a diode is named as a photodiode. Due to the electrical field of the reversely biased voltage, the depletion region between *p* and *n* layers is created. The size of the depletion region is dependent on the amount of the reversely biased voltage, and the doping concentrations of *p* and *n* layers. This depletion region incited by a photo can yield a pair of electron and hole which are drifted and guided by the electrical field to become a photocurrent. When the reversely biased voltage approaches the breakdown voltage, the depletion region becomes large to increase the incitation rate of electrons and holes, and the high electrical field expedites the drifting of electrons and holes and reduces the recombination chance of electrons and holes. This operation manner is called as the Geiger mode [5], which has a rapid response to incident light. Restated, slight light can induce a sufficient photocurrent for detection.

A 3D image can be captured by the time of flight technique where object depths are interpreted by the time of round trips associated with light that originates from a light source, like light emitting diode (LED), shining objects, and is reflected from objects. To effectively seize reflected light, photodiodes must quickly respond to photos owing to high speed of light. Hence, an avalanche photodiode which operates at the Geiger mode can be a good choice to accomplish TOF sensing. An avalanche photodiode has high sensitivity to precisely apprehend slight photos which are early birds. Under such physical characteristics, a 3D depth map can be successfully attained from an array of avalanche photodiodes.

Many avalanche complementary metal-oxide semiconductor (CMOS) photodiodes have been explored and proposed in the literature. For instance, Niclass *et al.* proposed a 3-D imager based on a 2-D array of single-photon avalanche diodes fabricated by the standard CMOS technology where sub-millimetric precision could be achieved under a depth-of-field of several meters [6]. Marshall *et al.* realized a CMOS 64×64 pixel array in which avalanche photodiodes and active pixel sensors were integrated [7]. Zappa *et al.* adopted a standard 0.8-μm CMOS technology to realize an integrated sensor consisting of photodiodes, input sensing circuit, photon-counting and control circuits where an active quenching and active reset circuit were implemented to drive a single-photon avalanche diode [8]. Faramazpour *et al.* developed avalanche photodiodes and their corresponding driving circuits under a standard 0.18-μm CMOS technology where the breakdown voltage was 10.2V and the dead time was 30ns [9]. Atef *et al.* implemented two photodiode structures using the 40-nm standard CMOS technology where one could perform like an avalanche photodiode and the other functioned as a regular photodiode [10]. Pancheri *et al.* presented a low-noise avalanche photodiode based on the graded junction in 0.15-μm CMOS technology [11]. The other studies focused on ameliorating the process, materials and doping concentrations in order to minimize noise, to lower dark currents and to enhance sensitivity [12]-[15].

By using the TSMC 0.35μm CMOS technology, P-diffusion_N-well_P-substrate, N-well_P-substrate and N-diffusion_P-substrate photodiodes are explored to understand the physical characteristics where their cross sections are shown in Fig. 2 [16]. Figures 3 and 4 depict breakdown voltages of P-diffusion_N-well_P-substrate and N-well_P-substrate photodiodes under the same area sizes with different perimeters, respectively. The measurement results reveal that the breakdown voltage becomes larger as the perimeter is decreased. Additionally, the breakdown voltage of a P-diffusion_N-well_P-substrate photodiode is smaller than that of an N-well_P-substrate photodiode. The photodiode with a deeper p-n junction, larger area and smaller perimeter has a higher breakdown voltage. Particularly, the lateral region of a photodiode may form a dead space which brings about the decrease of a fill factor, quantum efficiency and breakdown voltage [16]. From the abovementioned phenomena, a circular photodiode with the maximum ratio of area to perimeter is likely preferred to attain a high breakdown voltage.

As well as the boundary effect, the efficiency of induced photocurrents is examined where a light source adopts a laser at 850nm. Referring to [17], the current gain can be expressed as

$$G = (I_{photo\_Geiger} - I_{dark\_Geiger}) / (I_{photo\_typical} - I_{dark\_typical}) \tag{1}$$

$I_{photo\_typical}$ and $I_{dark\_typical}$ are the induced and dark currents of a photodiode at a typical reversely biased voltage, respectively. This typical voltage can be V_{dd}, which is a supply voltage of circuits at the currently used process technology. Restated, it is the maximum reversely biased voltage for 2D sensing of a photodiode. Additionally, I_{photo_Geiger} and I_{dark_Geiger} are the induced and dark currents of a photodiode at the Geiger mode, respectively. The measurement results, depicted in Fig. 5, reveal that the current gain of N-diffusion_P-substrate is the best, and the second is P-diffusion_N-well_P-substrate where each of three photodiodes has the area of 30μm×30μm, and V_{dd} is 3.3V. Due to high variations of currents, the logarithmic scale of 'log(fA)' at the left Y axis is adopted to clearly present the measured currents.

In addition to the breakdown voltage of a photodiode, the crosstalk effect, as shown in Fig. 6, need to be well investigated to minimize the interference of the neighboring pixels [18], [19]. Figure 7 displays a top view of 3×3 photodiodes, with and without an N-well guard ring, which were implemented by the TSMC 0.35μm CMOS technology. The photodiode in the center of a 3×3 photodiode array is measured when the neighboring eight photodiodes are biased at different voltages. The measured breakdown voltages of 3×3 P-diffusion_N-well_P-substrate and N-well_P-substrate photodiodes with and without an N-well guard ring are depicted in Fig. 8. The breakdown voltage of a photodiode with a guard ring is greater than that of a photodiode without a guard ring. Notably, the central pixel likely has a higher breakdown voltage as its neighboring pixels are reversely biased to a higher voltage. Figure 9 shows breakdown voltages of 3×3 P-diffusion_N-well_P-substrate, N-well_P-substrate and N-diffusion_P-substrate photodiodes with N-well guard rings while neighboring pixels are reversely biased at voltages from 0V to 9V. The N-well_P-substrate photodiode yields the largest breakdown voltage, the second is the P-diffusion_N-well_P-substrate photodiode and the last one is the N-diffusion_P-substrate photodiode.

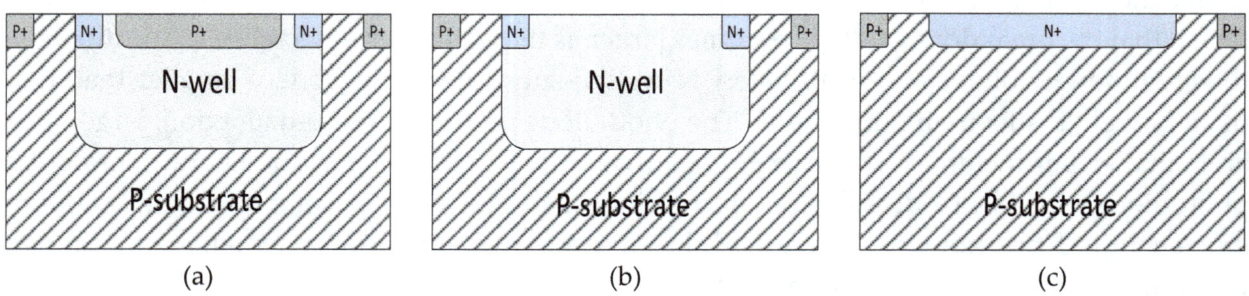

Figure 2. Cross sections of photodiodes. (a) P-diffusion_N-well_P-substrate. (b) N-well_P-substrate. (c) N-diffusion_P-substrate photodiodes.

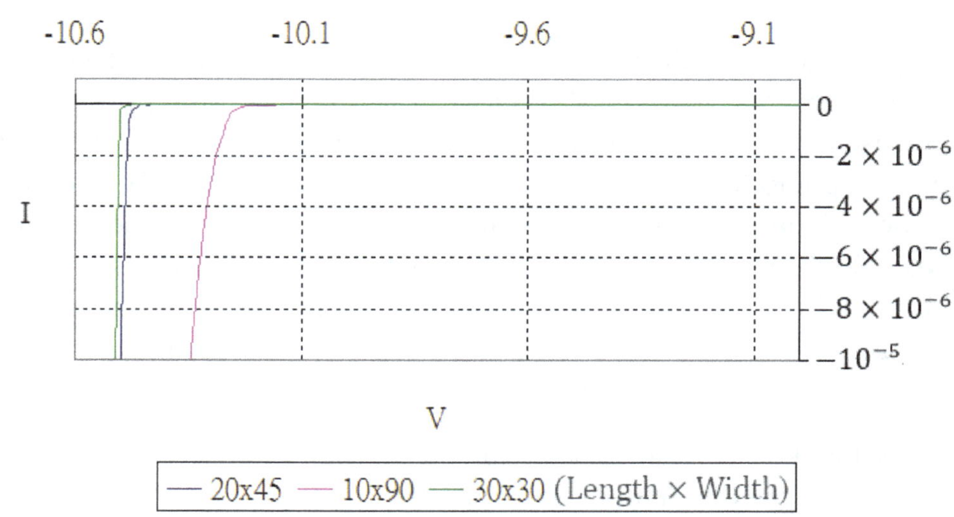

Figure 3. Breakdown voltages of P-diffusion_N-well_P-substrate photodiodes.

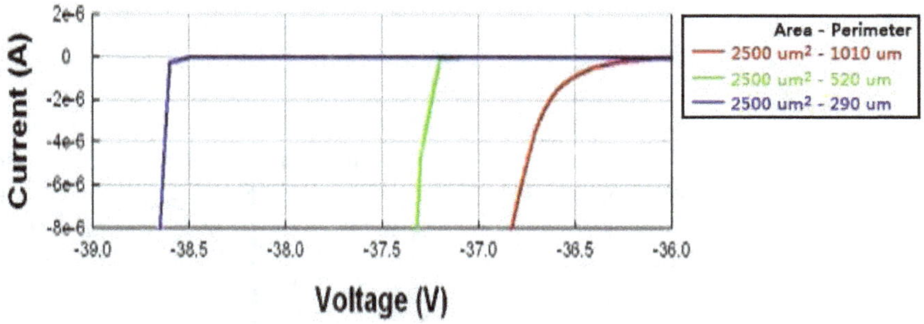

Figure 4. Breakdown voltages of N-well_P-substrate photodiodes.

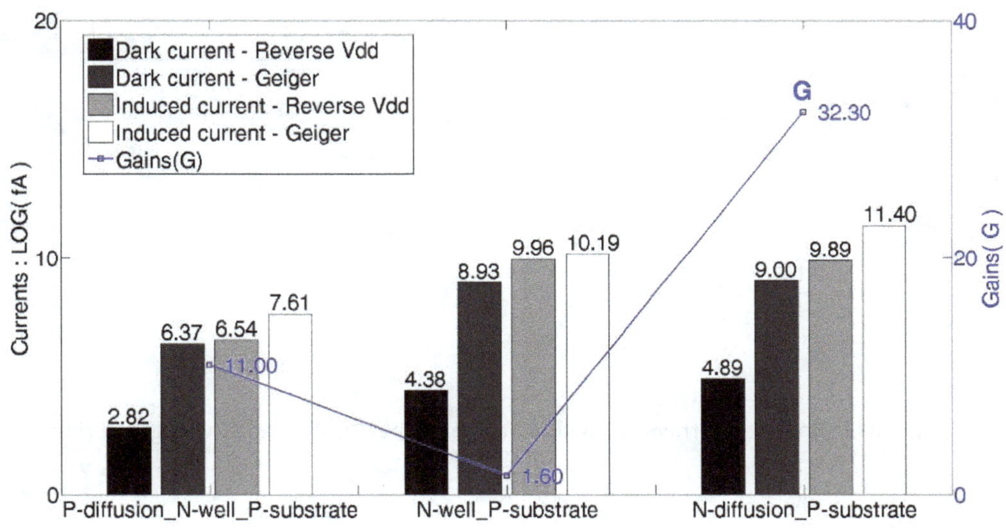

Figure 5. Induced currents, dark currents and gains of photodiodes at an 850nm laser.

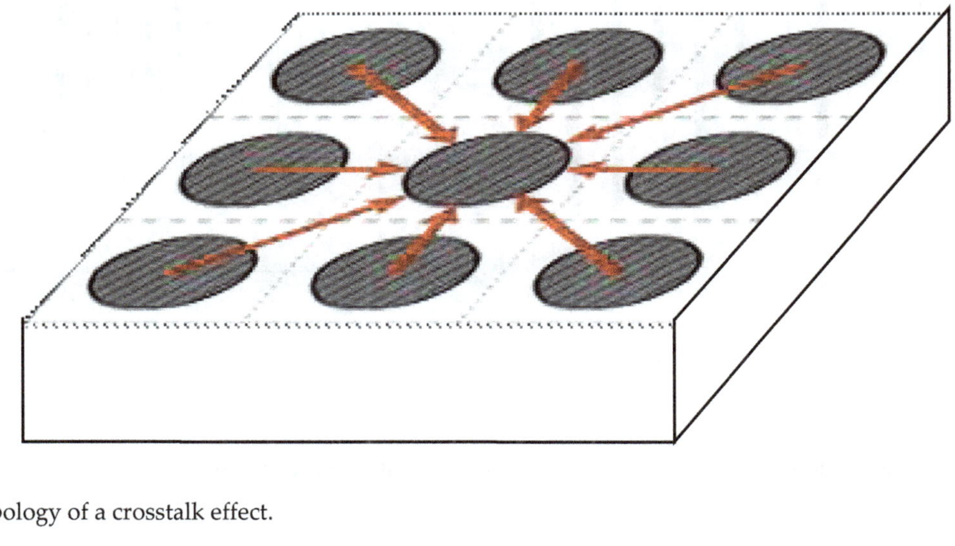

Figure 6. Topology of a crosstalk effect.

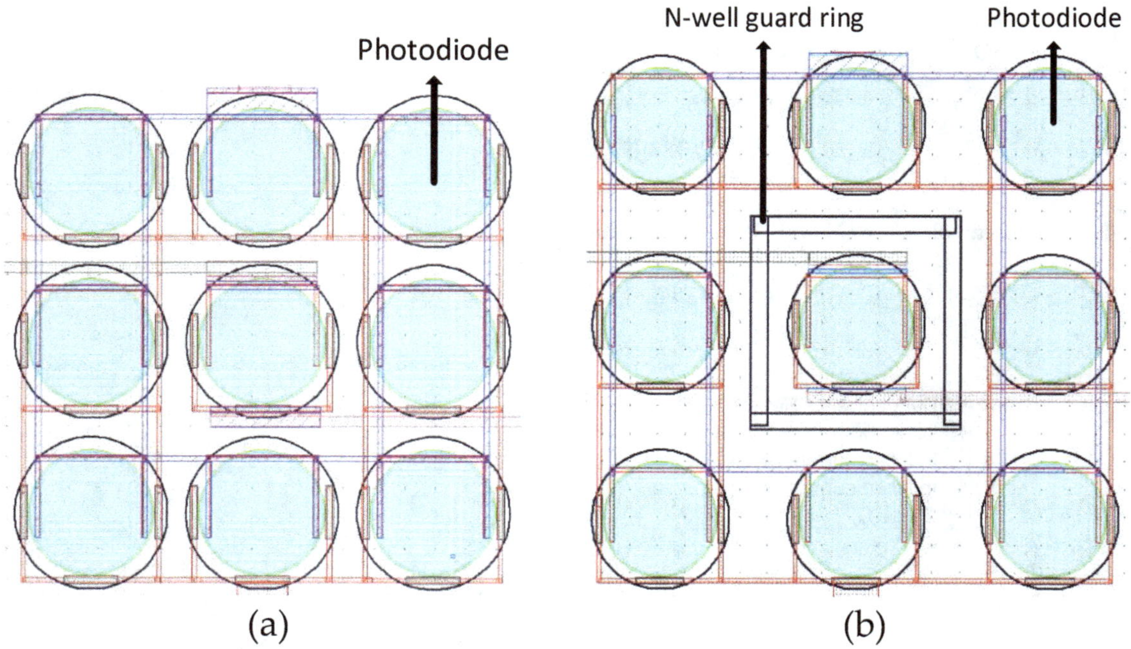

Figure 7. Top view of 3×3 photodiode arrays. (a) Without a guard ring. (b) With an N-well guard ring.

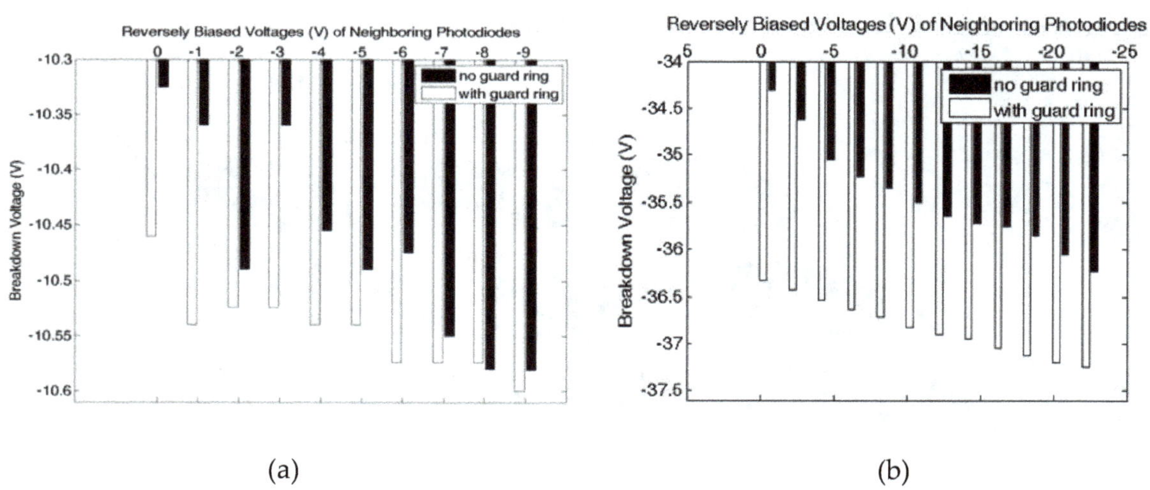

Figure 8. Measured breakdown voltages. (a) 3×3 P-diffusion_N-well_P-substrate photodiodes. (b) 3×3 N-well_P-substrate photodiodes.

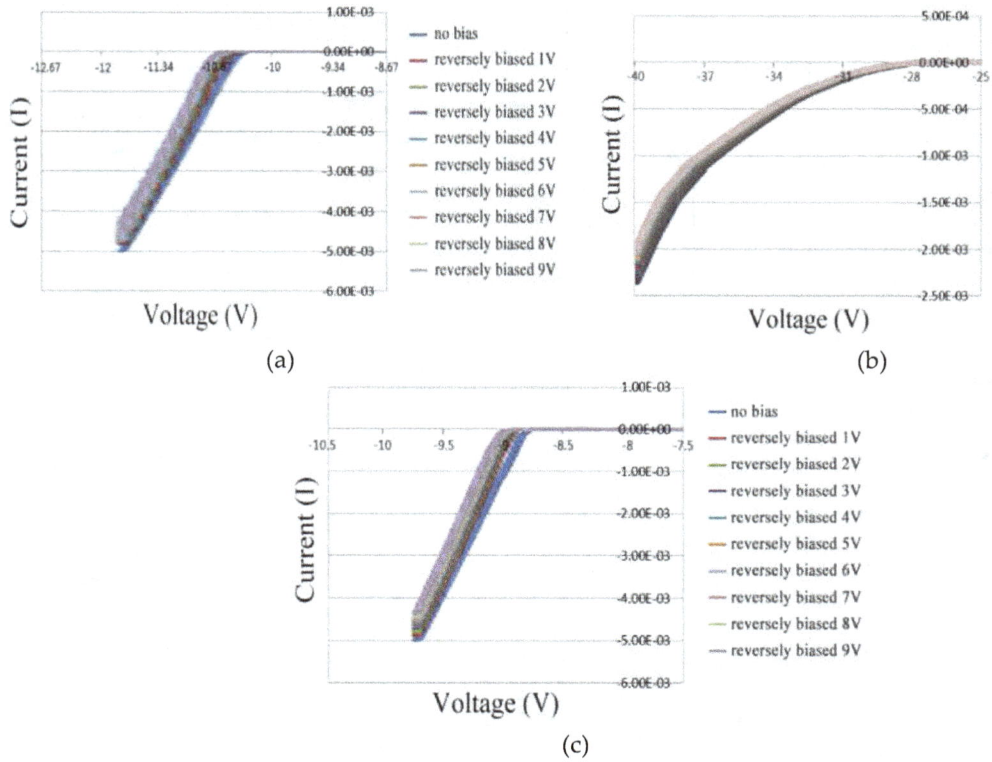

Figure 9. Measured breakdown voltages. (a) P-diffusion_N-well_P-substrate. (b) N-well_P-substrate. (c) N-diffusion_P-substrate.

3. 2D/3D-integrated pixel

A 3D depth sensor can yield a depth map of a scene but fails to provide delicate gray-scale information. However, a 2D image sensor can effectively interpret a pixel at a fairly fine gray scale rather than depth resolution. Accordingly, there is a demand for using multiple 2D or/and 3D sensors to capture a real-world scene and then to display a 3D picture to a watcher. For example, if two 2D image sensors are employed to mimic binocular vision, 3D perception adheres to a specific viewpoint. Nevertheless, if a 2D image sensor and a 3D depth sensor are used together to capture objects, the difference of viewpoints from these two sensors at different positions has to be amended. Additionally, the camera including two individual sensors takes relatively high hardware cost accompanied with great power consumption. Therefore, we propose to realize 2D luminance and 3D depth perceptions at one sensor of which 2D and 3D operations are alternately executed. The concept of how to design the proposed 2D/3D-integrated image sensor is illustrated in Fig. 10 [20].

The first step of integrating a 2D image sensor and a 3D depth sensor is to design a photodiode that can be shared by these two sensors. Figure 11 depicts the measured spectrum responses of P-diffusion_N-well_P-substrate, N-well_P-substrate and N-diffusion_P-substrate photodiodes using the TSMC 0.35μm CMOS technology where the reversely biased voltage is zero. From the measured results, N-well_P-substrate and P-diffusion_N-well_P-substrate photodi-

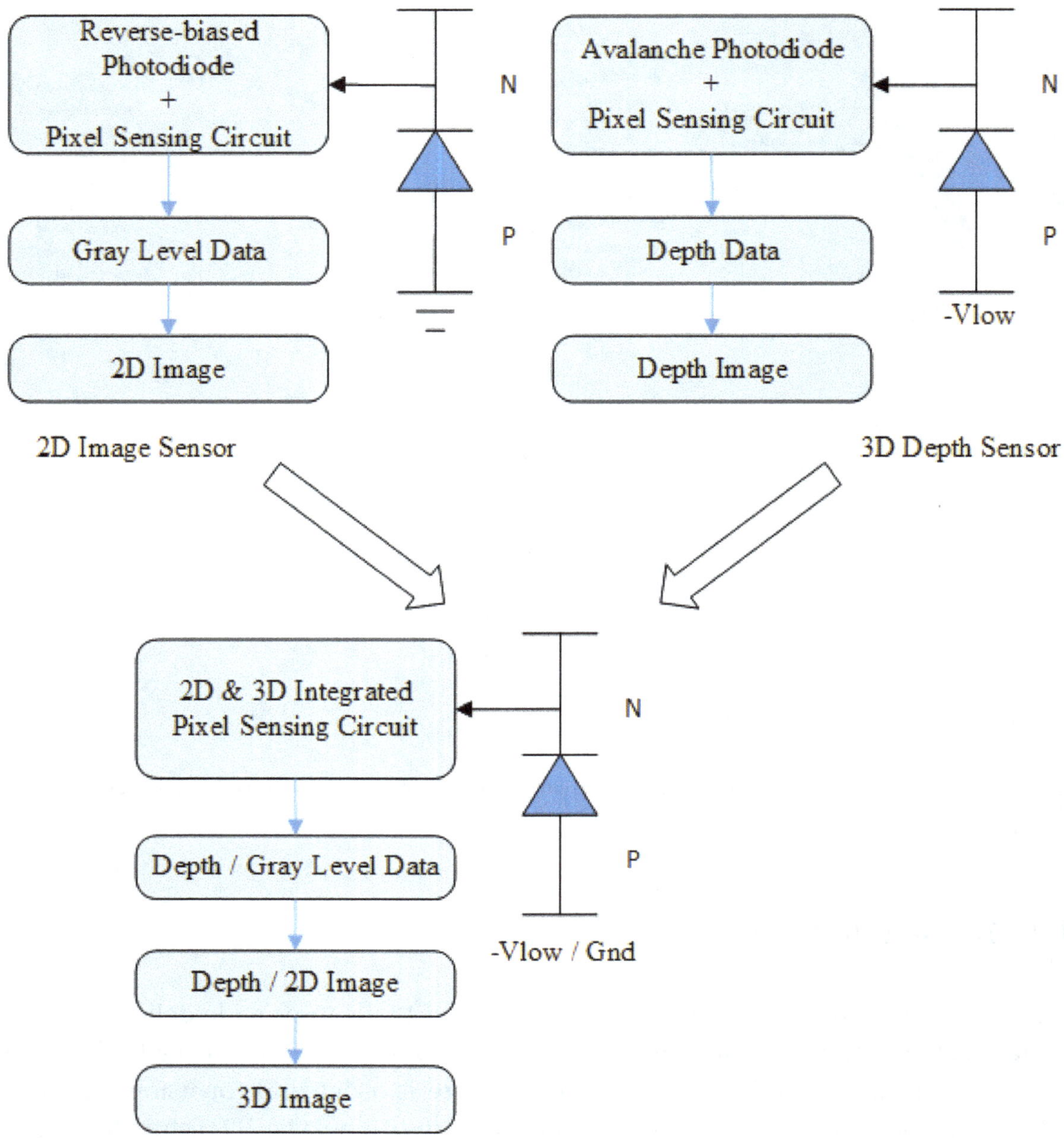

Figure 10. Design methodology of the 2D/3D-integrated sensor.

odes are the best and second best, respectively. In the following, the operations of photodiodes associated with 2D image and 3D depth sensors are studied to determine which photodiode structure is preferred.

3.1. Photodiodes of 2D and 3D sensors

When a PN photodiode is biased at a reverse voltage, incident light reaches the depletion region of this photodiode, and then incites numerous pairs of holes and electrons which create a photocurrent. The light intensity increases with the induced photocurrent. Based on an exposure period, the sensing circuit of a pixel converts an integrated photocurrent to an analogue voltage which represents gray-level luminance. Such a mechanism accomplishes a

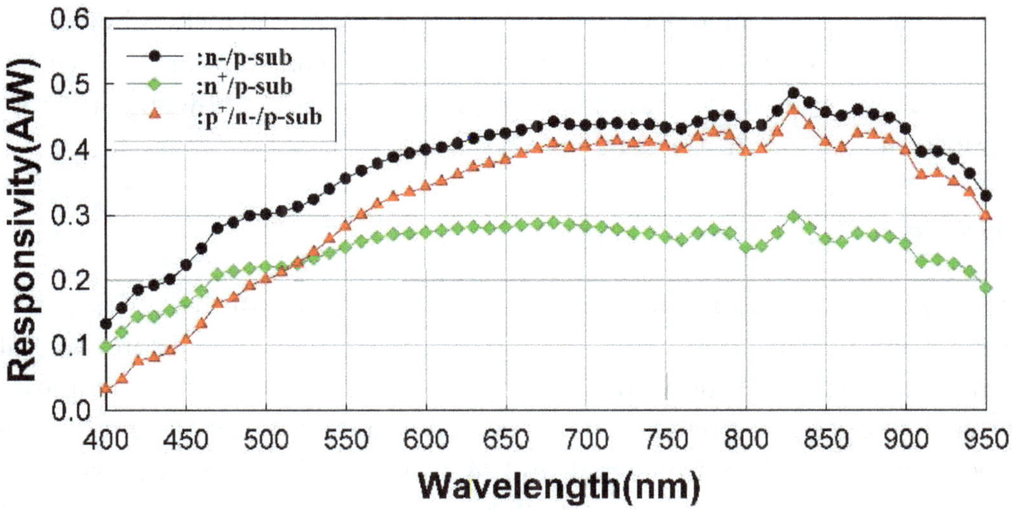

Figure 11. Measured spectral responses of three CMOS photodiodes.

2D image sensing. While the reversely biased voltage approaches the breakdown voltage of a photodiode, this photodiode has relatively large electrical field and depletion region in which pairs of holes and electrons are easily generated by few photons. This kind photodiode operating at the Geiger mode is to acquire a 3D depth map where the time of the flight trip associated with light emitted from a source, reaching objects, reflected from objects and arriving at photodiodes is computed.

Physical characteristics of photodiodes with different geometric and junction structures must be well understood in order to effectively integrate 2D and 3D photodiodes. From our previous study [16], the corner in the geometric shape of a photodiode easily leads to breakdown because of charged elements likely gathering at this point. Accordingly, the number of corners in a photodiode increases, resulting in the decrease of a breakdown voltage. Furthermore, the dark current is correspondingly enlarged as the number of corners increases. In the previous section, we conclude that the area and peripheral of a photodiode are somehow proportional and inversely proportional to the breakdown voltage, respectively. The increased number of pixels in an array lowers the breakdown voltage as well. Therefore, a photodiode with a circle shape and a guard ring is chosen to achieve high-efficiency sensing capability.

According to quantum efficiency of photodiodes in Fig. 11, N-well_P-substrate and P-diffusion_N-well_P-substrate photodiodes are good candidates. The reversely biased voltages of a photodiode used for 2D and 3D sensing are quite different. When the N-well_P-substrate photodiode is adopted for 2D and 3D sensors, P-substrate must be driven by two different voltages. However, P-substrate is a common base which is usually shared by transistors of all circuits. Under the standard CMOS technology, P-substrate is always connected to the ground. Hence, the N-well_P-substrate photodiode does not satisfy our demand. In the P-diffusion_N-well_P-substrate photodiode, P-diffusion and P-substrate are addressed by the ground to form two PN junctions for a 2D sensing where P-diffusion_N-well and N-well_P-substrate work for short-wavelength and long-wavelength light receiving, respectively. When a 3D sensing is conducted, P-diffusion and P-substrate are biased by a large negative voltage and 0V, respectively. Such a biasing manner makes P-diffusion_N-well an avalanche photodiode and N-well_P-substrate a photodiode under a normal reversely biased voltage. Once few photos reach the P-diffusion_N-well_P-substrate photodiode, P-diffusion_N-well reacts rapidly, and N-well_P-substrate may not respond too much.

3.2. Pixel circuits

To capture 2D gray-level information, the dynamic range is one of the key issues associated with a pixel design. A pixel circuit with a larger dynamic range can interpret a greater range of light luminance. Figure 12 shows the pixel circuit at a 2D mode, which employs an extra path to provide charges [21]. Such a path includes two transistors of M_2 and M_3 where M_2 is manipulated by an external control signal, and M_3 functions like an active resistor with gate-node sensing. $I_{charge\text{-}supply}$ goes through this path to compensate the current sink from $I_{photo,}$ and thus to delay the gate node of M_6 to become 0V. The simulations in Fig. 13 reveal that the pixel circuit with a charge supply mechanism can reach up to 110dB and greatly extend the dynamic range than without a charge supply mechanism.

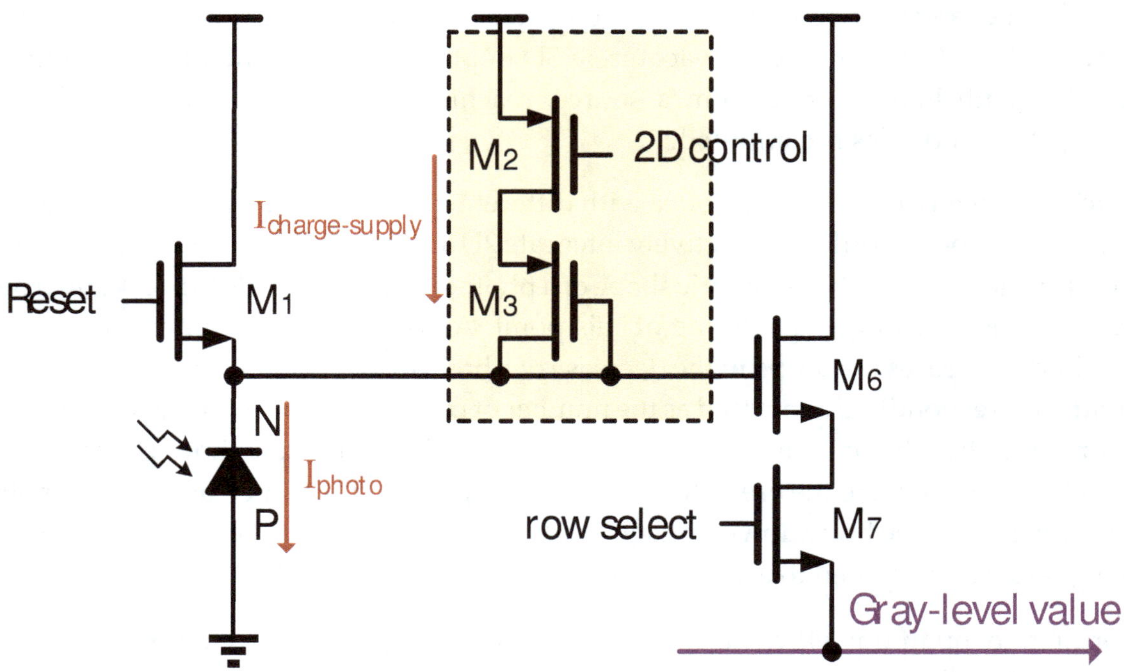

Figure 12. 2D pixel circuit.

The pixel circuit at a 3D mode is to discover objects associated with depth information. The photodiode in a pixel is reversely biased close to a breakdown voltage. Once the photodiode is triggered by few photons, a significant current is generated. Such an instant and great current accelerates the response of the pixel circuit. Referring to a simple passive quenching structure [22], we proposed a modified 3D pixel circuit, as depicted in Fig. 14. When a slight light reaches this 3D pixel, a great current is induced to make a voltage drop at the *N* node of a photodiode. This voltage drop is expedited by M_4, M_5 and an inverter as well. The timing diagram of the

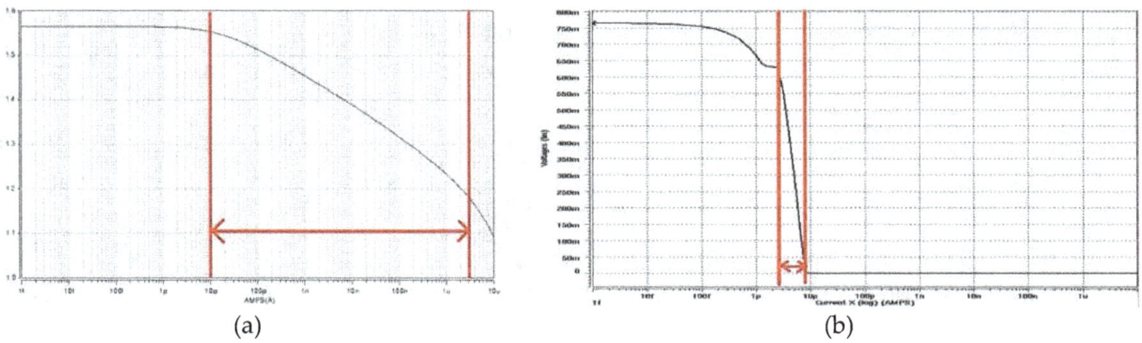

Figure 13. Simulated dynamic range of a 2D pixel circuit. (a) With a charge supply mechanism. (b) Without a charge supply mechanism.

proposed 3D pixel circuit is depicted in Fig. 15. During the reset, the *N* node of a photodiode is charged to a voltage, above $\frac{1}{2}V_{dd}$, which is within the input noise margin of an inverter outputting 0V. The source and gate nodes of M_4 are connected to the input and output of an inverter, respectively, to create a feedback loop. Additionally, M_5 is linked to the drain node of M_4, and is manipulated by a signal of '3D control' to build a pull-down path. After the reset, the photodiode begins to perceive photos. Once few photos are detected, a discharge action is taken. At the same time, M_4 and M_5 are activated to accelerate the inverter dropping to $1/2V_{dd}$. Such a pull-down path lessens the time interval associated with light detection, and thus raises the depth resolution. Figure 16 shows the circuit diagram and layout of the proposed 2D/3D-integrated pixel that is easily addressed by using the 2D and 3D control signals to switch its operation modes.

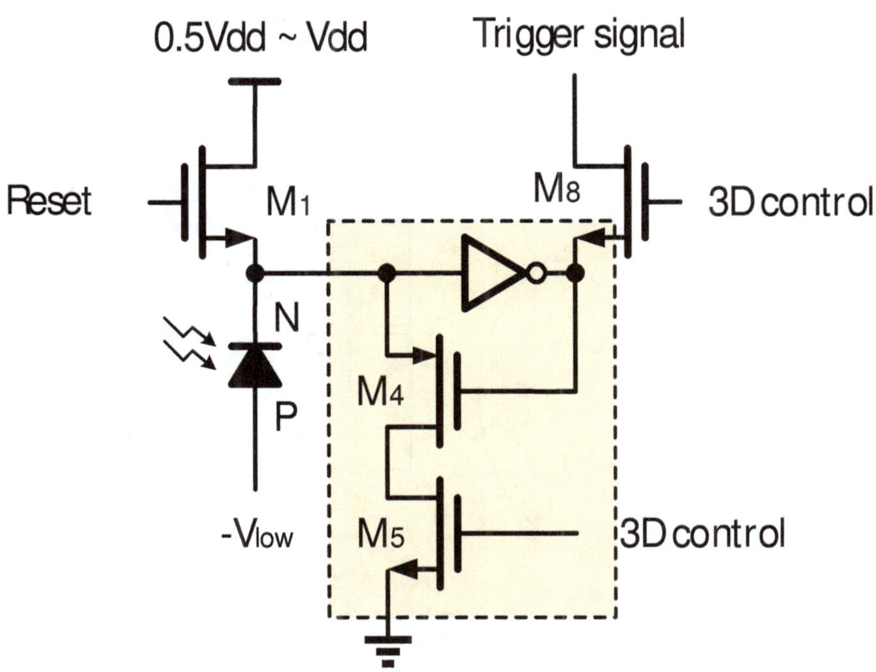

Figure 14. 3D pixel circuit.

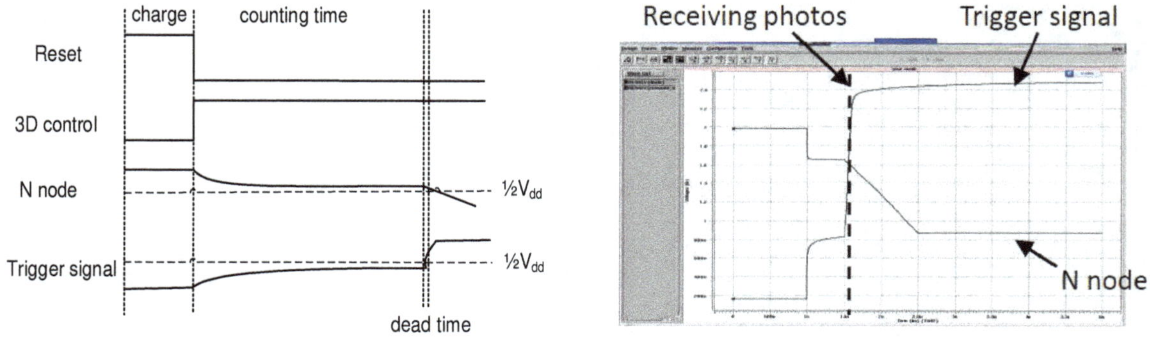

(a) (b)

Figure 15. Timing diagram of a 3D pixel circuit. (a) Timing chart. (b) Simulation chart.

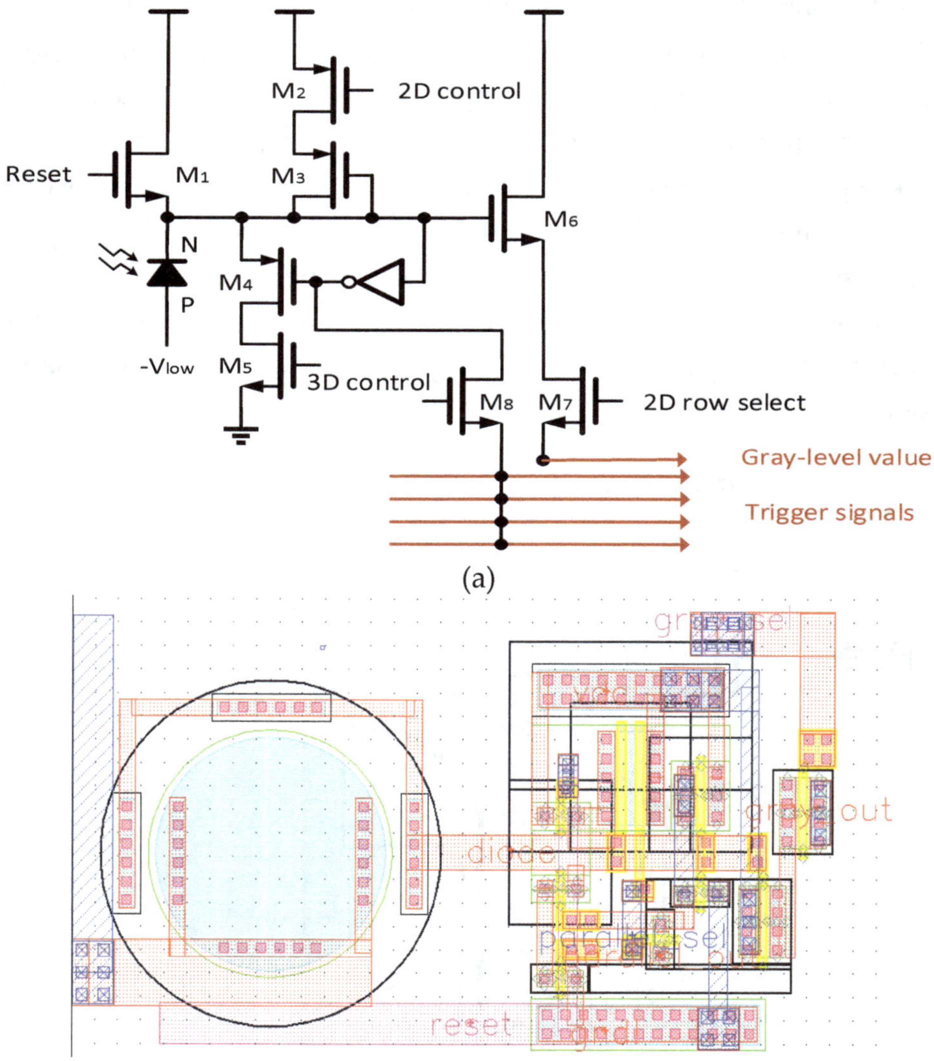

(a)

(b)

Figure 16. 2D/3D-integrated pixel circuit. (a) Circuit diagram. (b) Layout diagram.

4. Parallel reading using the bus-sharing mechanism

When the proposed 2D/3D-integrated pixel circuit is operated at a 3D mode, a pulse signal is triggered by the perceived photos. Afterwards, the time difference between light emitted and detected is calculated, and then the corresponding depth information is derived based on a velocity of light. An independent bus line from each pixel of a large pixel array may not be addressed at low hardware cost. Hence, there is a need for low-cost parallel reading and computation in time. Nevertheless, as the size of a pixel array goes up, the hardware complexity of parallel reading is extremely expanded. Not only is the area increased, but the coupling effect between photodiodes and transmission lines is likely induced as well. According to the conventional work [23], [24], the reading of trigger pulses is fulfilled at a time-multiplexed manner.

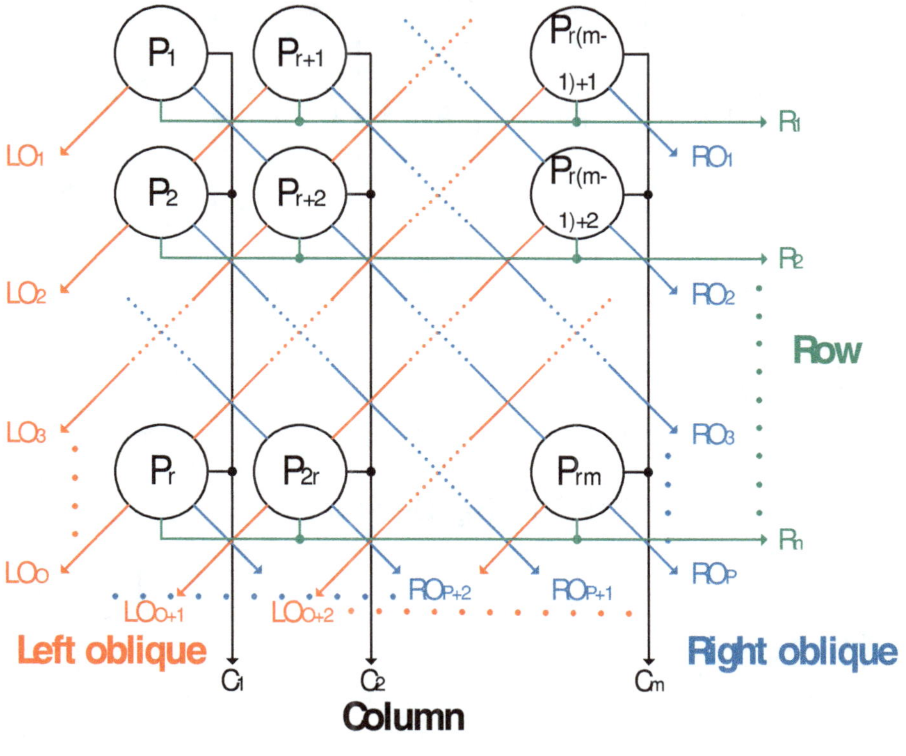

Figure 17. Parallel reading of the bus-sharing mechanism.

To overcome the abovementioned drawbacks, a bus-sharing mechanism is proposed to realize parallel reading at low hardware cost. This bus-sharing mechanism employs a connection topology in which each pixel connects to multiple shared buses, as displayed in Fig. 17. Since each bus is connected to multiple photodiodes, a decoder associated with the shared buses is demanded to determine which photodiode is activated. Restated, when buses are enabled by one or multiple photodiodes simultaneously, these buses become V_{dd}. Based on the pattern of the enabled buses, the locations of photodiodes are discovered. For instance, while light is sensed by a pixel, P_1, the buses of LO_1, C_1, R_1 and RO_p are pulled up to V_{dd}. As P_2 also observes light, it enables the buses of LO_2, C_1, R_2 and RO_{p1}. Although there is a bus of C_1 shared by P_1

and P_2, the other three buses are quite different, and can be used to distinguish P_1, P_2 and others. Therefore, according to the enabled buses and triggered time, when and which photodiode(s) are activated can be decoded and ascertained.

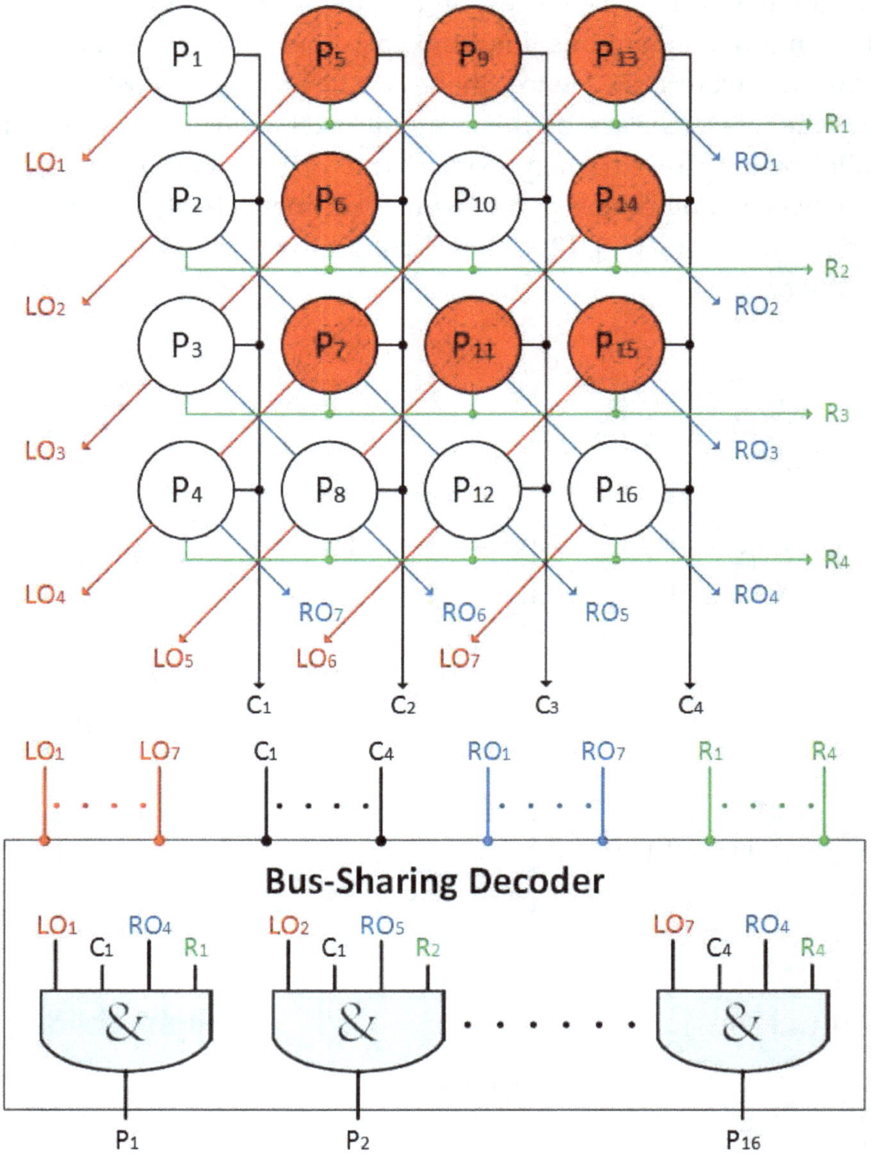

Figure 18. Special case of a photodiode mistaken.

Photodiodes triggered at different time can be discovered by the proposed bus-sharing mechanism at an effectual and low-cost manner. However, when many photodiodes receive photos simultaneously, a special condition must be considered. An un-triggered pixel completely encircled by triggered pixels that discover light at once is mistaken for a triggered one. A special case, depicted in Fig. 18, mistakes P_{10} where P_{10} does not capture any light, and photodiodes of $P_5 \sim P_7$, P_9, P_{11} and $P_{13} \sim P_{15}$ are activated by light. The buses of LO_4, C_3, RO_3 and R_2 addressed by P_{10} are enabled by (P_7, P_{13}), (P_9, P_{11}), (P_5, P_{15}) and (P_6, P_{14}), respectively. During decoding P_{10}, an AND gate with inputs of LO_4, C_3, RO_3 and R_2 yields logic-1, which is an error solution. Such a situation can be figured out while P_{10} is only enabled at an earlier or later time. Restated, to analyze the activated patterns of P_{10} and its neighboring pixels at different time intervals, the accurate triggering time associated with P_{10} can be attained.

4.1. Multi-channel time to digital converter

When an LED generates light projecting upon an object, the time to digital converter (TDC) begins to calculate the time of light travel from an LED to a photodiode. The derived travel time which is multiplied by a velocity of light can present a double distance from a sensor to an object. According to this distance information, the depth of an object apart from a sensor can be attained. Figure 19 shows the block diagram of multi-channel TDC which is an event-driven approach [25]. The counter in TDC is applied to obtain a timescale number which is correlated with time of light flight. Additionally, the delay line circuit is used to interpolate a fractional scale. To attenuate the influence of temperature and process, differential pairs are adopted to realize flip flops of a delay line.

A 3D depth map comes from a pixel array of which pixels demand timing information. If each pixel has the corresponding timing circuit to compute its depth, there are too many timing circuits which take great hardware cost and high power dissipation. To reach the TDC shared by multiple pixels, a multi-channel TDC composed of a ring TDC, a thermal encoder and a 4-bit counter, as displayed in Fig. 19, is adopted. Referring to [25], Fig. 20 shows a 15-stage ring TDC, which is a core of a multi-channel timing circuit. When the signal of 'Start' is active, a NAND gate and 14 inverters build a ring oscillation. This ring TDC produces 15 outputs of C_1, C_2,... and $C_{15,}$ which are compacted by a thermal encoder to give a 4-bit fine result. In the meantime, the counter yields a 4-bit coarse result as well. 4-bit fine and course results form 8-bit timing information which is stored in the latch array, and used to interpret the depth information.

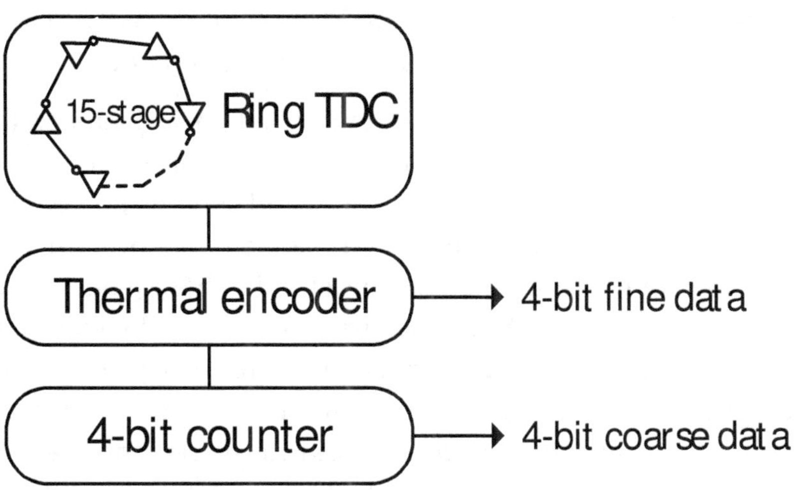

Figure 19. Block diagram of multi-channel TDC.

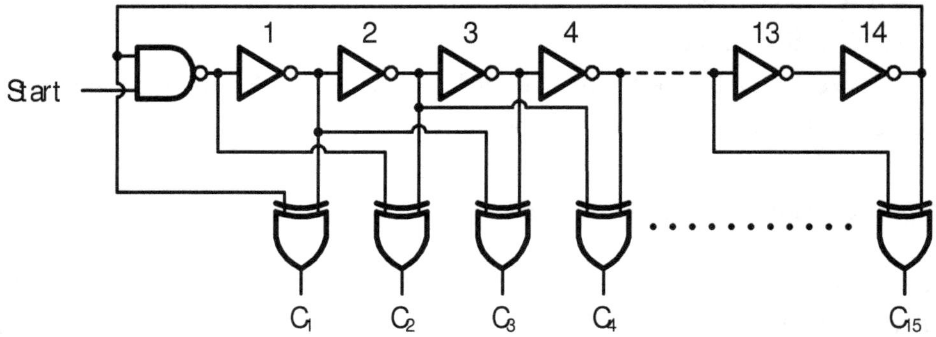

Figure 20. Ring TDC.

5. 2D/3D-integrated image sensor

The proposed 2D/3D-integrated image sensor employs P-diffusion_N-well_ P-substrate photodiode, which can be switched to different reversely biased voltages, and then operates at a 2D or 3D photo sensing mode. The Correlated Double Sampling (CDS) circuit and readout circuit used for a 2D mode, and Sense Amplifiers (SA), a multi-channel TDC and a readout circuit used for a 3D mode are implemented. Additionally, sequential and parallel reading is realized by using row and column decoders, and bus-sharing connections and decoders at 2D and 3D modes, respectively. The CDS circuit reduces the fixed pattern noise, and the SA boosts a trigger pulse generated from a pixel to lower the dead time. The block diagram of the proposed 2D/3D-integrated image sensor is depicted in Fig. 21. Since human visual perception has a good resolution in luminance rather than depth, the proposed 2D/3D-integrated image sensor adopts pixel dimensions of 352×288 and 88×72 pixels associated with 2D and 3D sensing, respectively. Restated, the pixel dimension at a 3D mode is one sixteenth of that at a 2D mode to lessen hardware cost. Particularly, to effectively decrease the overhead of parallel reading at a 3D mode, the bus-sharing mechanism can address 88×72 pixels by using 478 lines rather than 6,336 lines. After decoding 478 signal bits, 88×72 bits are stored in latches at every counting-time interval where each bit indicates if a pixel is triggered or not.

The TSMC 0.35μm 2P4M CMOS technology was employed to implement the proposed 2D/3D-integrated image sensor with a die size of 12mm×12mm [26]. The field factor is about 44% where a photodiode has a diameter of 10μm. The peak and average powers are 2.56W and 0.58W, at a supply voltage of 3.3V, respectively. The dynamic range of luminance detection can reach to 100dB. Each stage of the ring TDC was measured to have a delay of 234ps which can interpret a depth resolution of 4cm. When capturing an object, the external start signal resets the pixel circuit, enables the TDC and triggers a light source. The proposed sensor begins to receive the reflected light from an object and calculate the travel time using the TDC. The counted travel time is used to derive object depths. In our measurement, the FPGA board is programmed to manipulate an LED array and provide the timing control signals to read data from the proposed chip, as displayed in Fig. 22, where a cylindrical box is located in front of the sensor. At a light source from an 850nm LED array, Fig. 23 shows the depth map of a

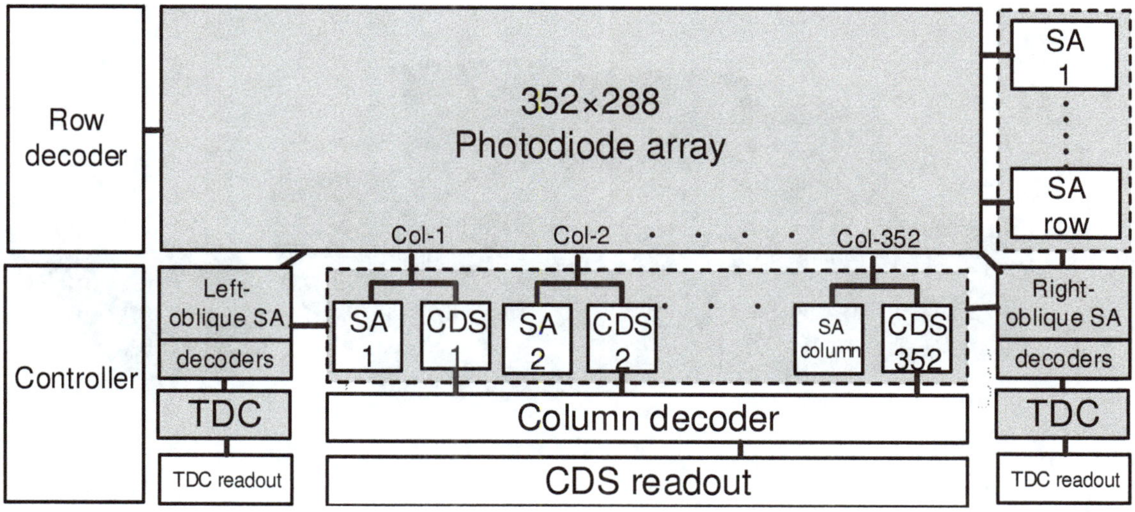

Figure 21. Block diagram of the proposed 2D/3D-integrated image sensor.

cylindrical box. This depth map exhibits a similar cylindrical shape but somehow includes little noise.

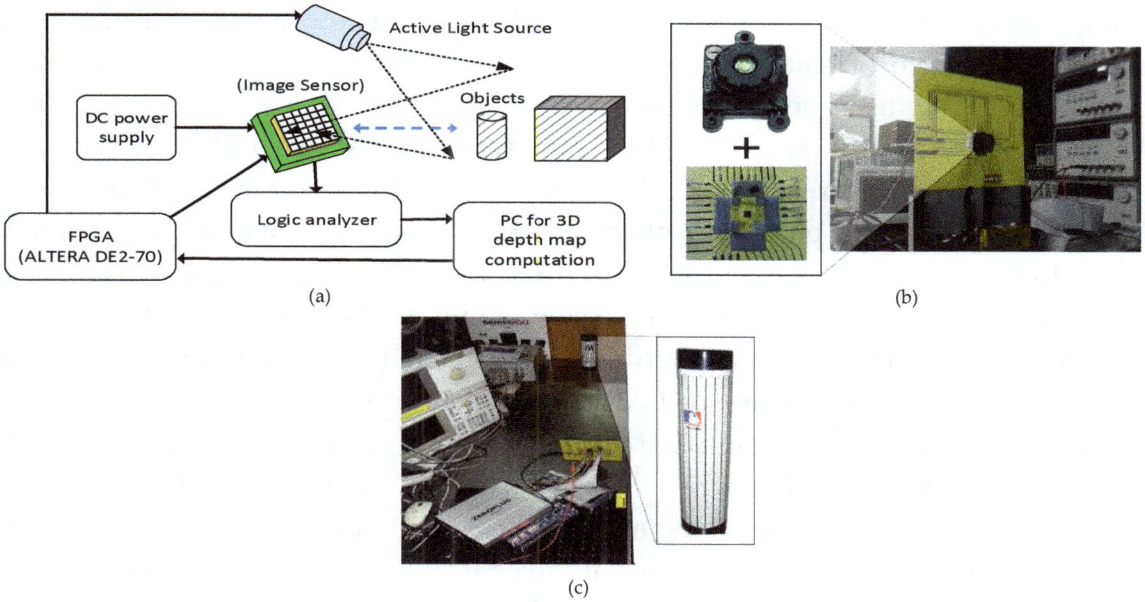

Figure 22. Measurement setup using the FPGA board. (a) Block diagram of measurements. (b) Proposed chip with on-board package and a convex lens. (c) Discovering a cylindrical box.

6. Currently available techniques of 2D and 3D image sensors

Nowadays, CMOS image sensors generally have extensive resolutions, high frame rates, large dynamic ranges and low power dissipation. To meet these demands, pixel circuits, read-out structures, analogue-to-digital converter (ADC) architectures and 3D integrated circuits (IC)

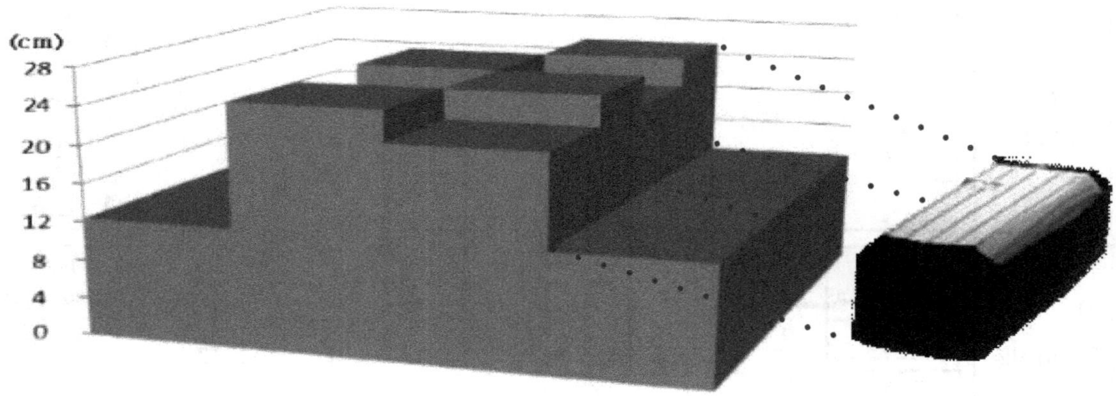

Figure 23. Measured depth map of a cylindrical box.

have been well explored as well as high-efficiency photodiodes. For example, Lim *et al.* adopted two column-shared cyclic ADCs arranged at a two-stage stack, to reduce the power consumption of the read-out circuit [27]. Seo *et al.* employed a column-parallel folding integration and cyclic ADC architecture to minimize the amount of noise, and reached to 13-bit or more resolution for each pixel [28]. To enhance the sensitivity of a pixel, Xu *et al.* designed a capacitive trans-impedance amplifier pixel with a tiny metal–oxide–metal (MOM) capacitor. To compensate the mismatch of small integration capacitors across the pixel array, an on-chip calibration scheme with in-pixel circuits was developed [29]. Afterwards, Xu *et al.* further implemented a small multi-layer MOM integration capacitor to achieve high sensitivity and low mismatches across a pixel array. Under such a design, the size of the previous capacitive trans-impedance amplifier pixel was effectively minimized [30]. Sakakibara *et al.* added a floating diffusion node and a storage node in a pixel circuit which supports two-channel read-outs for low and high intensities. Via a single-slope 12-bit ADCs, their pixel with dual storage structure made it possible to obtain up to 83dB dynamic range [31]. Chung *et al.* proposed a 0.5V operated pulse-width modulation CMOS image sensor with threshold-variation canceling and programmable current-controlled threshold schemes in which the fixed-pattern noise could be effectively minimized [32]. Yeh *et al.* used the 3D IC technology to stack a pixel array, an ADC array and an image processing array, which could operate in parallel, to achieve a high frame rate and high spatial resolution [33]. Sukegawa *et al.* combined the back-illuminated photosensitive layer and logic layer to become an image sensor. Such an approach could make the size of 8M pixels down to 1/4 inch. Additionally, the RGBW (Red, Green, Blue and White) coding was added in the color filter of the pixel to improve sensitivity, and binning-SVE was adopted to reach a high dynamic range [34]. The specifications of the abovementioned approaches are summarized in Table 1.

As well as 2D image sensors, there are some special approaches to implement 3D image sensors. For instance, Koyama *et al.* used a camera lens and a single sensor to realize the binocular images for left and right eyes [35]. Particularly, a lenticular lens separates incident light to become two beams for left-eye and right-eye viewing points, which were captured by

Authors	Xu *et al.* [30]	Sukegawal *et al.* [34]	Yeh *et al.* [33]	Chung *et al.* [32]	Sakakibara *et al.* [31]	Seo *et al.* [28]	Lim *et al.* [27]
Pixel architecture	CTIA	N/A	4T-APS PPD	N/A	N/A	4-TR PPD	2-T PPD
Shutter	Global	N/A	Rolling	N/A	Global	N/A	N/A
Process (CMOS)	0.18 μm	65nm (1P4M)	0.18 μm (1P6M)	0.18 μm (1P6M)	90nm (1P4M)	0.18 μm (1P4M)	0.13 μm (1P4M)
Array size	640×480	3280×2464	2048×1536	64×40	5M	1032×1024	1696×1212
Pixel size (μm^2)	8.7×8.22	1.12×1.12	2.8×2.8	10×10	5.86×5.86	7.5×7.5	2.25×2.25
Fill factor (%)	41	N/A	28	25.4	N/A	52	N/A
Frame rate (fps)	400	30	100	78.5	N/A	2.2	250
Dynamic range	50.1dB	60dB	N/A	82dB	83dB	78dB	59dB
FPN	0.55%	N/A	0.43 mV_{rms}	0.055%	N/A	N/A	0.1%
Sensitivity	1.89 $V/lux \cdot s$	6.7k $e^-/\ lux \cdot s$ (white pixel)	3.69 $V/lux \cdot s$	N/A	78k $e^-/\ lux \cdot s$	10 $V/lux \cdot s$	8.6k $e^-/\ lux \cdot s$
Temporal noise	15.6 e^-_{rms}	2.2 e^-_{rms}	16 e^-_{rms}	N/A	4.8e^-_{rms}	≤1e^-_{rms}	12.5 e^-_{rms}
ADC resolution (bit)	12	10	12	10	12	13-19	10
ADC type	SAR	N/A	Off-chip	Ramp	Column single-slope	Folding-integration/ Cyclic	Shared Cyclic
Power	N/A	185mW	19mW	29.6 μW	2W	450mW	300mW

Note: CTIA: Capacitive Trans-Impedance Amplifier; PPD: Pinned Photo-Diode

Table 1. Specifications of Currently Available 2D Image Sensors.

multiple pairs of neighboring pixels. However, to solve the energy losses due to the beam splitter and the crosstalk and so on, the polycyclic Si_3N_4 digital micro lenses were employed to make the beam focused at photodiodes. Wang *et al.* used a light-field image sensor to capture the intensity and angle of the incident light, and then derived depth information where angle sensitive pixels were adopted [36]. Based on the traditional pixel architecture and diffraction gratings, the Talbot effect could be addressed to capture the information of the incident light angle. By calculating the convergence and divergence angles, the depth information is attained. Kim *et al.* use single lens to capture 2D and 3D images via a time-multiplexed manner [37]. The pixel architecture consisted of a 4-transistor active pixel and two floating diffusion nodes where each pixel unit was adjacent to two transfer gates. Through a time-multiplexed manner, the transfer gates are used to switch the operational modes. At a normal mode, the capture manner is identical to that of a traditional CMOS image sensor, and at a 3D mode, all of the transfer gates are switched on to yield many 4×4− pixel units for capturing the depth information. Restated, they enlarged the sensing area of a photodiode to enhance the sensitivity at a 3D mode. However, the increase of the reversely biased voltage usually brings higher sensitivity improvement than the increase of the sensing area. The specifications of the abovementioned 3D image sensors and the proposed 2D/3D-integrated image sensors are listed in Table 2.

	Authors	Koyama *et al.* [35]	Wang *et al.* [36]	Kim *et al.* [37]	Proposed
Luminance	Pixel architecture	3-T APS	3-T APS	4-T APS (PPD)	5-T APS
	Process (CMOS)	0.11μm (1P3M)	0.18μm (1P6M)	0.11μm (1P4M)	0.35μm(2P4M)
	Array size	2.1M	400×384	1920×1080 (2D) 480×270 (3D)	352×288 (2D) 88×72 (3D)
	Pixel size (μm^2)	7.6	56.3	13.3 (2D) 213.2 (3D)	78.5
	Fill factor (%)	N/A	58	38.5	44
	Frame rate(fps)	60	200	30	30
Depth	Technique	Lenticular lens	Light-field	Time of flight	Time of flight
	Measured range (m)	0.2~1	1	0.75~4.5	1~3
	Non-linearity (%)	≤5	0.25	0.93	N/A
	Frame rate (fps)	N/A	N/A	11	N/A
	Resolution	N/A	2.5mm	38mm	40mm
	Calculation	Yes	Yes	No	No

Note: PPD: Pinned Photo-Diode

Table 2. Specifications of currently available 3D image sensors and proposed 2D/3D-integrated image sensors.

7. Conclusion

This chapter presents a 2D/3D-integrated image sensor which includes photodiodes, pixel circuits, CDS circuits, sense amplifiers, a multi-channel TDC, readout circuits, row/column decoders, bus-sharing connections/decoders and so on. The luminance and depth information of a scene can be captured by the same pixel at a time-multiplexed manner. Based on the standard CMOS technology, the circular photodiode of P-diffusion_ N-well_ P-substrate is utilized thanks to good quantum efficiency, fair breakdown voltage and easy integration of 2D and 3D pixels. Particularly, the proposed integrated pixel can yield a high dynamic range at a 2D mode using a charge supply mechanism and a high response speed at a 3D mode using a feedback pull-down mechanism. Additionally, the bus-sharing mechanism is employed to diminish hardware cost of parallel reading. By using the TSMC 0.35μm 2P4M CMOS technology, the 352×288-pixel 2D and 88×72-pixel 3D integrated image sensor was designed to have a dynamic range up to 100dB and a depth resolution of around 4cm. The measured results reveal very promising performance. Therefore, the 2D/3D-integrated image sensor proposed herein can be widely applied to various multimedia capturing applications under low hardware cost and low power dissipation.

Acknowledgements

This work was partially supported by National Science Council (NSC) and Ministry of Science and Technology (MOST), Taiwan, under the project numbers of NSC 97-2221-E-194-060-MY3, and MOST 102-2221-E-194 -047. The authors would like to thank Wei-Jean Liu, Meng-Lin Hsia, Zhe Ming Liu, Ming-Chih Lin, Chieh Ning Chan, Kuan-Hsien Lin, Shu Chun Wang and Hsiu-Fen Yeh who helped in designing, simulating, implementing and measuring the 2D and 3D image sensors. Additionally, Chip Implementation Center (CIC), Hsinchu, Taiwan, providing the services of VLSI CAD tools and chip fabrication is highly appreciated.

Author details

Oscal T.-C. Chen*, Yi-Yang Lee and Robin R.-B. Sheen

*Address all correspondence to: oscal@ee.ccu.edu.tw

National Chung Cheng University Chiayi, Taiwan

References

[1] Panasonic. Panasonic Product Support - DMC-3D1K. Available from: http://shop.panasonic.com/support-only/DMC-3D1K.html.

[2] Andersen MR, Jensen T, Lisouski P, Mortensen AK, Hansen MK, Gregersen T. Kinect depth sensor evaluation for computer vision applications. Aarhus University, Department of Engineering, 2012.

[3] Microsoft. Kinect for Windows features. Available from: http://www.microsoft.com/en-us/kinectforwindows/meetkinect/features.aspx.

[4] Charbon E. Introduction to time-of-flight imaging. In: Proc. of IEEE Sensors Conference; 2-5 Nov. 2014; Valencia. p. 610-613.

[5] Aull BF, Loomis AH, Young DJ, Heinrichs RM, Felton BJ, Daniels PJ. Geiger-mode avalanche photodiodes for three-dimensional imaging. Lincoln Laboratory Journal. 2002; 13(2):335-349.

[6] Niclass C, Rochas A, Besse P-A, Charbon E. Toward a 3-D camera based on single photon avalanche diodes. IEEE Journal of Selected Topics in Quantum Electronics. 2004; 10(4):796-802. DOI: 10.1109/JSTQE.2004.833886.

[7] Marshall GF, Jackson JC, Denton J, Hurley PK, Braddell O, Mathewson A. Avalanche photodiode-based active pixel imager. IEEE Trans. on Electron Devices. 2004; 51(3): 509-511. DOI: 10.1109/TED.2003.823051.

[8] Zappa F, Lotito A, Tisa S. Photon-counting chip for avalanche detectors. IEEE Photon Technology Letters. 2005;17(1):184-186. DOI: 10.1109/LPT.2004.838136.

[9] Faramarzpour N, Deen MJ, Shirani S, Qiyin F. Fully integrated single photon avalanche diode detector in standard CMOS 0.18-μm technology. IEEE Trans. on Electron Devices. 2008; 55(3):760-767. DOI: 10.1109/TED.2007.914839.

[10] Atef M, Polzer A, Zimmermann H. Avalanche double photodiode in 40-nm standard CMOS technology. IEEE Journal of Quantum Electronics. 2013; 49(3):350-356. DOI: 10.1109/JQE.2013.2246546.

[11] Pancheri L, Dalla Betta GF, Stoppa D. Low-noise avalanche photodiode with graded junction in 0.15-μm CMOS technology. IEEE Electron Device Letters. 2014; 35(5): 566-568. DOI: 10.1109/LED.2014.2312751.

[12] Kang Y, Mages P, Clawson AR, Yu PKL, Bitter M, Pan Z. Fused InGaAs-Si avalanche photodiodes with low-noise performances. IEEE Photonics Technology Letters. 2002; 14(11):1593-1595. DOI: 10.1109/LPT.2002.803370.

[13] Ong DSG, Jo Shien N, Yu Ling G, Chee Hing T, Shiyong Z, David JPR. InAlAs avalanche photodiode with type-II superlattice absorber for detection beyond 2-μm. IEEE Trans on Electron Devices. 2011; 58(2):486-489. DOI: 10.1109/TED.2010.2090352.

[14] Jun H, Banerjee K, Ghosh S, Hayat MM. Dual-carrier high-gain low-noise superlattice avalanche photodiodes. IEEE Transactions on Electron Devices. 2013; 60(7): 2296-2301. DOI: 10.1109/TED.2013.2264315.

[15] Zhen Guang S, Dun Jun C, Hai L, Rong Z, Da Peng C, Wen Jun L. High-gain AlGaN solar-blind avalanche photodiodes. IEEE Electron Device Letters. 2014; 35(3):372-374. DOI: 10.1109/LED.2013.2296658.

[16] Chan CN, Chen OTC. Physical effects of avalanche CMOS photodiodes. In: Proc. of OptoElectronics and Communications Conference; 5-9 July 2010. p. 824-825.

[17] He J, Xi X, Chan M, Hu C, Li Y, Zhang X. Equivalent junction method to predict 3-D effect of curved-abrupt p-n junctions. IEEE Transactions on Electron Devices. 2002; 4 9(7):1322-1325.

[18] Hsia M-L, Liu ZM, Chan CN, Chen OTC. Crosstalk effects of avalanche CMOS photodiodes. In: Proc. of IEEE Conference on Sensors; 28-31 Oct. 2011; Limerick, Ireland: IEEE; 2011. p. 1689-1692.

[19] Chen OTC, Lin K-H, Liu ZM. High-efficiency 3D CMOS image sensor. In: Proc. of 18th OptoElectronics and Communications Conference; June 30 - July 4, 2013; Kyoto, Japan: Optical Society of America; 2013.

[20] Liu ZM, Lin M-C, Chan CN, Chen OTC. 2D and 3D-integrated image sensor. In: Proc. of IEEE 53rd Midwest Symposium on Circuits and Systems; 1-4 Aug. 2010; Seattle, USA: IEEE; 2010. p. 292-295.

[21] Liu W-J, Yeh H-F, Chen OTC. A high dynamic-range CMOS image sensor with locally adjusting charge supply mechanism. In: Proc. of IEEE 48th Midwest Symposium on Circuits and Systems; 7-10 Aug. 2005: IEEE; 2005. p. 384-387.

[22] Cova S, Ghioni M, Lacaita A, Samori C, Zappa F. Avalanche photodiodes and quenching circuits for single-photon detection. Applied Optics. 1996; 35(12): 1956-1976.

[23] Niclass C, Favi C, Kluter T, Gersbach M, Charbon E. A 128 128 single-photon image sensor with column-level 10-bit time-to-digital converter array. IEEE Journal of Solid-State Circuits. 2008; 43(12):2977-2989.

[24] Gersbach M, Maruyama Y, Labonne E, Richardson J, Walker R, Grant L. A parallel 32× 32 time-to-digital converter array fabricated in a 130 nm imaging CMOS technology. In: Proc. of IEEE European Conference on Solid-State Circuits; 14-18 Sept. 2009; Athens: IEEE; 2009. p. 196-199.

[25] Jianjun Y, Fosterdai F, Jaeger R. 12-Bit vernier ring time-to-digital converter in 0.13 μm CMOS technology. IEEE Journal of Solid-State Circuits. 2010; 45(4):830-842.

[26] Chen OTC, Lin K-H, Liu ZM, Wang SC, Hsia M-L. 2D and 3D integrated image sensor with a bus-sharing mechanism. In: Proc. of IEEE 55th Midwest Symposium on Circuits and Systems; 5-8 Aug. 2012; Boise, Idaho, USA: IEEE; 2012. p. 138-141.

[27] Lim S, Cheon J, Chae Y, Jung W, Lee D-H, Kwon M. A 240-frames/s 2.1-Mpixel CMOS image sensor with column-shared cyclic ADCs. IEEE Journal of Solid-State Circuits. 2011; 46(9):2073-2083.

[28] Seo M-W, Suh S-H, Iida T, Takasawa T, Isobe K, Watanabe T. A low-noise high intrascene dynamic range CMOS image sensor with a 13 to 19b variable-resolution column-parallel folding-integration/cyclic ADC. IEEE Journal of Solid-State Circuits. 2012; 47(1):272-283.

[29] Xu R, Liu B, Yuan J. A 1500 fps highly sensitive 256 256 CMOS imaging sensor with in-pixel calibration. IEEE Journal of Solid-State Circuits. 2012; 47(6):1408-1418.

[30] Xu R, Ng WC, Yuan J, Yin S, Wei S. A 1/2.5 inch VGA 400 fps CMOS image sensor with high sensitivity for machine vision. IEEE Journal of Solid-State Circuits. 2014; 49(10):2342-2351.

[31] Sakakibara M, Oike Y, Takatsuka T, Kato A, Honda K, Taura T. An 83dB-dynamic-range single-exposure global-shutter CMOS image sensor with in-pixel dual storage. In: Technical Digest of IEEE International Solid-State Circuits Conference; 19-23 Feb. 2012; San Francisco, CA: IEEE; 2012. p. 380-382.

[32] Chung M-T, Lee C-L, Yin C, Hsieh C-C. A 0.5 V PWM CMOS imager with 82 dB dynamic range and 0.055% fixed-pattern-noise. IEEE Journal of Solid-State Circuits. 2013; 48(10):2522-2530.

[33] Yeh S-F, Hsieh C-C, Yeh K-Y. A 3 Megapixel 100 Fps 2.8 m pixel pitch CMOS image sensor layer with built-in self-test for 3D integrated magers. IEEE Journal of Solid-State Circuits. 2013; 48(3):839-849.

[34] Sukegawa S, Umebayashi T, Nakajima T, Kawanobe H, Koseki K, Hirota I. A 1/4-inch 8Mpixel back-illuminated stacked CMOS image sensor. In: Technical Digest of IEEE International Solid-State Circuits Conference; 17-21 Feb. 2013; San Francisco, CA: IEEE; 2013. p. 484-485.

[35] Koyama S, Onozawa K, Tanaka K, Kato Y. A 3D vision 2.1 Mpixel image sensor for single-lens camera systems. In: Technical Digest of IEEE International Solid-State Circuits Conference; 17-21 Feb. 2013; San Francisco, CA: IEEE; 2013. p. 492-493.

[36] Wang A, Molnar A. A light-field image sensor in 180 nm CMOS. IEEE Journal of Solid-State Circuits. 2012; 47(1):257-271.

[37] Kim S-J, Kang B, Kim JD, Lee K, Kim C-Y, Kim K. A 1920× 1080 3.65μm-pixel 2D/3D image sensor with split and binning pixel structure in 0.11 pm standard CMOS. In: Technical Digest of IEEE International Solid-State Circuits Conference; 19-23 Feb. 2012; San Francisco, CA: IEEE; 2012. p. 396-398.

4

Colorless and Transparent High-Temperature-Resistant Polymer Optical Films: Current Status and Potential Applications in Optoelectronic Fabrications

Jin-gang Liu, Hong-jiang Ni, Zhen-he Wang,
Shi-yong Yang and Wei-feng Zhou

Abstract

Recent research and development of colorless and transparent high-temperature-resistant polymer optical films (CHTPFs) have been reviewed. CHTPF films possess the merits of both common polymer optical film and aromatic high-temperature-resistant polymer films and thus have been widely investigated as components for microelectronic and optoelectronic fabrications. The current paper reviews the latest research and development for CHTPF films, including their synthesis chemistry, manufacturing process, and engineering applications. Especially, this review focuses on the applications of CHTPF films as flexible substrates for optoelectrical devices, such as flexible active matrix organic light-emitting display devices (AMOLEDs), flexible printing circuit boards (FPCBs), and flexible solar cells.

Keywords: colorless polymer films, high temperature, synthesis, flexible substrates

1. Introduction

Various polymer optical films have been widely applied in the fabrication of optoelectronic devices [1]. Recently, with the ever-increasing demands of high reliability, high integration, high wiring density, and high signal transmission speed for optoelectronic fabrications, the service temperatures of polymer optical films have dramatically increased [2, 3]. For instance, in the fabrication of new-generation flexible thin-film transistor-driven active matrix liquid crystal display devices (TFT-LCDs) or active matrix organic light-emitting display devices (AMOLEDs), the processing temperature on the flexible plastic substrates might be higher than 300°C [4-6]. Most of the common polymer optical films would lose their optical and mechanical properties at such high processing temperatures. Thus, colorless and transparent high-temperature-resistant polymer optical films (CHTPFs) have attracted increasing attentions from both the academic and engineering aspects in the past decades.

According to the different servicing temperatures or glass transition temperatures (T_g), the polymer optical films could be roughly classified into three types, including conventional optical films ($T_g < 100°C$), common high-temperature optical films ($100 \leq T_g < 200°C$), and high-temperature optical films ($T_g \geq 200°C$), as shown in Figure 1. The typical chemical structures for the polymer optical films are illustrated in Figure 2. Main physical and chemical characteristics for the typical optical polymers were tabulated in Table 1 [7, 8]. It can be clearly seen that conventional polymer optical films, such as polyethylene terephthalate (PET, T_g: ~78°C) or polyethylene naphthalate (PEN, T_g: ~123°C), possess excellent optical transparency. However, they are facing great challenges in advanced optoelectronic fabrication due to their limited service temperatures. On the other hand, high-temperature-resistant polymer films such as wholly aromatic polyimide films (PI) exhibit excellent thermal stability up to 300°C. However, they suffer from deep colors and poor optical transmittance in optoelectronic applications. Thus, achieving a compromise between the pale color and high thermal stability for the polymer optical films is one of the most challenging projects for optoelectronic polymeric films development.

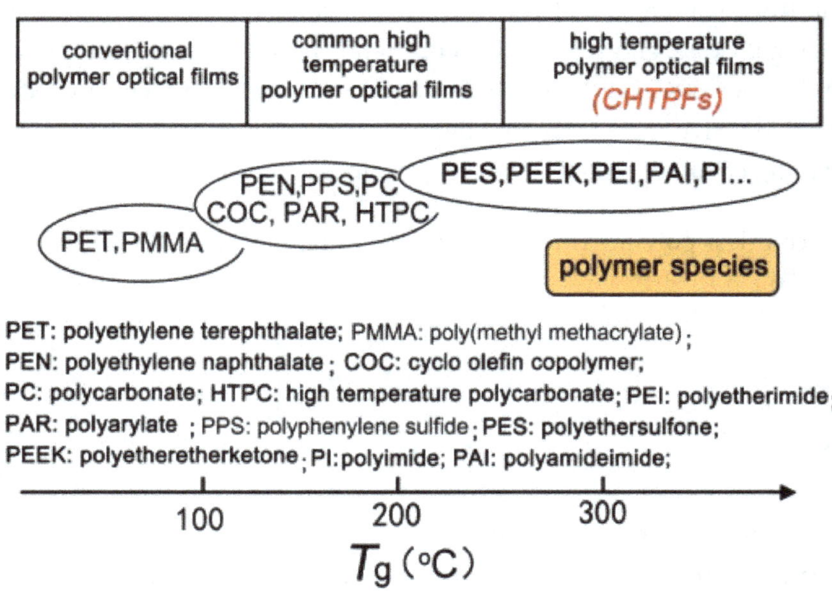

Figure 1. Classification of polymer optical films

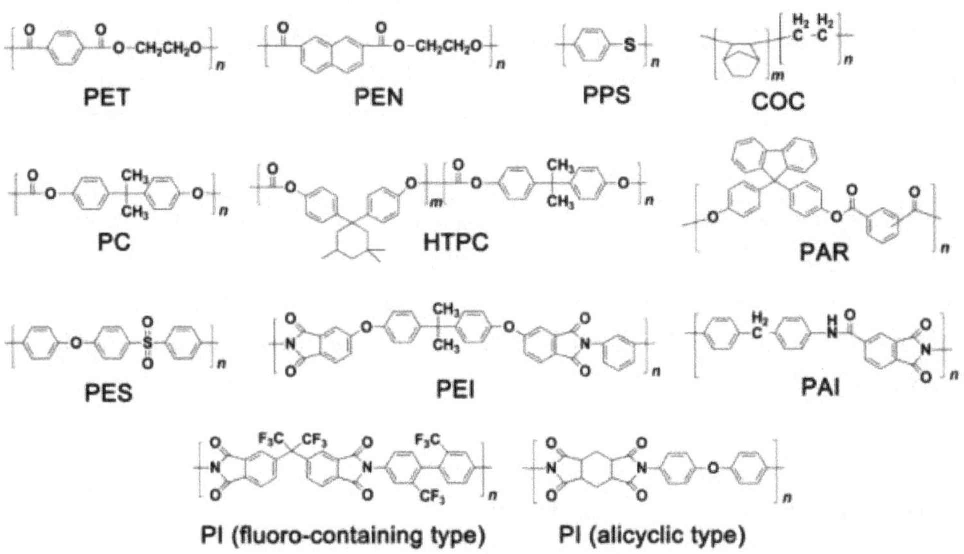

Figure 2. Typical chemical structures for polymer optical films

Item[1]	Unit	PET	PEN	PC	PPS	PEI	PES	PI	CPI[2]
Density	g/cm^3	1.40	1.36	1.20	1.35	1.27	1.37	1.43	1.23
Transmittance	%	90	87	92	85	80	89	30-60	90
T_m	°C	256	266	240	285	365	380	NA[3]	NA
T_g	°C	78	123	150	90	217	223	>300	303
WVTR	g/m^2 day	21	6.9	60	8	43.5	73	64	93
OTR	cm^3/m^2 day	6	2	300	6	220	235	22	NA
Water uptake	%	0.3	0.4	0.2	0.05	1	0.5	1.3	2.1
σ	MPa	225	275	98	250	130	95	274	112
E_b	%	120	90	140	50	70	70	90	12
D.S.	V/μm	280	300	250	250	250	260	280	NA
ε	-	3.2	3.0	3.0	3.0	3.5	4.0	3.3	2.9

[1] T_m: melting point; T_g: glass transition temperature; WVTR: water vapor transmission rate; OTR: oxygen transmission rate; σ: tensile strength; E_b : elongation at break; D.S.: dielectric strength; ε: dielectric constant;. [2] Data from colorless PI film Neopulim® L-3430 developed by MGC, Japan; [3] Not available.

Table 1. Typical properties of polymer optical films

In the past decades, considerable progress has been achieved in both the academic development and commercialization for novel CHTPFs. According to the classification in Figure 1, the leading materials, in terms of comprehensive properties and potential market volume, include colorless polyimide (PI) films, polyethersulfone (PES) films, polyetheretherketone (PEEK) films, polyamide (PA), and polyamideimide (PAI) films. They have been the main components for CHTPF families. According to the statistics from Techno Create Corp. (TCC, an authoritative consulting agency in Japan), the market of CHTPFs in 2011 has been close to 1 billion Japanese yen and the market will see a rapid increase higher than 15% per year in the following years [9].

In this review, the state of art and future development of CHTPFs in optoelectronic fabrications has been reviewed. The molecular design, synthesis chemistry, and film fabrication techniques for CHTPFs were introduced first. Then, the applications of CHTPFs in several important optoelectronic fields including flexible display, flexible printing circuit boards (FPCBs), and flexible solar cells were presented.

2. CHTPF manufacturing technology

Generally, the overall production process for CHTPF products consists of several steps, including monomers synthesis, polymer resin preparation, and the film preparation. Technically, these three steps all have their own core technologies and are usually interwinded and interrelated. The reactivity and purity of monomers will definitely affect the physical and chemical properties of the derived polymer resins, including molecular weights and their distribution, inherent viscosities, solubility in organic solvents, appearance, color, and so on. The features of the resins have great effects on the properties of the final polymer films, including their color, optical transparency, mechanical strength, thermal stability, and dielectric properties. Meanwhile, the preparing technologies for the films, including casting procedure, the uniaxial or biaxial stretching process, high-temperature curing program, and even the final winding and rewinding process will also affect the features of the CHTPF products. Thus, the manufacture of CHTPFs is usually a multidisciplinary technology.

The manufacturing techniques for polymer films usually include several types, such as casting, melting extrusion, and blowing procedures. Extrusion is the process of forming a film continuously through an opening. Most extruders do this by rotating a screw inside a stationary heated cylindrical barrel, to melt the polymer resins and pump the melt through a suitably shaped slit. This is used for direct manufacture of finished film products. It may also be used to feed a second process such as injection molding, blow molding, coating, laminating, or thermo-forming process. Blowing procedure is usually performed from a single-screw extruder by extruding polymer resins, cooling it with external and/or internal air streams, stretching it in the machine direction (MD) by pulling it away from the die; stretching it in the transverse direction (TD) by internal air pressure, flattening it by passing through nip rolls, and winding it onto a cylindrical roll. Optional post-stretching operations may include flame or corona surface treatment for wettability, adhesion, and sealing.

Many factors influence the choice of suitable procedures for polymer film manufacture, including physical and chemical properties of the polymer resins, color and appearance demands, and the current abilities of film-producing equipment, and so on. For example, as shown in Figure 3, for crystalline polymer resins that have clear melting points, such as PET and PEN, nonsolvent melting extrusion technique is mainly used. However, for amorphous polymers with low to moderate T_g values, such as PC and PES, both melting extrusion and solution casting techniques can be used. As for high-T_g amorphous polymers, such as PIs, solvent-casting procedure is usually the optimal choice.

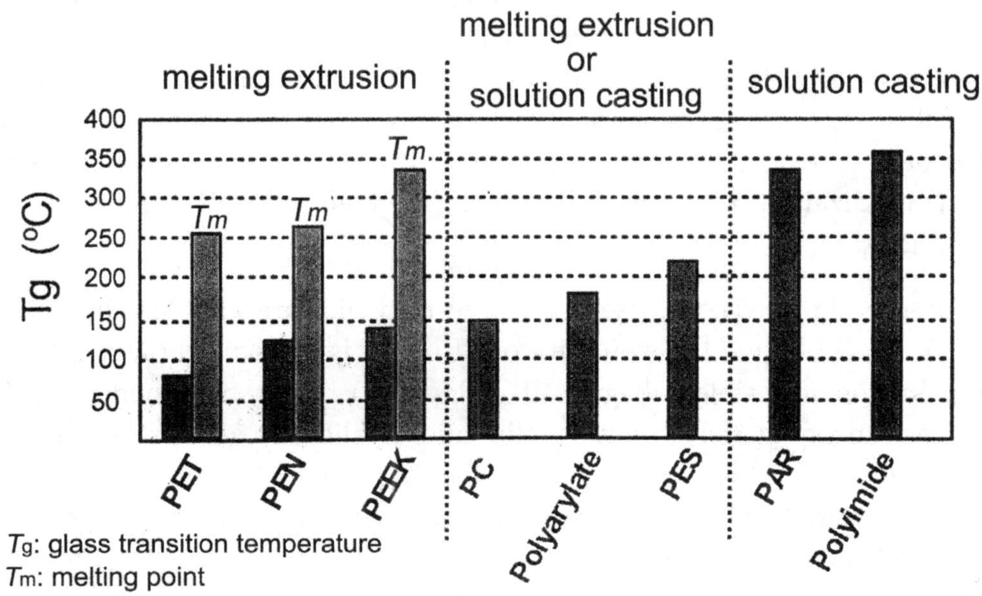

Figure 3. Processing methods for polymer films

The manufacturing techniques for CHTPF films have the similarities with the common optical films; however, they have their own uniqueness at the same time. This is mainly due to their relatively higher T_g values (or melting points) and lower solubility in common solvents caused by the more rigid molecular skeletons compared with the common optical polymers. Thus, for CHTPF films, solvent-casting procedure is most commonly used, especially in laboratory. For the solvent-casting procedure, it can be classified into two approaches: uniaxial stretching (machine direction, MD) and biaxial stretching (transverse direction, TD, and machine direction, MD) techniques. Biaxial stretching at temperatures above the T_g values of the CHTPF resins can usually improve the high-temperature dimensional stability of the obtained films.

Then, in the present paper, PI films are taken as examples to illustrate the development of lab-scale and industrial-scale manufacturing techniques for CHTPF films.

2.1. Laboratory preparation of PI films

Before industrial-scale manufacturing for CHTPF optical films, it is quite necessary to make a film prototype in laboratory so as to determine the optimal processing parameters. For PI films, the common fabrication techniques include two pathways: standard route via poly(amic acid) (PAA) and new route from organosoluble PIs, as illustrated in Figure 4. Both routes have their advantages and drawbacks. Generally speaking, the first standard PAA route is suitable to all kinds of PI films. In this procedure, dianhydride and diamine monomers will first polymerize in *N,N*-dimethylacetamide (DMAc) to afford PAA solution. The obtained PAA solution is sensitive to heat and moisture, which is easily degrading when stored at room temperature. Thus, it had been better using the newly synthesized PAA for preparing PI films. The PAA solution is cast onto clean glass or stainless steel substrates, followed by thermally curing from room temperature to elevated temperatures. This curing process consists of not only the physical course of solvent evaporation but also the chemical course of imidization or cyclization with the elimination of water. It has been well established that the imidization temperatures as high as 300-350°C is necessary to finish the transition from PAA to PI. Such a high temperature will definitely affect the color of the produced PI films. On the other hand, during the elimination of water from the system, microscopic defects such as pinholes and crack might occur. Thus, the imidization condition should be deliberately controlled in order to produce high-quality PI films.

Technically, the second route is only useful for PI resins which are soluble in organic solvents (mainly DMAc). As we know, the solubility of PI resins is particularly associated with its structure. Introduction of flexible linkages (-O-, $-CH_2-$, etc.), bulky substituents (alkyl groups, phenyl, etc.), and unconjugated structure (aliphatic or alicyclic groups) are all beneficial increasing the solubility of PI resins in organic solvents. From this point of view, this route is quite useful for colorless PI films production, because most of the PI resins for colorless PI films are soluble in polar solvents. In addition, the curing procedure for preimidized PI solution is nearly pure physical course of solvent evaporation. Thus, the PI films can be produced at relatively low temperature and exhibit good surface smoothness. This is undoubtedly beneficial for the production of colorless PI films.

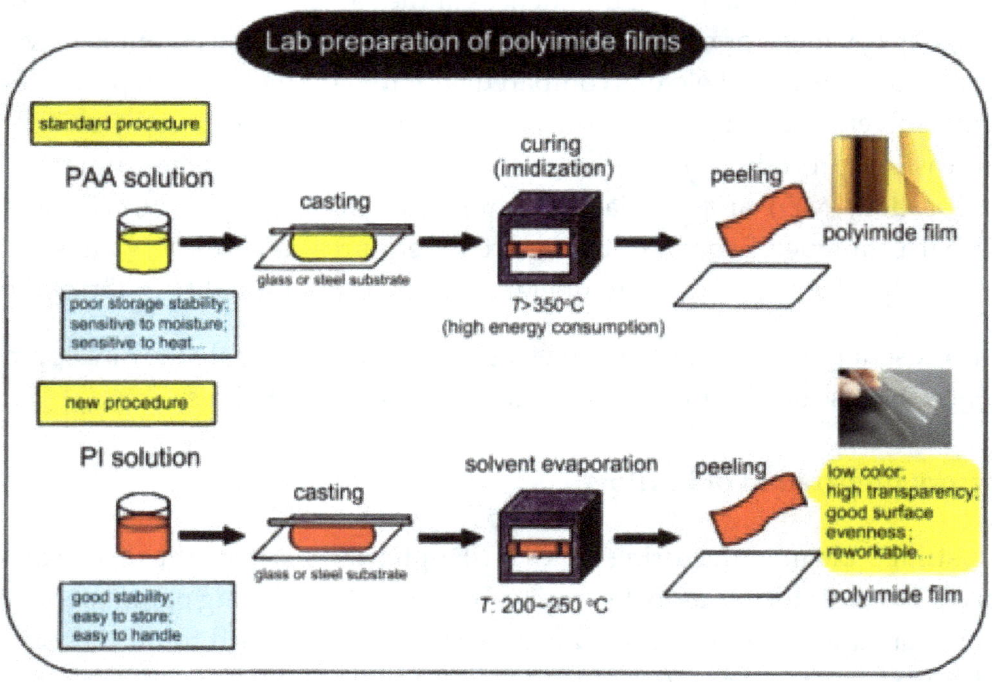

Figure 4. Lab-scale preparation of PI films

2.2. Industrial preparation of PI films

Ever since the commercialization of PI films with the trademark of Kapton® [poly(pyromellitic anhydride-oxydianiline), PMDA-ODA] in 1960s by DuPont Corporation in the USA, they have been becoming one of the most important basic materials for modern industry [10]. PI films have found various applications in civil and military high-tech fields. In 2011, the worldwide consumption for PI films is more than 8000 metric tons and this consumption is estimated to reach 13000 tons in 2016. The three major markets for PI film are flexible printed circuit substrates, high-temperature wire and cable wrapping, and magnetic wire insulation. The wide applications of PI films are mainly attributed to their excellent properties, including extreme servicing temperatures (-296-400°C for Kapton), high mechanical properties, excellent dielectric features, and good environmental stability. The superior property for PI films, on one hand, is associated with their heteroaromatic molecular structures, and on the other hand, owes to their unique producing techniques.

Compared with the lab-scale preparation, the greatest difference for industrial manufacturing of PI films is the stretching process [11]. Stretching process, either uniaxial or biaxial stretching of the gel-like PAA films, will result in the full orientation and extension for the PI molecular chains. The gelation of PAA films can be achieved either by partially evaporating the solvent or by chemical treatment with a dehydrating agent (acetic anhydride, dicyclohexylcarbodiimide, etc.) and its catalyst (pyridine). From a viewpoint of polymer physics, stretching will greatly enhance the mechanical properties of the obtained PI films. For example, the values of elongations at break for the lab-making PI films without any stretching treatment are usually below 20%. However, this value can be increased several times after stretching treatments.

A diagram of the biaxial stretching production line for PI films is shown in Figure 5. In this procedure, the monomers are first feed into the polymerization reactor containing fully dried DMAc solvent. After polycondensation, the obtained PAA solution is deaerated and cast from the slit die in the form of a continuous film onto the surface of a heated rotating stainless steel drum. The solvent in PAA is partially evaporated and a portion of imidization reaction occurs in PAA at the same time. Thus, a self-supporting PAA film is formed. Alternatively, the PAA solution on the rotating drum can pass through a bath containing dehydrating agent and cyclization catalyst to afford a gel-like PAA film. Then, the gel-like PAA film is peeled from the metal drum and first stretched in the machine direction (MD) while controlling the stretching rate using nip rolls. The stretching ratio can usually be regulated by the drive source and a speed regulator. The gel film stretched in the machine direction is subsequently introduced into a tenter frame where it is gripped at both transverse edges. Various means may be employed to grip the film, including pins, clips, clamps, and rollers. The gel film is then stretched in the transverse direction due to outward movement of the tenter clips, the volatile organic solvent is removed by evaporation, and the film is heat-treated by means of hot air or radiant heat from an electrical heater to give a biaxial oriented polyimide film (BOPI). The transverse stretching is carried out at temperatures around 350°C to facilitate the imidization of PAA. Such a procedure has been widely used for PI film production and there has been significant patent activity in the past half century since the commercialization of PI films in 1960s. Up to now, most of the commercially available wholly aromatic PI films have been produced by such kind of procedure.

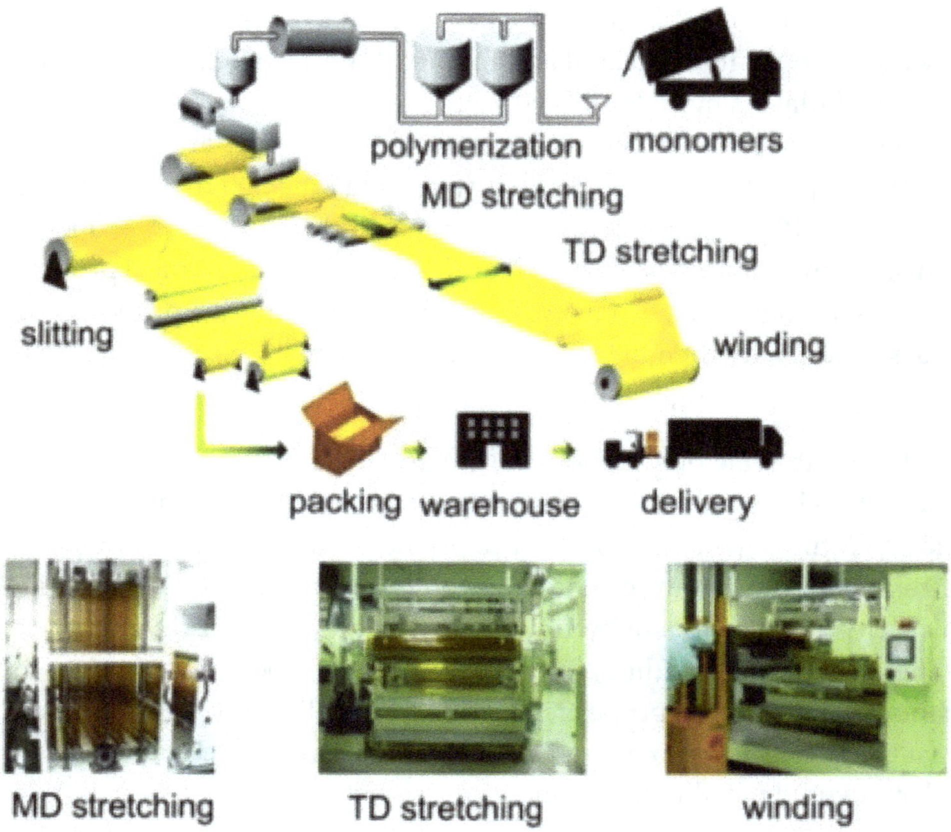

Figure 5. Industrial-scale manufacturing for PI films via PAA precursors

Because the full imidization temperature for PAA is usually higher than 350°C, the manufacturing procedure mentioned above might be difficult for colorless PI film production, whose color and transparency is highly sensitive to high temperatures. Thus, a new manufacturing technique has been developed in recent years [12].

As illustrated in Figure 6, the new procedure uses soluble PI resins as the starting materials instead of PAAs. The key elements for this procedure include: (a) the PI resin must be soluble in a volatile solvent; (b) a stable PI solution with a reasonable solid content and viscosity should be formed; and (c) formation of a homogeneous film and release from the casting support must be possible. In the procedure, PI resins are first dissolved in polar solvents to afford the PI solution, which are purified by filtration through screen mesh. Then, the PI solution is cast onto stainless steel belt, followed by thermally drying at high temperatures to remove the solvent. This drying procedure is only to remove the solvent in the PI solution. Thus, the temperature is usually lower than the common imidization temperature (300-350°C). Similarly, the PI films can also be stretched at an appropriate solvent content. For instance, NASA (National Aeronautics and Space Administration, USA) Langley research center investigated the molecularly oriented colorless PI films for space applications [13]. In large space structures with designed lifetimes to be 10-30 years, there exists a need for high-temperature (200-300°C) stable, flexible polymer films that have high optical transparency in the visible light region. For this purpose, a colorless and transparent PI film, LaRC-CP1, derived from 6FDA and fluoro-containing diamine, 2,2-bis[4-(4-aminophenoxy)phenyl]hexafluoropropane (BDAF), has been developed in NASA. This film is prepared from soluble PI resin. The LaRC-CP1 film was uniaxially stretched at 1.5, 1.75, and 2 times the original length of the film. Table 2 shows the influence of stretching on the physical and mechanical properties of LaRC-CP1 film. Apparently, the tensile properties of the film increased with increased stretching ratio. For example, the tensile strength of 2.0× stretched film increased from 93.0 to 145.4 MPa after stretching treatment; and the elongations increased from 16% to 65%. After stretching treatment, the dimensional stability, stiffness, elongation, and strength of the film were greatly enhanced, which are crucial for the applications in space environments.

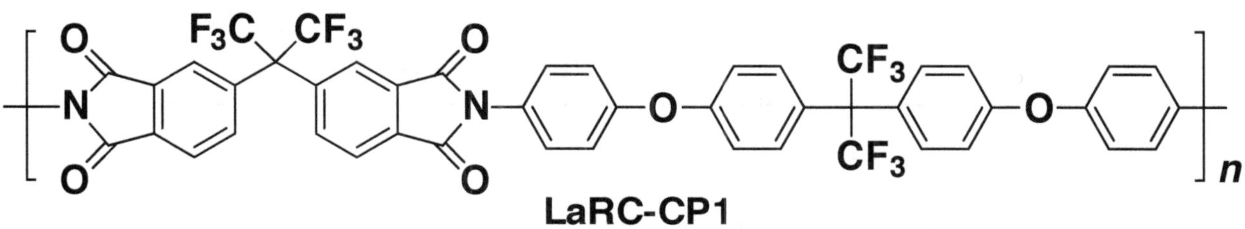

Stretch ratio	CTE (ppm/°C)	Tensile strength (MPa)	Modulus (GPa)	Elongation (%)
None	50	93.0	2.0	16
1.5×	42	88.9	1.7	20
1.75×	44	107.5	1.8	49
2.0×	46	145.4	2.1	65

Table 2. Characterization of stretched LaRC-CP1 film

US Patent 8357322 assigned to Mitsubishi Gas Chemical Company describes a method for producing colorless and transparent PI films by a solution casting procedure. The biaxially stretched colorless PI films exhibit excellent optical transparency, heat resistance, and reduced dimensional changes [14]. The films were produced with the soluble PI resin as the starting materials, which were derived from 1,2,4,5-cyclohexanetetracarboxylic dianhydride and aromatic diamines by one-step high-temperature polycondensation route. The PI film was biaxially stretched in the machine direction by 1.01 times and in the transverse direction by 1.03 times at 250°C for 11 min under a stream of nitrogen. Then, the PI film was dried by blowing nitrogen containing 1000 ppm oxygen at a flow rate of 3.3 m/sec at 280°C for 45 min. The obtained PI film had a thickness of 200 μm, a total light transmittance of 89.8%, a yellow index of 1.9, and a haze of 0.74%. The solvent residual ratio in the film was 0.5% by weight. By virtue of these properties, the colorless PI films might find various applications in optoelectronic applications, such as transparent conductive film, transparent substrates for flexible

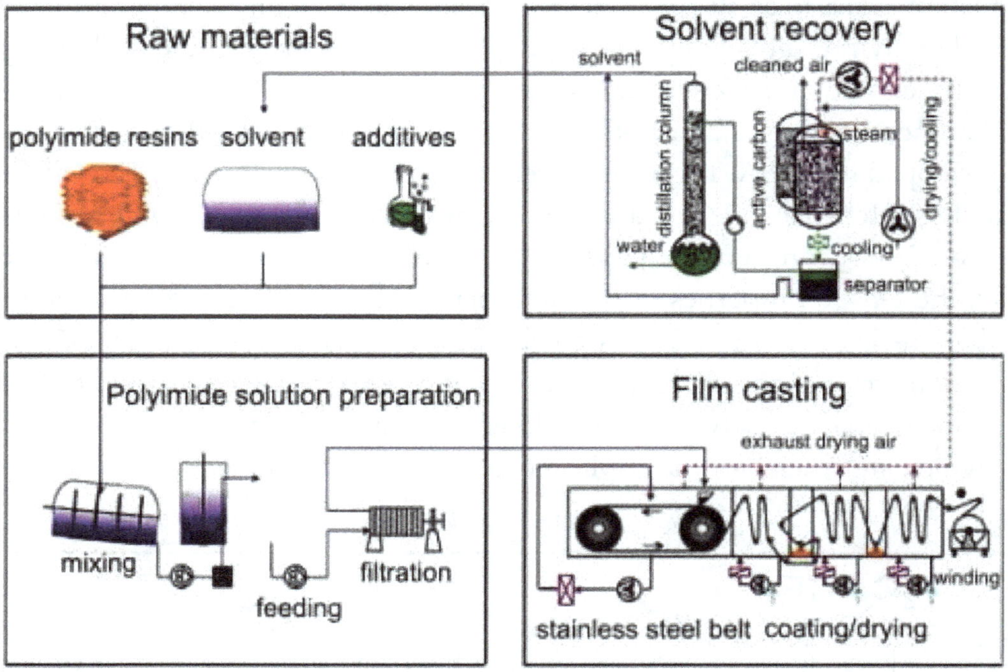

adapted, modified, and reprinted from ref. [12]

Figure 6. Industrial-scale manufacturing for PI films via soluble PI resins

display, flexible solar cells, and flexible printing circuit board (FPCB). Similar procedures were also reported by the company [15].

In addition, US Patent 7550194 assigned to DuPont Company [16] and US patent 8846852 to Kolon Industries [17] report the low-color PI films derived from the copolymers of fluoro-containing dianhydride, 2,2-bis(3,4-dicarboxyphenyl)hexafluoropropane dianhydride (6FDA), 3,3′,4,4′-biphenyl-tetracarboxylic dianhydride (BPDA), and fluoro-containing diamine, 2,2′-bis(trifluoromethyl)-benzidine (TFMB). The copolymers were prepared via PAA precursors, followed by chemical imidization of the PAAs to afford the gel-like PAA films or soluble PI resins. Then, the PI films were produced from these intermediums at high temperature up to 300°C. Flexible and tough PI films with low color and high transparency were obtained.

2.3. CHTPF films analysis and evaluation techniques

In practical applications for CTHTP optical films, various properties have usually been analyzed and evaluated. For different applications, specific properties might be specially emphasized. For example, in the fabrication of AMOLEDs, the water vapor transmission rate (WVTR) and oxygen transmission rate (OTR) of the flexible substrates are severely limited to be below 10^{-4} cm^3/m^2 day and 10^{-6} g/m^2/day, respectively, because the penetration of water and oxygen through the substrates might poison the emitting components, resulting in the reduced operating life of the devices [18]. In flexible solar cells, optical transmittance and yellowness might be the most concerned parameters because yellowness of the polymer substrates might decrease the conversion efficiency of solar light to electricity. Thus, it would be helpful to understand the analysis and evaluation techniques for CHTPF films.

2.3.1. Optical properties

Common optical properties, such as yellow index (YI), haze, optical transmittance at specific wavelength, ultraviolet-visible cutoff wavelength, refractive index are usually needed to be evaluated for CHTPFs.

Yellowness index (YI) indicates the degree of departure of an object color from colorless or from a preferred white toward yellow. Haze value indicates the degree of cloudiness in a film. The YI and haze value of a film can usually be measured by a colorimeter and can be computed by a given procedure from colorimetric or spectrophotometric data [19]. Optical transmittance of a film indicates the percentage of incident light that is transmitted by the film. The reciprocal of optical transmission is the haze value, which increases as the percent of transmission decreases. As a general rule, 0% haze relates to complete transparency, up to 30% is translucent, and more than 30% haze is considered opaque. Optical transmittance of a film can usually be measured with an ultraviolet-visible light spectrophotometer.

Refractive index of a film indicates the ratio of the velocity of light in vacuum to that in a film. It is the ratio of the sine of the angle of incidence to the sine of the angle of refraction. Refractive index values can be measured with a prism coupler. Birefringence of a film indicates the difference in the refractive indices of two perpendicular directions in a film. When the refractive indices measured along three mutually perpendicular axes are identical, they are classified as optically isotropic. When the film is stretched, providing molecular orientation, and the refractive index parallel to the direction of stretching is altered so that it is no longer identical to what is perpendicular to this direction, the film displays birefringence. The common apparatus for optical parameters measurement of CHTPF films are shown in Figure 7.

2.3.2. Thermal properties

As mentioned before, the thermal stability of optical films is becoming increasingly important for their applications in optoelectronic fabrications. The thermal properties of an optical film include thermal decomposition temperature (T_d), glass transition temperature (T_g), coefficient of thermal expansion (CTE), and high-temperature dimensional stability. Generally, thermal analysis for an optical film indicates any analysis of physical or thermodynamic properties of the film in which heat is directly involved, with the heat either being added or removed. Different methods are used with each method providing certain useful data or information.

Thermogravimetric analysis (TGA) is an analysis by the measurement of weight changes of an optical film as a function of increasing temperature with time. Properties measured include thermal decomposition temperature and relative thermal stability. Dimensional stability of a polymer optical film indicates its ability to retain the precise initial shape and size. It is the temperature above which the films lose their dimensional stability. For most films, the main determinant of dimensional stability is their glass transition temperature.

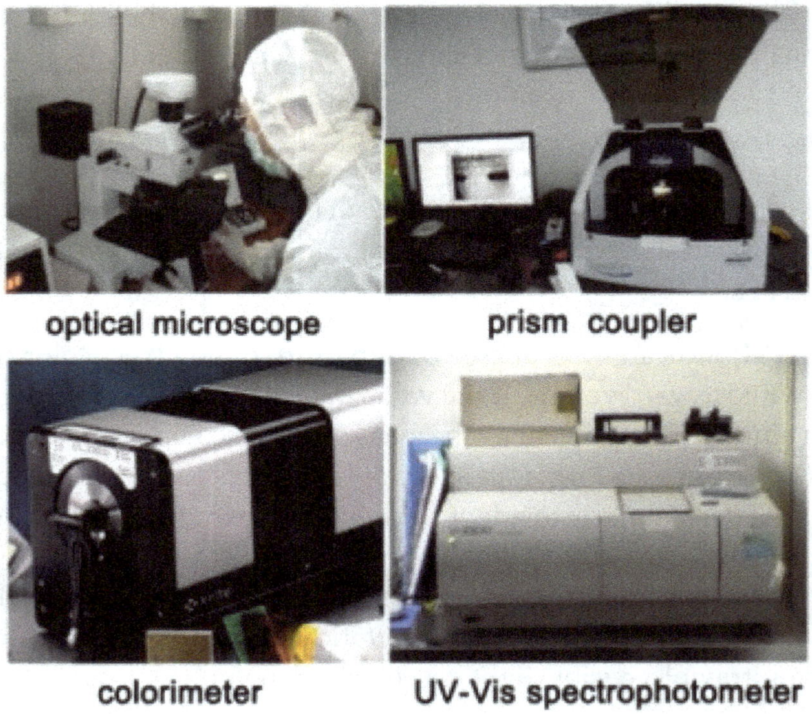

Figure 7. Optical properties measurement system

Below a certain temperature, polymer optical films will behave as hard glass-like substance. When heated above this temperature, individual segments of the polymer films will achieve large mobility; as a result the films become soft and elastic. The temperature at which this change happens is called the glass transition temperature (T_g). In other words, T_g indicates the reversible change in phase of a film from a brittle glassy state to viscous or rubbery state. At T_g, the film's volume or length increases, and above it, the properties of the film decrease. The T_g value of a film can be determined with specific equipment, such as differential scanning calorimetry (DSC), dynamic mechanical analysis (DMA), and thermal mechanical analysis (TMA) and the obtained values depend on the method used. For crystalline polymer films, such as PET and PEN, the crystalline melting points are usually above T_g.

In a typical DSC measurement, two pans are placed on a pair of identically positioned platforms connected to a furnace by a common heat flow path. One pan contains the polymer film, the other one is empty (reference pan). Then, the two pans are heated up at a specific rate. The computer guarantees that the two pans heat at exactly the same rate, despite the fact that one pan contains polymer and the other one is empty. The polymer film sample will take more heat to keep the temperature of the sample pan increasing at the same rate as the reference pan. A plot is created where the difference in heat flow between the sample and the reference is plotted as a function of temperature. The inflection point in the heat flow plot is recorded as the T_g value for the film.

DMA indicates a technique in which either the modulus or the damping of a polymer film under oscillatory load or displacement is measured as a function of temperature, frequency, time, or other combinations. TMA indicates a test that measures the dimensional changes as a function of temperature. The dimensional behavior of a film material can be determined precisely. Measurements made include coefficient of linear thermal expansion (CTE), T_g, and softening characteristic. CTE value of a film reflects the change in volume per unit volume resulting from a change in temperature of the material. The mean coefficient is commonly referenced to room temperature and expressed in mm/mm °C. CTE value is quite important for polymer optical films which are used with other heterogeneous materials, such as metal, glass, or ceramic. The unmatched CTE values between the polymer films with the other materials are thought to be one of the most important reasons for delamination, cracking, and other failures in the devices.

2.3.3. Gas permission properties

When a plastic substrate is used for the flexible OLED application, the water-vapor transmission rate (WVTR) and oxygen transmission rate (OTR) feature of the plastic substrate become critical because most high-performance semiconductor organic compounds show degraded performance when exposed to environmental moisture [20]. As mentioned before, WVTR and OTR of the flexible substrates are severely limited to be below 10^{-4} cm^3/m^2 day and 10^{-6} g/m^2 day, respectively, for AMOLED and organic solar cells [21]. Unlike glass, plastic substrates usually cannot provide sufficient protection to the permeants. For example, general PI films have WVTR values of 10^0-10^2 g/m^2/day dependent on the aggregation structures of their molecular chains. Addition of some specific additives, such as graphene [22], might improve their moisture barrier properties to a limited extent. Thus, inorganic thin films with extremely higher barrier properties have to be used on the substrate in practical applications. In order to evaluate the WVTR and OTR features of one polymer film, it is necessary to understand these two parameters. WVTR and OTR can now be measured with water vapor or oxygen gas permeation measurement systems produced by Mocon Corp., USA. The Mocon test has a measurement limit of ~10^{-4} g/m^2 day. A lower WVTR measurement has to be measured by calcium test, which is able to measure up to ~10^{-6} g/m^2 day.

3. Applications of CHTPF films in optoelectronics

3.1. Commercialization of CHTPF films

It is safe to say that the commercialization of CHTPFs is highly promoted by their potential applications for flexible optoelectronic devices, such as flexible light-emitting diodes (F-LED), flexible solar cells or photovoltaic cells (PV), flexible thin-film transistors (F-TFT), flexible printing circuit boards (FPCB), and so on. Many present and future applications of optical films make greater demands for higher properties, and especially combinations of properties, than are available from the commodity materials. To satisfy these requirements, organic polymer chemists and chemical engineers have developed and commercialized many types of polymers, offering improved properties. Table 3 briefly summarizes the commercially available and R&D CHTPF optical films in the world. Optical polymers containing various thermal-stable units, such as PI, PAI, PA, PES, and PS, have been extensively investigated and commercialized. In addition, some kind of inorganic-organic hybrid optical films have also been developed. As mentioned before, a statistical date from Techno Create Corp shows that the market of CHTPF optical films in 2011 has been close to 1 billion Japanese yen and the market will see a rapid increase higher than 15% per year in the following years [9]. Some of the typical CHTPF optical films are shown in Figure 8.

Up to now, CHTPF optical films have found various applications as plastic substrate candidates for flexible optoelectronic devices, including FPCB, flexible display (TFT-LCDs or AMOLEDs, etc.), touch panel, electronic paper, and thin photovoltaic cells. Plastic substrates with both optical transparency and high-temperature resistance have great potential applications in these areas due to the superior flexibility, lightness, cost-effectiveness, and processability to their fragile and expensive glass analogs. For instance, in the fabrication of flexible bottom-emission AMOLED devices, the processing temperatures of light emitting-components on the flexible substrates usually precede 300°C. Under such processing conditions, only CTHTP optical films such as colorless PI films could meet the severe demands.

Company	Product name	Resin	Transmission %	T_g °C
Mitsubishi Gas Chemical	Neopulim®	PI	89–90	>300
DuPont-Toray	Colorless Kapton®	PI	87	>300
Kolon	NA[1]	PI	88	>300
Japan Synthetic Rubber	Lucera®	NA	88	280
Toyobo	HM type	Polyamideimide (PAI)	91	225
Nippon Steel Chemical	Sillplus®	Resin+glass	91–92	NA
Toray	Aramid®	Polyamide (PA)	NA	315
Sumitomo Bakelite	Sumilite® FS-1300	Polyethersulfone (PES)	89	223
Showa Electricity	Shorayal®	NA	92	250
Tosoh	OPS film	Polysulfone (PS)	93	220
Kurabo	Examid®	Polyamide (PA)	NA	220

[1] Not available.

Table 3. Commercialization of CHTPF optical films in the world

3.2. Applications of CHTPF films

There has been growing interest in the use of plastic film substrates in the fabrication of future electronic devices, such as flexible displays, photovoltaics, batteries, sensors, and antennas [23]. This developing trend provides great opportunities for the development of CHTPF optical films. As shown in Figure 9, CHTPFs have found widespread applications in optoelectronics as various substrates for flexible display devices, flexible solar cells, FPCBs, touch panels, and so on.

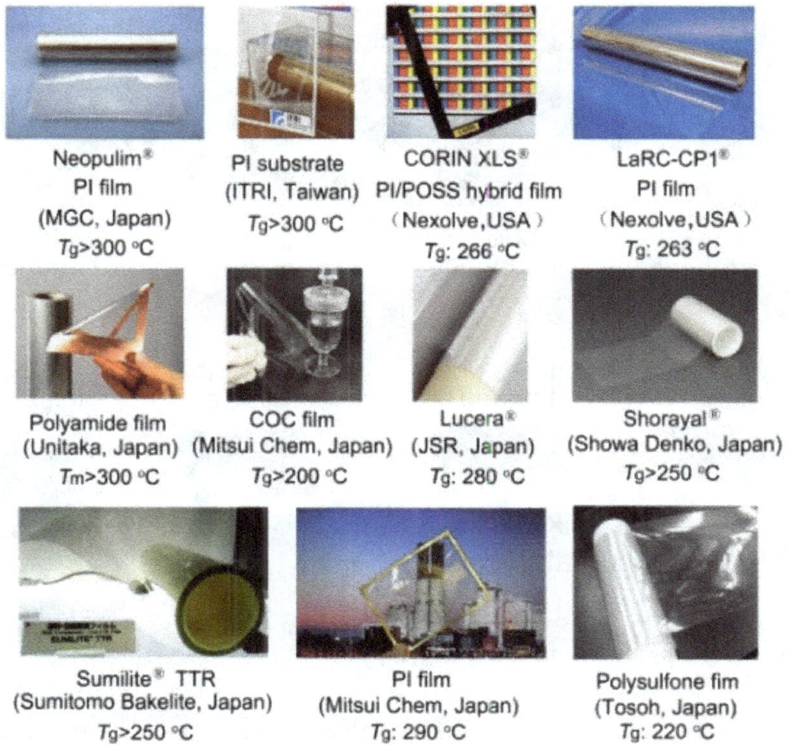

Figure 8. Commercially available or R&D CHTPF products in the literature

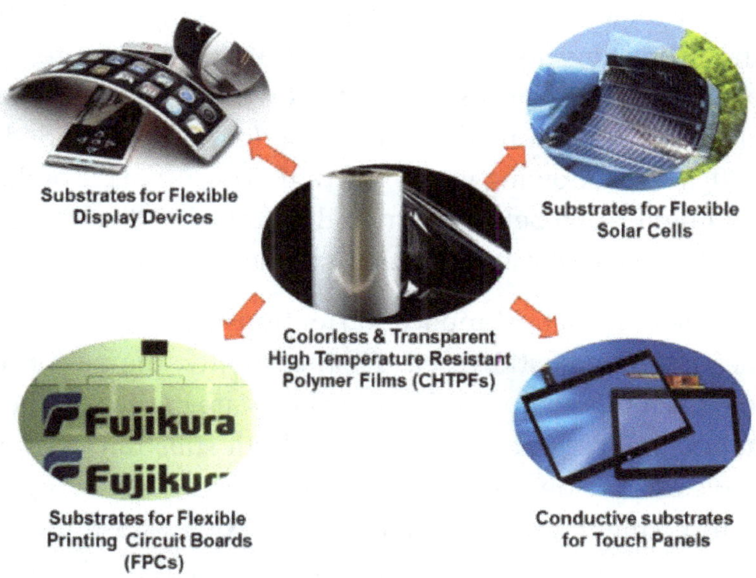

Figure 9. Potential applications of CHTPFs in optoelectronics

3.2.1. Substrates for advanced flexible display devices

As the structural support and optical signal transmission pathway and medium, flexible substrates are playing ever-increasing important roles in advanced optoelectronic display devices [24], the characteristics and functionalities of flexible substrates have been becoming

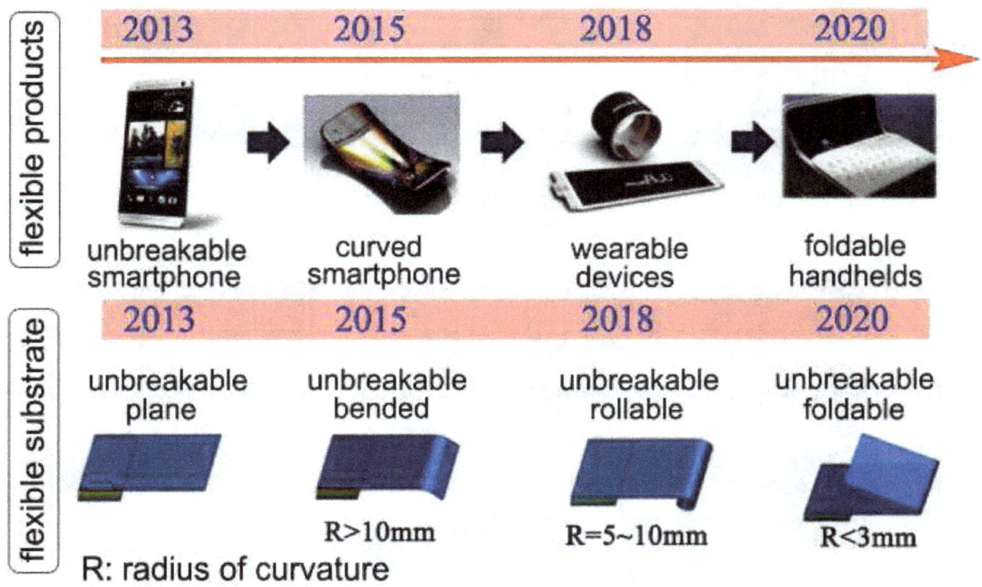

Figure 10. Roadmap for flexible products and substrates [24]

the important factors that affect the quality of flexible devices. Currently, there are mainly three types of substrates for flexible displays: thin glass, transparent plastic (polymer), and metal foil. Transparent plastic substrates possess good optical transmittance similar to that of thin glass; meanwhile the good flexibility and toughness comparable to those of metal foils. Thus, they are ideal for flexible display. A flexible display using a plastic substrate is considered to be one of the promising displays because of attractive features, such as thinness, lightweight, and good flexibility. For instance, as shown in Figure 10, the development of flexible substrates is experiencing a roadmap of plane (current)→ bended (2015)→ rollable (2018)→ foldable (2020) in the following years. The radius of curvature of the highly transparent flexible substrates might reach below 3 mm in the year of 2020. At that time, transparent plastic substrates might be the best candidate that can meet the demands.

However, in order to achieve a practical application for transparent plastic substrates in flexible display, several issues have to be addressed. First, currently, the performance of thin-film transistors (TFTs) built on common optical films or sheets are limited by the low-temperature process caused by the low thermal stability of current plastic substrates, typically below 250°C. For instance, for flexible display devices, such as active matrix-driven organic light-emitting diodes (AMOLED) processing, fabrication of TFTs on flexible substrates is one of the most important procedure. Up to now, there have been four types of production technologies for TFT fabrications in AMOLED, including amorphous silicon (a-Si) TFTs, low-temperature polysilicon (LTPS) TFTs, oxide TFTs, and organic TFTs (OTFTs). The key features for the current TFTs are summarized in Table 4 [25].

It can be seen that LTPS TFTs technique exhibits the highest field-effect mobility and stable electrical performance. However, the procedure requires a high process temperature of about 500°C during silicon crystallization. Conventional polymer optical film substrates cannot meet the application. For a-Si TFTs process, it has been widely used for AMOLED devices owing to uniform electrical characteristics over large areas, reasonable field-effect mobility, low-temperature process (< 300°C), and low cost compared to the other techniques.

	a-Si TFTs	LTPS TFTs	OTFTs	IGZO TFTs
Field-effect mobility (cm^2/V-s)	< 1	50–100	0.1–1	10–30
Process temperature (°C)	<300	300–500	<300	<300
Device stability	Challenging	Good	Challenging	OK
uniformity	Good	Challenging	OK	OK
Manufacturability	Excellent	Maturing	Developing	Developing
cost	Low	Medium	To be determined	

Table 4. Key features for a-Si TFTs, LTPS TFTs, OTFTs, and IGZO TFTs [25]

ITRI (Industrial Technology Research Institute, Taiwan) developed a unique flexible-universal-plane (FlexUP) solution for flexible display applications [26]. This new technique relies on two key innovations: flexible substrate and a debonding layer (DBL). As for the flexible substrate, ITRI developed a colorless PI substrate, which exhibits good optical transmittance (90%), high T_g (>300 °C), low CTE (28 ppm/°C), and good chemical resistance. In addition, the PI substrate with barrier treatment shows a WVTR value less than 4×10^{-5} g/m^2/day. Moreover, this barrier property suffered only to a minor drop, to 8×10^{-5} g/m^2/day, after the flexible panel had been bent 1000 times at a radius of 5 cm. The substrate used a hybrid technique, which contains a high content (>60 wt%) of inorganic silica particles in the PI matrix. A 6-inch flexible color AMOLED display device was successfully fabricated using this substrate. By using this colorless PI substrate, flexible touch panel was also successfully prepared.

A 7-inch flexible VGA transmissive-type active matrix TFT-LCD display with a-Si TFT was successfully fabricated on the colorless PI substrate developed by ITRI [27]. The colorless PI substrate has the features of high T_g (>350°C) and high light transmittance (>90%), which ensure the successful fabrication of 200°C a-Si:H TFT in the flexible device, as shown in Figure 11. The flexible panel showed resolution of 640×RGB×480, pixel pitch of 75×225 mm, and brightness of 100 nit. This technique is fully a-Si TFT backplane compatible, which makes it attractive for applications in high-performance flexible display. Similarly, a-Si TFTs deposited on clear plastic substrates (from DuPont) at 250-280°C was reported [28]. The free-standing clear plastic substrate has a T_g value higher than 315°C and a CTE value below 10 ppm/°C. The maximum process temperature of 280°C has been close to the temperature used in industrial a-Si TFT production on glass substrates (300-350°C).

Toshiba Corp., Japan, successfully fabricated a flexible 10.2-inch WUXGA (1920×1200) bottom-emission AMOLED display device driven by amorphous indium gallium zinc oxide (IGZO) TFTs on a colorless and transparent PI film substrate, as shown in Figure 12 [29]. Firstly, a transparent PI film was formed on a glass substrate and then a barrier layer was deposited to prevent the permeation of water. Then, the gate insulator, IGZO thin film, source-drain metal,

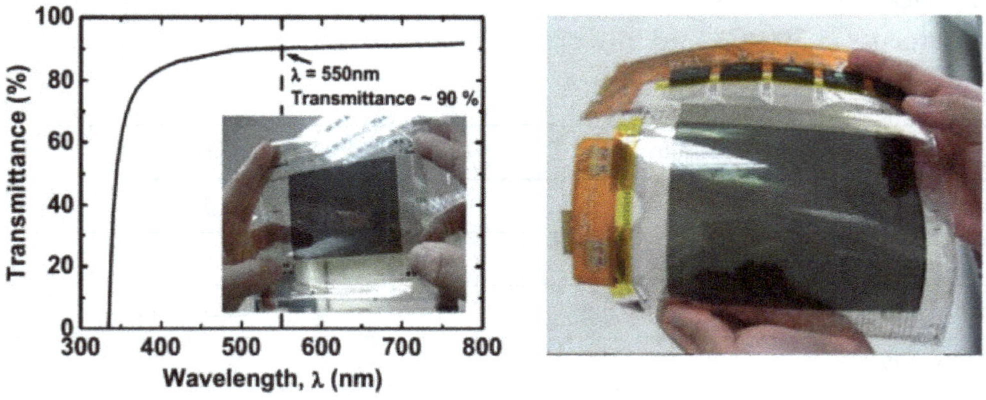

adapted, modified, and reprinted from ref. [27]

Figure 11. Colorless PI substrate and the color VGA flexible TFT-LCD

and passivation layer were successively deposited to afford the IGZO TFT. Secondly, the flexible AMOLED panel was fabricated using the IGZO TFT, color filter, white OLED, and encapsulation layer. Finally, the OLED panel was debonded from the glass substrate to afford the final AMOLED panel. The threshold voltage shifts of amorphous IGZO TFTs on the PI substrates under bias-temperature stress have been successfully decreased to less than 0.03 V, which is equivalent to those on glass substrates. ITRI also reported high-performance flexible amorphous IGZO TFTs on transparent PI-based nanocomposites substrates [30].

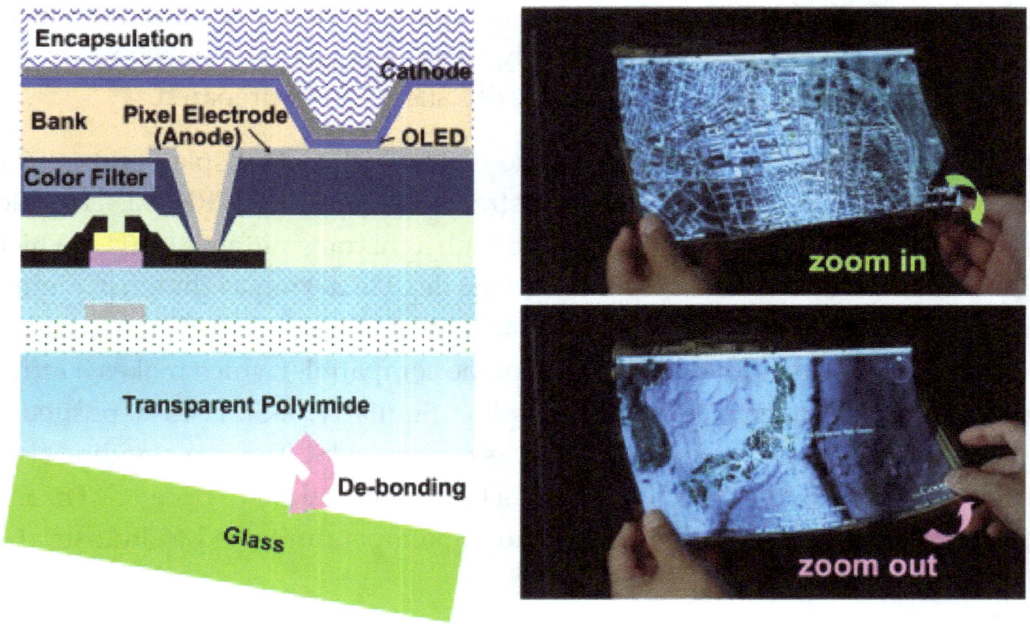

Figure 12. Flexible 10.2-inch AMOLED devices on transparent PI substrates [29]

Besides PI flexible substrates, other CHTPF substrates have also been developed. For instance, Teijin Ltd, Japan, developed novel high-temperature polycarbonate (PC) substrates for flexible displays [31]. The PC base film was obtained by a solvent casting process from dichloromethane solution and exhibited high optical transmittance (91%), high T_g value (215 °C), ultra-low intrinsic birefringence and low retardation (1 nm), good elastic and dimensional stability, and an extremely smooth surface. The new substrates consisted of the high-temperature PC base film, silicon oxide gas barrier layer, and transparent indium zinc oxide (IZO) conductive film which showed promise in overcoming the obstacles in producing many kinds of thin, lightweight, and flexible display devices. Similarly, a high heat resistance PC film with the T_g of 240°C and optical transmittance higher than 90% in the visible light region has been reported by General Electric [32]. A transparent, high barrier, and high heat substrate for organic electronics was successfully prepared by the film.

In summary, with the development of CHTPF optical films, the fabricated TFTs have showed similar characteristics to those of industry-standard a-Si TFTs fabricated on glass in the 300-350°C range. This result represents an important step toward a generic TFT backplane on flexible and optically clear film substrates.

3.2.2. *Substrates for transparent Flexible Printing Circuit Boards (FPCBs)*

Over the years, the FPCB applications have always been the largest market for high-temperature polymer films, such as PI, polyamideimide, and polyetherimide films. The flexible nature of FPCBs allows their convenient use in compact electronic equipment such as portable computer, digital cameras, watches, and panel boards. Generally, the traditional FPCB is mainly prepared from flexible copper-clad laminates (FCCLs), as shown in Figure 13. FCCLs consist of a layer of PI film bonded to copper foil. Depending on the intended use of the laminate, copper may be applied to one (single-sided) or both sides (double-sided) of the PI film. PI film almost completely dominates the portion of FCCL market in which heat resistance is needed to withstand the soldering temperatures. Recently, with the development of flexible displays, necessity for a transparent film substrate in place of glass substrate is increasing. Correspondingly, a transparent film substrate for FCCLs is increasingly desired. However, most of the all-aromatic PI films currently used in FCCLs show colors from yellow to deep brown, and thus cannot be used in transparent FCCLs.

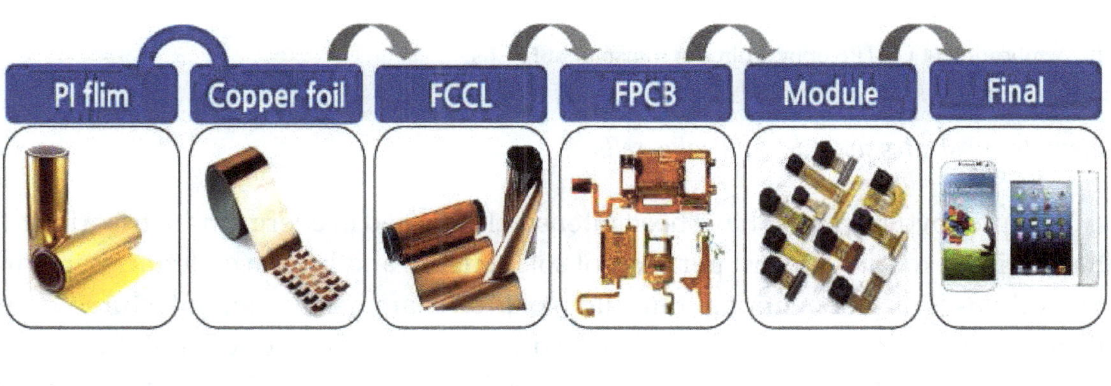

Figure 13. FPCB industry chains from FCCL to final products

Toyobo Corp., Japan, recently patented a colorless and transparent FCCL and the derived FPCB based on a PAI film [33]. The PAI film was synthesized from 1,2,4-cyclohexanetricarboxylic anhydride (HTA) (Table 2) and aromatic diisocyanate monomers (Figure 6) and the curing procedure was 200°C/1 h, 250°C/1 h, and 300°C/ 30 min under nitrogen. The film exhibited good thermal stability with T_g of 300°C, light transmittance of 89%, good tensile properties with tensile strength of 140 MPa, elongation at break of 30%, tensile modulus of 3.9 GPa, and low CTE of 33 ppm/K. The single-side FCCL from the PAI film and copper coil showed good soldering resistance, high bonding strength (10.6 N/cm), and good dimensional stability under the condition of 150°C for 30 min. In addition, the FCCL showed good optical transparency with a transmittance of 75% at the wavelength of 500 nm.

Very recently, there has been vigorous activity in developing and commercializing transparent FPCB products in the world. This is mainly driven by the urgent needs of such products for mobile communication optoelectronics. Typical products reported by multiple manufacturers in public are summarized in Figure 14. Various optical films including PEN, PAI, and PI films have been used as the substrates in these new products. It can be anticipated that CHTPF optical films will play an increasingly important role for the future development of transparent FPCBs.

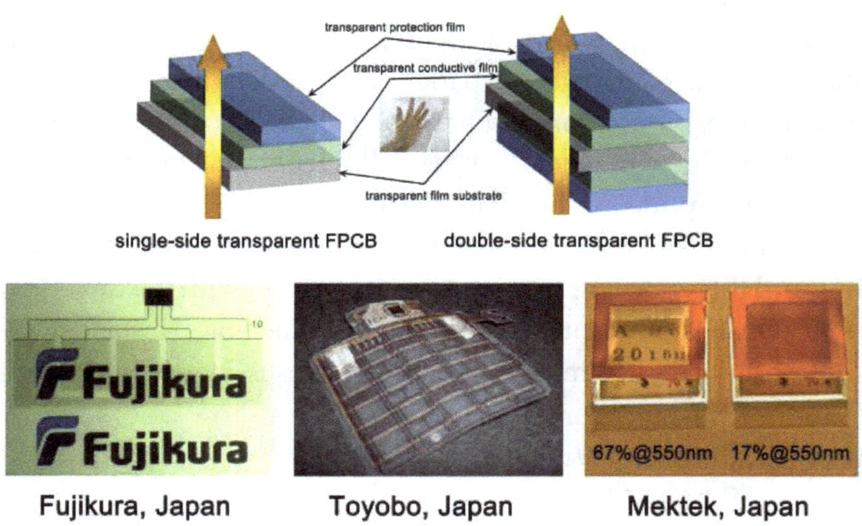

Figure 14. Application of CHTPF optical films in transparent FPCBs

3.2.3. Flexible substrates for thin-film solar cells

Solar cells or photovoltaics (PV) have been intensively studied in energy industries due to their potential ability to reduce the cost per Watt of solar energy and improve lifetime performance of solar modules [34]. Conventional thin film solar cells are usually manufactured on transparent conducting oxide coated 3-5 mm thick soda-lime glass substrates and offer no weight advantage or shape adaptability for curved surfaces. Fabricating thin-film solar cells on flexible polymer substrates seems to offer several advantages in practical applications, such as weight saving, cost saving, and easy fabrication. The polymer substrates for thin-film solar cell fabrications should be optically transparent and should withstand the high processing temperatures. For example, for the current cadmium telluride (CdTe) cell fabrication techniques, the processing temperatures are in the range of 450-500°C. Most of the transparent polymers will degrade at such a high temperature. Undoubtedly, the lack of a transparent polymer which is stable at the high processing temperature of solar cells is one of the biggest obstacles for the applications of polymer substrates in flexible solar cells.

Wholly aromatic PI films, such as Kapton® (DuPont, USA) and Upilex® (Ube, Japan) can withstand a high temperature round 450°C. However, they exhibit deep colors and strongly absorb visible light. CdTe solar cells on such PI substrates will yield only low current due to large optical absorption [35]. The development of colorless PI film with good high-temperature stability makes it possible to produce high-efficiency solar cells. One of the most promising reports on the successful applications of colorless PI films in flexible solar cells fabrication might be the work carried out in Swiss Federal Laboratories for Materials Science and Technology (Empa) [36]. As one of the Empa's continuous work on developing high-efficiency thin-film solar cells aiming at enhancing their performance and simplifying the fabrication processes, they utilized colorless PI film (developed by DuPont) as the flexible substrate for CdTe thin-film PV modulus in 2011 (Figure 15). A conversion efficiency of 13.8% using the new substrates was achieved, which was the new record among this type of solar cells at that time.

Figure 15. CdTe solar cells on colorless PI substrate (Source: Empa)

4. Conclusions

Undoubtedly, CHTPFs represent a class of new materials with both high technological contents and high additional value. High comprehensive properties make them good candidates for advanced optoelectronic devices. It can be anticipated that, with the ever-increasing demands of optoelectronic fabrication, CHTPFs will attract more attentions from both the academia and the industry. For example, demand will continue to grow for displays of smart phones, tablet PCs, and other types of mobile electronic devices. Furthermore, these displays will be continuously improved in terms of visibility, flexibility, durability, and lightweight. In this context, CHTPF optical films are facing great developing chance. However, up to now, these still have several obstacles that should be overcome for the wide applications of CHTPFs in advanced fields. First, very limited commercially available CHTPF products greatly increase their cost, which lead to a very limited application only in high-end optoelectronic products. Low-cost CHTPFs are highly desired for their wide applications. Secondly, the combined properties of current CHTPFs should be further enhanced, such as further improving their optical transmittances at elevated temperatures, improving their mechanical and gas barrier properties. Thirdly, the manufacturing technology for CHTPFs should be further perfected in order to increase their uniformity, colorlessness, and dimensional stability at high temperatures.

Acknowledgements

Financial support from National Basic Research Program (973 Program) of China (2014CB643605), National Natural Science Foundation of China (51173188) and Beijing Science and Technology Project (D141100003314002) are gratefully acknowledged.

Author details

Jin-gang Liu[1*], Hong-jiang Ni[1], Zhen-he Wang[1], Shi-yong Yang[1] and Wei-feng Zhou[2]

*Address all correspondence to: liujg@iccas.ac.cn

1 Laboratory of Advanced Polymer Materials, Institute of Chemistry, Chinese Academy of Sciences, Beijing, People's Republic of China

2 BOE Technology Group Co. Ltd., Beijing, People's Republic of China

References

[1] MacDonald WA. Engineered films for display technologies. J Mater Chem 2004;14:4-10.

[2] Choi MC, Kim YK, Ha CS. Polymers for flexible displays: from material selection to device applications. Prog Polymer Sci 2008;33:581-630.

[3] Logothetidis S. Flexible organic electronic devices: materials, process and applications. Mater Sci Eng B 2008;152:96-104.

[4] Huang JJ, Chen YP, Lien SY, Weng KW, Chao CH. High mechanical and electrical reliability of bottom-gate microcrystalline silicon thin film transistors on polyimide substrate. Curr Appl Phys 2011;11:S266-70.

[5] Nakano S, Saito N, Miura K, Sakano T, Ueda T, Sugi K, Yamaguchi H, Amemiya I, Hiramatsu M, Ishida A. Highly reliable a-IGZO TFTs on a plastic substrate for flexible AMOLED displays. J Soc Inform Display 2012;20(9):493-8.

[6] Yamaguchi H, Ueda T, Miura K, Saito N, Nakano S, Sakano T, Sugi K, Amemiya I, Hiramatsu H, Ishida A. 11.7-inch flexible AMOLED display driven by a-IGZO TFTs on plastic substrate. SID Sympos Digest Tech Pap 2012;43(1):1002-5.

[7] Ei-chi S. The thermal resistant films for electronics. Kogyo Chosakai Publishing: Tokyo, Japan; 2006 (in Japanese).

[8] Ito H, Oka W, Goto H, Umeda H. Plastic substrates for flexible display. Japan J Appl Phys 2006;45(5B):4325-9.

[9] Tsukidate K. Current Status and Prospect of Heat-resistant transparent film market. Techno-Create Month J 2012;2:1-3.

[10] Ghosh MK, Mittal KL. Polyimides. New York: Marcel Dekker; 1996.

[11] DeMeuse MT. ed. Biaxial stretching of film. Principles and applications. Woodhead: Cambridge; 2011.

[12] Siemann U. Solvent cast technology - a versatile tool for thin film production. Prog Colloid Polymer Sci 2005;130:1-14.

[13] Fay CC, Stoakley DM, St Clair AK. Molecularly oriented polymeric thin films for space applications. High Perform Polymers 1999;11(1):145-56.

[14] Oishi J, Hiramatsu S, Kihara S. Process and apparatus for production of colorless transparent resin film. US Patent 8357322, Jan 22, 2013.

[15] Oguro H, Kihara S, Bito T. Process for producing solvent-soluble polyimide. US Patent 7078477, Jan 18, 2006.

[16] Simone CD, Auman BC, Carcia PF. Low color polyimide compositions useful in optical type applications and methods and compositions relating thereto. US Patent 7550194, Jun 23, 2009.

[17] Jeong Y H, Cho HM, Park H J. Polyimide film, US Patent 8846852, Sep. 30, 2014.

[18] Lewis J. Material challenge for flexible organic devices [J]. Mater Today 2006;9(4): 38-45.

[19] Rosato DV. Concise Encyclopedia of Plastics. New York: Springer; 2000.

[20] Erlat AG, Yan M, Duggal AR. Substrates and thin-film barrier technology for flexible electronics. In: Wong WS, Salleo A (eds.) Flexible Electronics: Materials and Applications. New York: Springer Science + Business Media; 2009.

[21] Park JS, Chae H, Chung HK, Lee S I. Thin film encapsulation for flexible AM-OLED: a review [J]. Semiconduct Sci Technol 2011;26:034001.

[22] Tsai MH, Tseng IH, Liao YF, Chiang JC. Transparent polyimide nanocomposites with improved moisture barrier using graphene. Polymer Int 2013;62(9):1302-9.

[23] Crawford GP. Flexible Flat Panel Displays. Chichester: John Wiley & Sons Ltd; 2005.

[24] Chen JL, Liu CT. Technology advances in flexible displays and substrates. IEEE Access 2013;1:150-8.

[25] Pang HQ, Tajan K, Silvernail J. Recent progress of flexible AMOLED displays. Proc SPIE 2011;7956:79560J-1-10.

[26] Chen J L, Ho J C. A flexible universal plane for displays. J Soc Inform Display 2011;2(11):6-9.

[27] Yeh YH, Cheng CC, Ho KY, Chen PC, Lee MH, Huang JJ, Tyan HL, Leu CM, Chang CS, Lee KC, Fang Sy, Chen TH, Pan CY. 7-inch color VGA flexible TFT LCD on colorless polyimide substrate with 200°C a-Si:H TFTs. SID Sympos Digest Tech Pap 2007;38(1):1677-9.

[28] Long K, Kattamis AZ, Cheng IC, Gleskova H, Wagner S, Sturm JC. Stability of amorphous-silicon TFTs deposited on clear plastic substrates at 250°C to 280°C. IEEE Electron Device Lett 2006;27(2):111-3.

[29] Miura K, Ueda T, Saito N, Nakano S, Sakano T, Sugi K, Yamaguchi H, Amemiya I. Flexible AMOLED display driven by amorphous InGaZnO TFTs. The 20th International Workshop on Active-matrix Flat panel Displays and Device: Conference proceedings, July 2-5, 2013, Kyoto, Japan.

[30] Chien CW, Wu CH, Tsai YT, Kung YC, Lin CY, Hsu PC, Hsieh HH, Wu CC, Yeh YH, Leu CM, Lee TM. High-performance flexible a-IGZO TFTs adopting stacked electrodes and transparent polyimide-based nanocomposites substrates. IEEE Trans Electronic Devices 2011;58(5):1440-6.

[31] Hanada T, Shiroshi I, Negishi T, Shiro T. Plastic substrate technologies for flexible displays. Proc SPIE 2010;7618:76180Q-1~8.

[32] Yan M, Kim T W, Erlat AG, Pellow M, Foust DF, Liu J, Schaepkens M, Heller CM, McConnelee PA, Feist TP, Duggal AR. A transparent, high barrier, and high heat substrate for organic electronics. Proc IEEE 2005;93(8):1468-77.

[33] Shimeno K, Ito T, Aoyama T, Nishimoto A, Nagata S, Kurita T. Polyamideimide resin, as well as a colorless and transparent flexible metal-clad laminate and circuit board obtained therefrom. US Patent 82222365, Jul. 17, 2012.

[34] Fraas I, Partain L. Solar Cells and their Applications, Second Edition. New Jersey: Wiley; 2010.

[35] Mathew X, Enriquez JP, Romeo A, Tiwari AN. CdTe/CdS solar cells on flexible substrates. Solar Energy 2004;77:831-8.

[36] Aliyu MM, Islam MA, Hamzah NR, Karin MR, Martin MA, Sopain K, Amin N. Recent developments of flexible CdTe solar cells on metallic substrates: Issues and prospects. Int J Photoenergy 2012; 2012:1-10, doi:10.1155/2012/351381.

5

Optoelectronic Biomedical Systems for Noninvasive Treatment and Control with Informated Support in Solutions

Barylo Hryhoriy, Gotra Zenon, Ivakh Mariya, Kozhukhar Oleksandr, Makara Ivanna and Virt Volodymyr

Abstract

The chapter presents the development of an optical-electronic system with informated support in doctor solutions for its application in hematology, otorhinolaryngology, and dermatology. With the purpose of improving the quality of treatment by the perfection of the technology, the introduction of new contact lenses controls the experimental researches of spectrums of different objects and uses an optical-electronic system for the control of blood during and after the process of photopheresis. Optoelectronic systems for the treatment of ear noise by method of sedation with the use of optical bio-impact on the human psycho-physical condition through visual receptors with a special light medical information program. The proposed specialized optical-electronic system for dermatology allowed to significantly reduce the length of time on the process of diagnosis, provide a rapid process and expand diagnostic capabilities for the identification of detected pathologies.

Keywords: optical-electronic system, informated support, contact lenses controls, hematology, otorhinolaryngology, dermatology

1. Introduction

The development of optical and optoelectronic non-invasive means of diagnosis and research of opportunities for their effective use, in particular, to assess the structural condition of the tissue area in normal and pathologically is an actual problem of modern biomedicine [1, 2]. A variety of pathological tissue area neoplasms forced to seek new approaches to conditioning, differentiation, and diagnosis. Such possibilities in dermatology, otolaryngology, hematology, and other fields of medicine have become a reality with the beginning of the application of the new principles of optoelectronic systems, highly efficient miniature light sources, and the design of new technologies using promising microcontrollers and flexible algorithms based on probabilistic methods approach (Bayesian method) and consistent statistical analysis (Wald method). The functional, pathophysiological, and psychophysical conditions of the patient, as well as changes in the conditions during illness, treatment, and individual treatments affect the optical properties of tissues. In particular, the express diagnosis of surface and near-surface layers of the skin on the principle analysis of signals reflected traversed, and its own electromagnetic radiation output information of the simulated process is to analyze the spectra of the light reflection of the scanned tissue.

For example, there have been new specialized optoelectronic and optical-electronic systems developed for [3]:

- automating the analysis of peripheral parameters and psychophysical conditions in the patient;
- activating the control blood system during photopheresis treatment procedure and (in the research study) optoelectronic active control of micro- and nano structures of biological fluids for new bacterial implantens technologies;
- physiotherapeutic procedures on the photo treatment of ear noises, tonsillitis, and pharyngitis; and
- express diagnostics and authentication of skin illnesses.

2. Optical-electronic system for the photopheresis treatment in hematology

2.1. Photospheresis

Photospheresis is a modern nanotechnology of blood, which is used for the treatment of heavily cured illnesses such as a T-cells lymph adenoma of the skin and psoriasis, considered as the precancerous stages. Due to the division of aluminous factions of blood with the introduction of a sensitizer under the action of electromagnetic radiation of area of a fellow creature of ultraviolet range and the electronic absorption of radiation of certain wavelengths, it is possible to carry out the treatment of the higher mentioned illnesses. Thus, the indexes of the key-in of optical radiation blood change in the different areas of spectrum [4].

2.2. Experiment and results

With the purpose of improving the quality of treatment by the perfection of technology due to the introduction of new contact lens controls in the experimental researches of spectrums of different objects that touch control of blood after and in the process of photopheresis were conducted. Conducting of these researches is extremely a necessity for the determination of prospective and possibility of method. The capacity of the structure consonant with the existing technology of photopheresis and the subsequent development of the checking system are offered below [5, 6].

The structure of the optoelectronic element of the research (Fig. 1) consists of the radiate 2 and sensory parts 5. The radiate part shows, by itself, the matrix of varicolored light-emitting diodes 1 with the chart of the electric feed. The light streams are from each of which, passing through the contactor from blood, is disposed between the radiate and sensory parts 3, gets on a photo detector sensory part 4. The sensory part includes itself, except for a photo detector or photo detectors electronic charts of the previous strengthening, feed, and tuning of the intensity of radiation. Such structure enables the comparison of the noted optical descriptions of blood with a marked higher resolution photo in the medical technology.

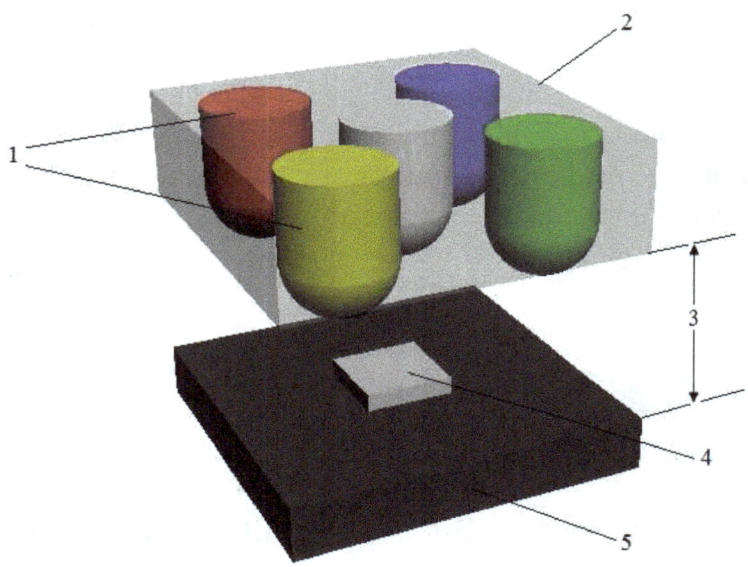

Figure 1. Structure of optoelectronic element. 1 - LED, 2 - radiative part, 3. Place of location of contactor, 4 – photo detector of sensory part, 5 - sensory part.

On the basis that we got and published earlier, recommendations in relation to principles and constituents of the checking system of the chart of experiment (Fig. 2) were developed, and the ground of choice of its components is carried out. A chart allowed to conduct the research of the dependencies of the intensity of electric signals, which were derived from the sensory part by the passing of a light stream from every light-emitting diode through the objects set in an experiment. By them were standard capacity with the bio-pierce of blood, empty capacity, layer of air with the sizes of capacity for blood and others like that.

For the choice of the constituents of an irradiate part - LED block 1, experimental researches that are spectral were conducted with the descriptions of light-emitting diodes in different areas within a visible range. The results of the researches of the light-emitting diodes are shown in Fig. 3. These results enabled us to choose certain types of light-emitting diodes according to some previous works pertaining to spectrum bars. There are also foreseen changes of light transmittance as a result of photopheresis. For the exception of the unforeseen influencing of

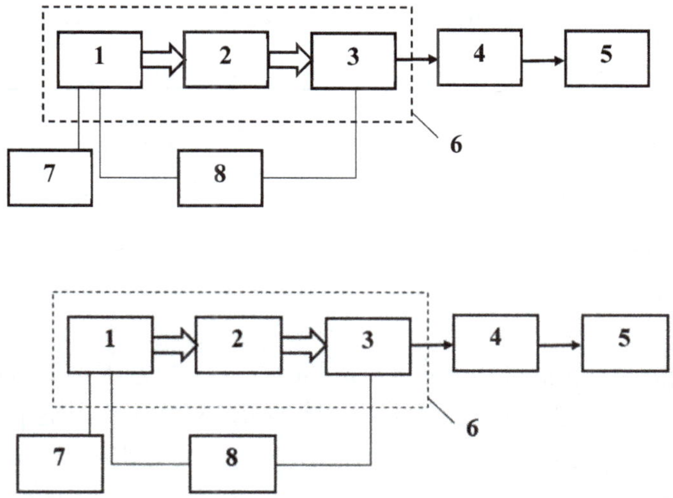

Figure 2. Chart of experiment: 1 – LED block, 2 – bio-sample, 3 – photo detector block, 4 – strengthener of photo-stream, 5 – measuring device of photo-stream, 6 – optoelectronic element, 7 – switchboard, 8 – feed.

extraneous radiates optoelectronic element 6 disposed in a light protective shell. For the influence on the results of the spectral descriptions of light transmittance and light reflection cuvette, the probed blood took place. And air around a cuvette is observed for the research of the passing of radiation of the chosen light-emitting diodes through this material, and the layer of air is chosen after its sizes. The intensity of radiation on spectral descriptions (Fig. 3) is resulted through the values of intensities and the maximal values for this description:

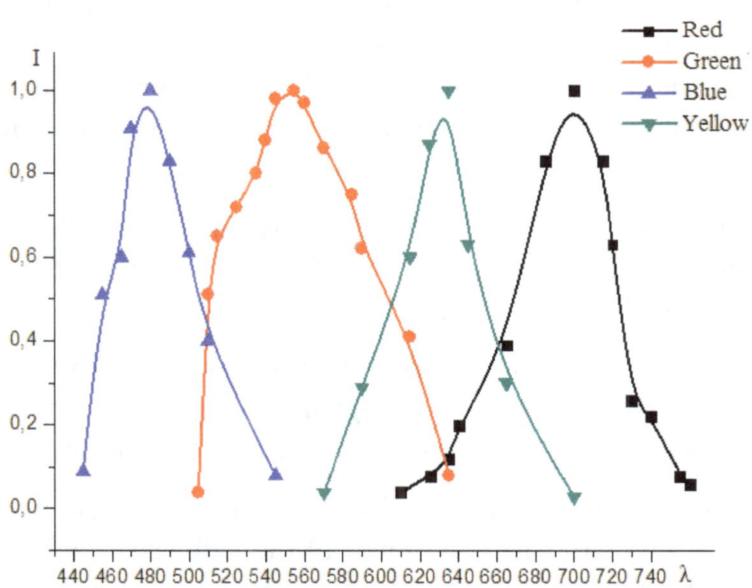

Figure 3. Spectral descriptions of separate light-emitting diodes I(λ).

On the basis of the derived results, the coefficients expected higher admissions of the objects chosen, the value of which is shown in Table 1. It is possible to correct the spectral description, which allows the exactness and authenticity of experiment.

Color of LED Object	Yellow	Blue	Green	Red
Air	1,7	5,8	8,7	10
Cuvette	1,4	5,4	6	6
Cuvette with blood	1,2	2,2	4,3	5,8

Table 1. Expected coefficients of admission.

To the sensory part, silicious photodiodes were applied as FD-24Ê with a spectral sensitiveness in the area of wave lengths of 590-980nm. The type and location of the photodiodes in a photo detector block enabled to get signals about the changes of radiation intensity after its transmission through the chosen object from each of the noted light-emitting diodes.

Both parts of the flat, thin-walled cuvette, which was filled the bio-pierce of blood, were set. The standard medical cuvette or, as it is named in hematological technologies, the contactor, was made from a nontoxic optically transparent polyethylene with a 2mm thickness of walls. The volume of blood simultaneously contained in the cuvette is 140x23x1mm.

After the known values of the volume of blood in a cuvette and its ultraviolet display in accordance with technology of photopheresis, the energy of excitation that blood gets during irradiation is estimated.

The radiation of the visible range of wave lengths, which is fixed with the basis of the control of changes in the spectral descriptions of blood as a result of OOH displays, must go through the walls of the cuvette twice. The developments of the controls and needs of the experimental researches with its own spectral descriptions must be taken into account. Given all these, the research for the different areas of the spectrum is done for every bio-sample and its absence on the basis of the value of the admission coefficient of optical radiation in the different areas of the spectrum.

The light-emitting diodes feed was carried out through the managed ballast resistors with the consistently included current measuring devices. For the switching of light-emitting diodes, an electro-mechanic switchboard was applied.

For the control of diets, light-emitting diodes and photo detectors were foreseen as the proper measuring devices of current and tension. The signals of the photo detector after strengthening by an electronic chart was seen in the measuring device through photo stream 5.

A research method included light-emitting diodes and measuring each of them to the photo stream. Results achieved with the use of a radiate block on the basis of 4 light-emitting diodes for different objects, which apply in photopheresis, are presented in Fig. 4.

Through the relation of the photo stream attained for the different light-emitting diodes in the different areas of spectrum for the objects chosen, the value of coefficients of light transmittance was higher. With the account of changes in the spectrum as a result of the cooperation of light with the cuvette and air, the expected values of the coefficients for the chosen light-emitting diodes are presented in the light transmittance of blood shown in Table 2.

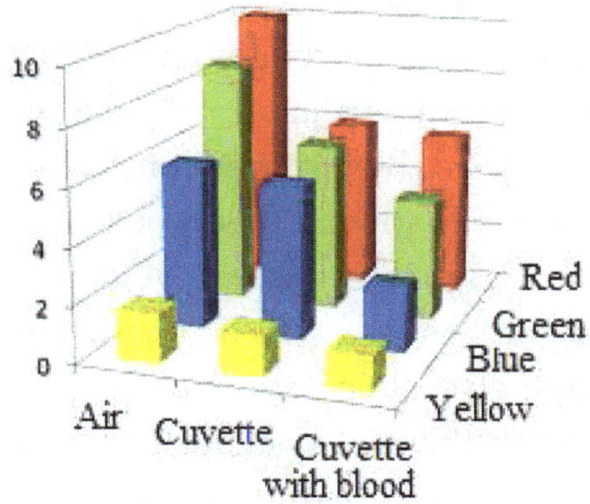

Figure 4. Dependencies of photo streams are for the chosen light-emitting diodes and different objects which apply in photopheresis.

Colour of LED	Blue	Green	Yellow	Red
Coefficients of transmittance	0,4	0,72	0,86	0,97

Table 2. The value of coefficients of light transmittance for blood.

These results are illustrated below in Fig. 5.

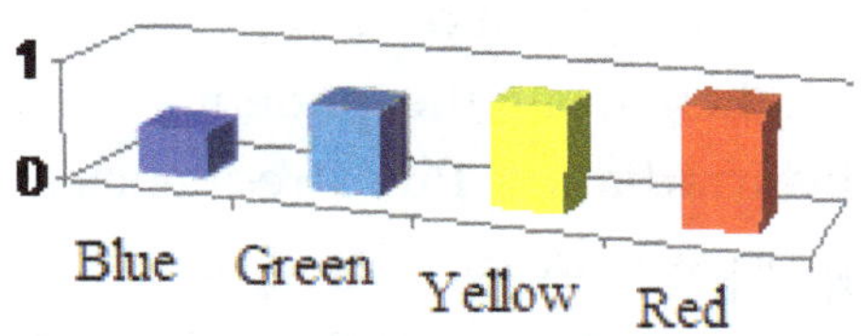

Figure 5. Coefficients of light transmittance for blood and the radiative part from four chosen light-emitting diodes.

Such set of bars can be applied to assess the facilities of the electro-optical research of blood as a result of its photo technological treatment.

On the developed operating optoelectronic prototype of the research of blood system during photopheresis or other photo-medic technologies, experimentally probed changes of coefficient admission of exposed to the rays at photopheresis are seen in four basic areas of spectrum.

It is shown that most spectral changes from photopheresis are tested by red, green, and dark blue areas of spectrum, which can be applied in the development of the optoelectronic checking system as a constituent of equipment for conducting of photopheresis.

After the results are acquired, it is possible to consider that the application of the offered parameters of the optoelectronic checking system will enable the development not only to control the technology of photopheresis but also for its perfection.

2.3. Model device for the study of optical parameters of blood during photopheresis

There is an urgent need to improve the efficiency of photomedical technology (PMT) for many medical areas such as hematology, otorhinolaryngology, dermatology, surgery, etc. Also, there is the problem of creating devices on the principles of programmable photo stimulation and information feedback from bio-objects (BO), minimal interference to BO, and no damage to the patient. One of prospective ways of solving this problem is to formulate the basic medical and technical requirements and the principles of development and application in medical practice and photo medicines, the basis of which would be assigned to new advances in research in the direction of optoelectronics and the design specialized for specific PMT optoelectronic systems (OES). Such thermal systems, discharge plasma, and semiconductor sources aiming incoherent radiation to allow the creation of BO stimulating irradiation with software controllable (according to the spatial frequency characteristics of the irradiated BO) dynamics and continuous noninvasive research of medicinal sessions for the analysis of changes in the optical parameters of the BO. This analysis can be based on the comparison of optical characteristics that are relevant to a particular medical technology (photo or any other technology) because, like the patient's blood, blood-filled organs, such as healthy, infected, or operated tissues, etc., due to differences in their BO radiation, reflection, and transmission BO test light streams [7-9].

Thus, a specialized OES has been developed for hematological PMT, including photopheresis technology as one of the most promising treatment for difficult and incurable illnesses such as psoriasis and T-cell lymphoma skin. The results of theoretical and experimental research regularities are also revealed. A clinical research device was created and implemented to provide a certain intensity and wavelengths of certain parts of the spectrum of ultraviolet and visible medical test irradiation taken from the patient's blood samples. To get the best stimulation effect of therapeutic irradiation flux, frequencies must be modulated, which corresponds to the frequency processes in irradiated samples of blood or blood fractions. Considering that there are only some accurately known ranges of frequencies, as they may be different for various patients, conditions should be provided for resonance frequency scanning. Contactless sensors are used as the correction modes for session therapeutic irradiation and for further treatment, based on the feedback provided, which is informative about the progress and effects of the treatment session. With the device, the volume of samples taken from the patients irradiated the treatment stream in the UV-A (320-400 nm) with specific power of 0.1... 0.4 mW/cm^2.

Upon the completion of the session, which had a duration of 5-10 minutes, the irradiated sample is returned to the patient's bloodstream. Information support based on continuous control changes, with the help of the optical characteristics of a subsidiary doctor who conducted the photopheresis, may be an important condition to contribute to the success of the treatment components of the information.

A similar treatment on the basis of a developed specialized OES with a dynamic radiation capacity of 10-10^2 mW (Fig. 6) allows therapeutic irradiation, control, and correction [10, 11].

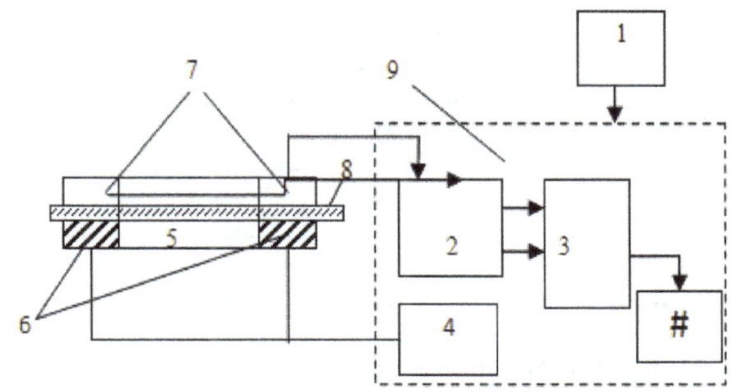

Figure 6. Block diagram of the online device for the analytical research of photopheresis efficiency: 1 - power supply 2 - the Amplifier, 3 - comparator, 4 - switch 5 – UV, 6 – LED, 7 - photodetector, 8 - sample BO, 9 – measurement unit, # – display.

OES is guided by a microprocessor, which laid irradiation modes and continuous control that are transmitted to a separate reflector and to a computer, via a USB-port, processing and display.

Before the start of the treatment, the doctor calibrates the OES, selects the mode of irradiation, puts the cell with the biosample in the device, and turns on the device. He watches the changes in the optical parameters of the biological test on the display. Upon reaching changes that correspond to those obtained in the clinical experiments' generalized value (Table 3), the doctor makes decisions on the sufficiency of the session or if a repeat session is necessary. At the same time, the doctor considers receiving understated indicators, the inexpediency for the further treatment, and in some cases, about unacceptable treatments for photopheresis of the patient.

Based on our research, the proposed model of the device is made for substantiated test light streams in spectral wavelengths of 642nm, 590nm, 505nm, and 465nm, which are also based on the changes in the intensity (Fig. 7) and colour temperature (Table 3) of the light streams through their passage in the BO.

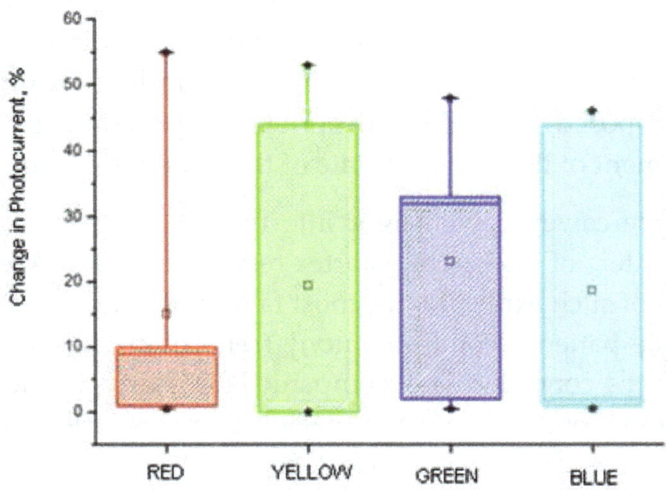

Figure 7. Changes in the photocurrents test light streams passed through the sample BO in a given spectral region.

Group of patients	Befor irradiation	After irradiation
	T_K, [K]	T_K, [K]
1	5300	4900
2	5550	4900
3	5350	5000

Table 3. Generalized changes in the optical parameters of biosample color temperature of test light flux after the irradiation and treatment process.

3. Optical-electronic system in otorhinolaryngology

3.1. Photomedical technology with light stimulation

On the basis of scientific and practical researches established, rhythmic light stimulation (light stimuli) can influence brain activity, and the dependence of the activity of the cerebral cortex on the stimulation frequency is associated with frequencies of biorhythms [12-13]. Thus, β-waves - over 13 Hz - support and activate the wakeful state, whereas α-waves - 8... 13 Hz debilitate the wakeful state, and τ and δ-waves - less than 8 Hz cause a slight or profound trance, or relaxation. Light-stimulation modes including those with light pulse repetition frequencies in accordance with the frequencies of space-time processes in biomedical objects (BO) can cause the bio-resonance effect in these objects. One can reach the effect of profound relaxation using an optical programmable trigger for the brain waves in the direction of low-frequency τ and δ-waves.

The obtained reactions can be explained by the changes in the activity of the somatosensory areas of the cerebral hemispheres' cortex under the influence of the photostimulation program. The moment of stabilization of these changes indicates that the cerebral cortex centers that are responsible for the relaxation are stimulated sufficiently. These changes are reflected by influencing stem vegetative and somatic nerve centers, and by reducing the tone of skeletal muscles and peripheral vessels. This results to the registered optical and thermal effects, and this fact agrees with physiology of processes in a human body during the stimulation of photoreceptors. It is proposed to use the appropriate contactless sensors for getting information about the achievement of the required state of the patient's organism.

Visual receptors that perceive a given light information program on the communication channels are excited centers of the cerebral cortex responsible for the psycho-emotional state of a person. As a result of such exposure, the most favorable conditions for treatment is when it is administered in the patient's condition atony (relaxation). It was determined that such influence on the patient's condition is accompanied by increased heat production, which caused an increase in the blood flow of small vessels and capillaries. In response, the system activates the regulation of temperature homeostasis.

These processes change the optical properties of tissue blood filled authority (BFA) [14], in particular, the detected variation of transmittance and reflectance flux at its interaction with a certain connective tissue, especially tissue BFA. Change in the first parameter causes, above all, an increase of the blood flow to the cells, and the second has an impact to the heat production on the surface of epithelial tissue changes (cf. deformation, change shape, sweating, etc.). Moment stabilizing changes in these parameters over time indicates the achievement of the required excitation centers of the cerebral cortex responsible for relaxing the body.

The optoelectronic system for the treatment of ear noise by method of sedation with the use optical bio-impact on the human psychophysical condition through visual receptors (Fig. 8) was equipped with a special light medical information program (Fig. 9). The program provides the appropriate frequency to change photos' stimulus on visual receptors of patients within 30 minutes of treatment [15].

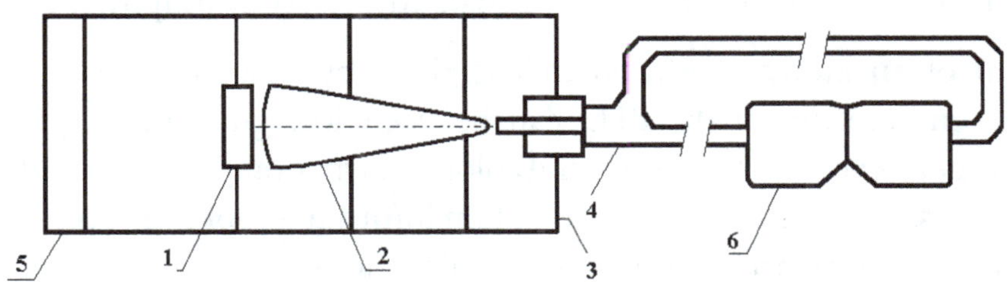

Figure 8. Model of photo stimulation device: 1 - LED terminal, 2 - focusing cone, 3 - case, 4 - lightguide, 5 - control unit, 6 - dimming glasses patient.

The measurement of change in the optical and thermal parameters of blood-filled object (BFO) in this experiment was carried out during the term of the medical program photo stimulation of bio influence.

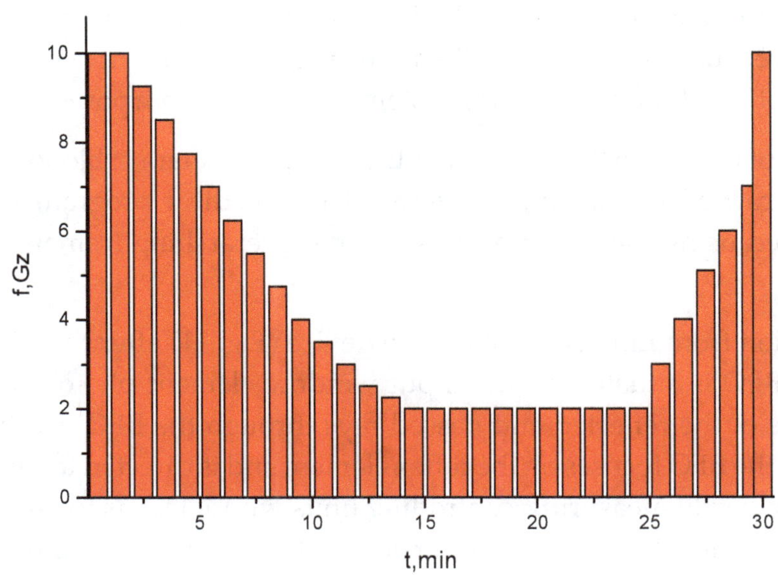

Figure 9. Special medical light information program of dependence time of the frequency photo stimulus for ear noise treatment by sedation [13].

In response, the system activates the regulation of the temperature of homeostasis. These processes change the optical properties of the blood-filled authority tissue. In particular, it is possible to fix the variation of transmittance and reflectance of the light flux in the interaction with certain connective tissues. The change of the first parameter causes, above all, an increase in the blood flow to the cells, while the second cause an impact on heat generation on the surface changes of the epithelial tissue (pores deformation, change shape, sweating, etc.). Moment stabilizing changes in these parameters achieves the required excitation centers of the cerebral cortex responsible for relaxing the body over time.

Based on these physiological processes developed and implemented for monitoring (changes of optical parameters of peripheral organ), the effectiveness of the procedures of the photo treatment of people who suffer from the disease otosclerotic is identified. The assessment of the impact on subjective indicators requires highly skilled medical staff to work during a treatment session and the use of complex and costly diagnostic equipment.

The achievement of an atony condition that is fixed by assessing changes in the optical parameters and temperature of the BFO can be used to signal the beginning of medical technologies by other factors, for example, the color information program on the specified area of organ and acupressure points. In order to obtain information of the necessary conditions of the patient organism at the start of the next medical technology, a contactless sensory actuator element, as an example, should be provided. Its objective is to obtain temperature and changes in the optical characteristics of BFO, and furthermore, to use hardware to determine the original set of changes. This task can be implemented, for example, by means of LED and photodiode optoelectronic components receiver and thermal radiation, which allows determining its temperature settings at the time. Information on optical performance and the temperature of the BFO is also assessed at the time when the original state of these changes is close to zero. This means that the system signals the biological response. The advantage of the proposed approach over the other is to carry out remedial action, capturing the individual signals of the bioresonance response to the impact. However, information on experimental research in this direction is not enough. Therefore, there is an urgent need for further studies on the implementation of modern control systems for medical purposes.

The measurement of change in the optical and thermal parameters of blood-filled object in this experiment were carried out during the term of the medical program photo stimulation bioeffects suggested using the exercise shown in the overarching framework of Fig. 10 and Fig. 11 [16-18].

The optical radiation from LED elements of the device (Fig. 11) after the interaction with the soft tissue of the BFO is situated on the input window element photo diode device that is located on the axis of the reflected and passed rays through the BFO. At the same time, the actual radiation of the BFO focuses on the reflex element and is directed to the photo convertor. These elements are partly measuring-controlling units whose task is to obtain optical signals after the interaction with the CA that are carriers of the reviews on the body healing effect. Subsequently processed opto-electronic units are converted from optical to electrical signals and after further analysis and comparison of test signals, the medical personnel are informed about the results on a computer monitor. The results of the described device provide a whole

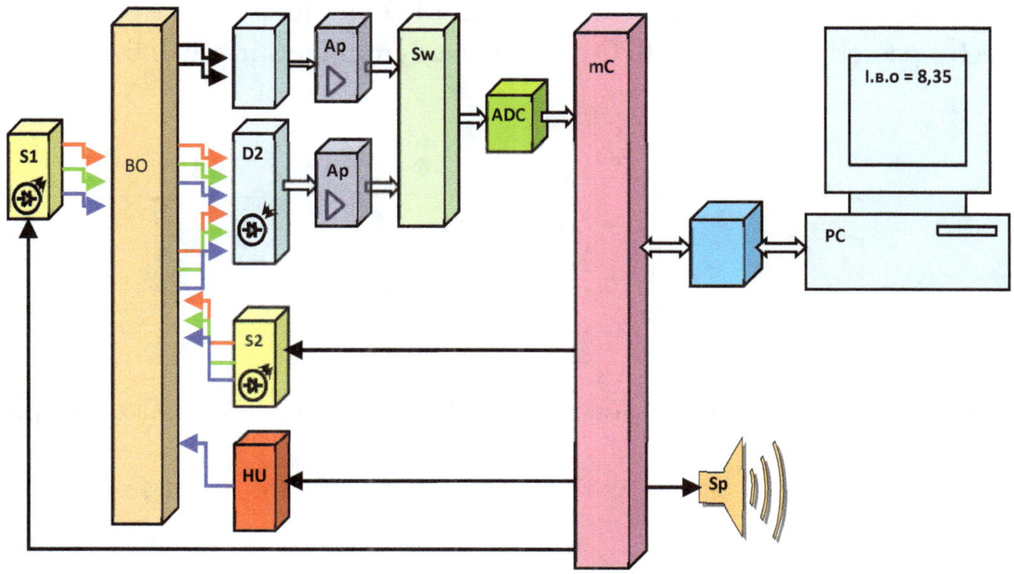

Figure 10. Generalized structure of optoelectronic photo diagnostic systems: BO - biological object, S1, S2 - light source, D1, D2 - transducers, Ap - signal amplifier, Sw - switch, ADC - analog - digital converter, mC - microcontroller, HU - medical device, TR - circuit connection, Sp - signaling device, PC - personal computer.

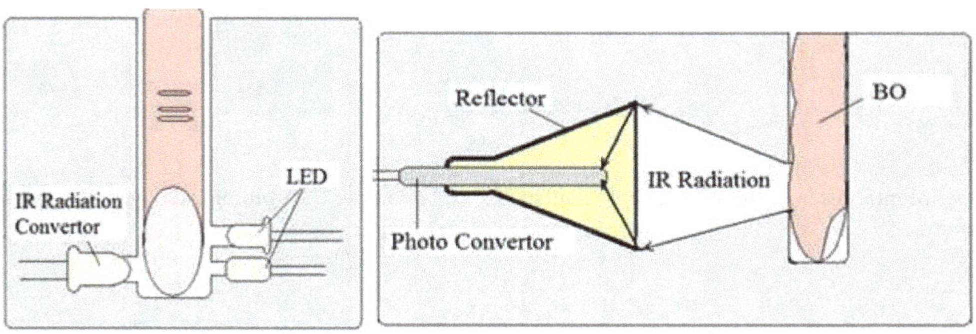

Figure 11. Converter circuit of the opto-electronic information system of evaluating the effectiveness of treatment procedures.

body response to the therapeutic effect of photostimul devices and systems, used for future decisions.

At the time of achieving biological resonance, the proposed fixed equipment during the photostimul program, the system allows to apply the proposed method of OES treatment. With the use of computer technology or online help from a physician-operator, the method for further treatment is determined by choosing the light program information and referral optical radiation with the settings specified in the scope.

The experimental studies applied some research, which are shown in Figs. 12, 13, 14. The transducer (TD) is placed on the optical axis, which focuses radiation on the TD BO, two LEDs, and photodetectors. The non-contact measurement of temperature and changes in the optical parameters of the experiments were carried out with the BO. The TD was placed on the focal length of the direct and reflected heat rays. The LED emitters are placed so that the fixed photodetector passed through the BFO and reflected from the surface of the light output.

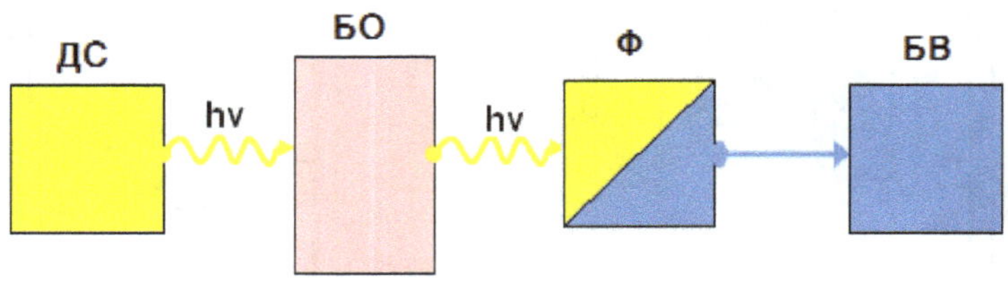

Figure 12. Experimental research of the transmission coefficients: ДС - light source, БО - biological object, Ф - hotodetector, БВ - unit measurement.

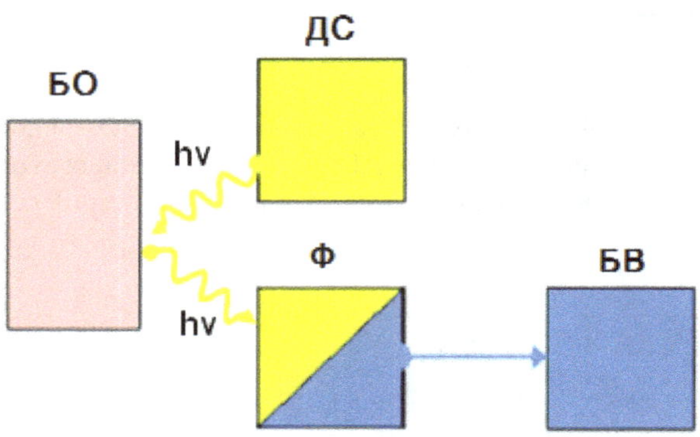

Figure 13. Experimental research reflecting coefficients: ДС - light source, БО - biological object, Ф - hotodetector, БВ - unit measurement.

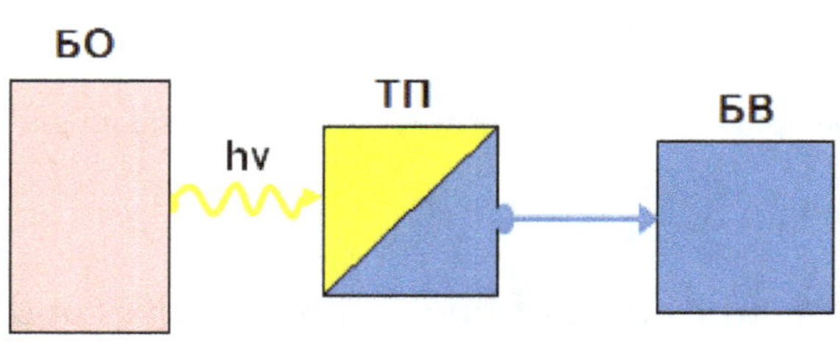

Figure 14. Experimental research showing temperature change: БО - biological object, ТП- thermo receiver, БВ- unit measurement.

The choice of LEDs to conduct research according to Fig. 12 and Fig. 13 was carried out based on the obtained spectral characteristics of the BO, which in this case is the phalanx of the finger (Fig.15).

Thus, the choice of the optical fiber array to conduct research to determine the coefficient of light transmission is made within the spectral characteristics of the BO. To determine the coefficient of the reflection elected LED emission spectrum, which does not fall within a specified range transmission BO.

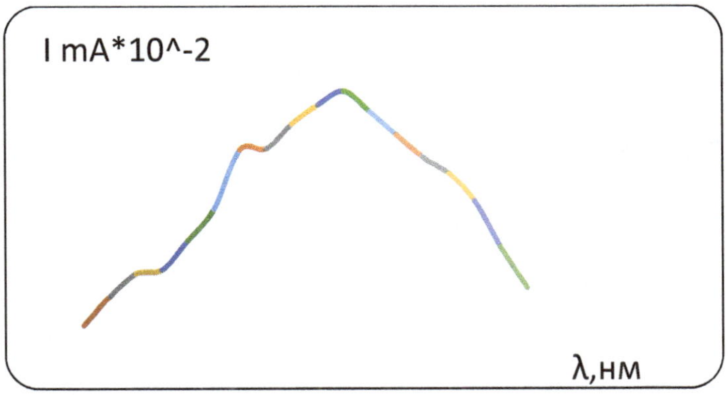

Figure 15. Spectral characteristics of BO.

For research to determine the dynamics of change in the transmission coefficient, the BO used the LED red area of the spectrum (Fig. 12). For research according to Fig. 13, the white LED spectral characteristics were used.

The results of the experimental research for one of the 20 patients are shown in Fig. 16 and Fig. 17.

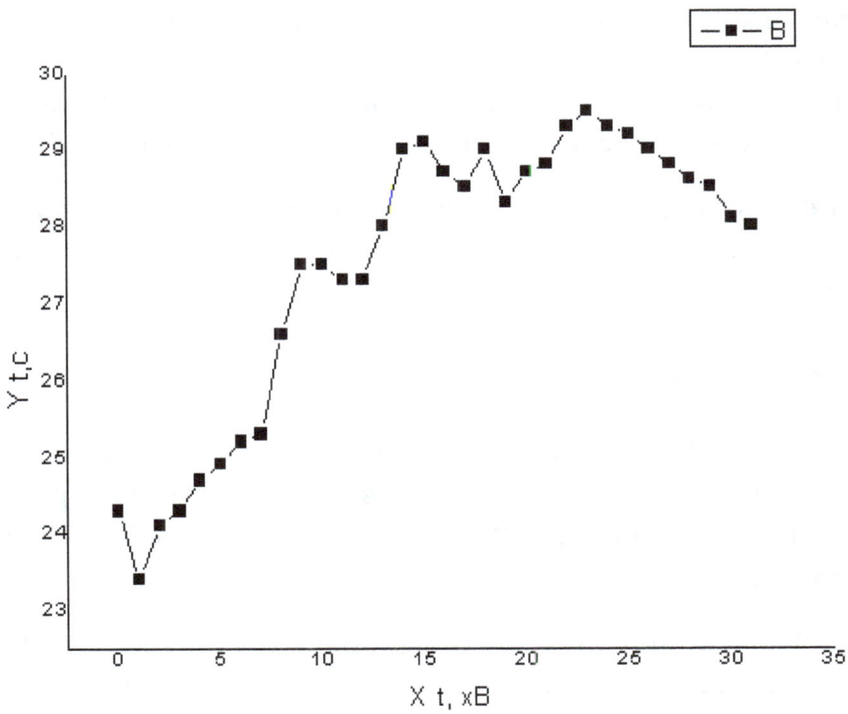

Figure 16. The change in temperature over time photo stimulation.

However, in all cases for the patients studied, the recurrence of dynamic changes of the measured parameters was confirmed. The difference in the time change of these settings for each patient was within 10-15 minutes of the actual photostimul medical applications. This is

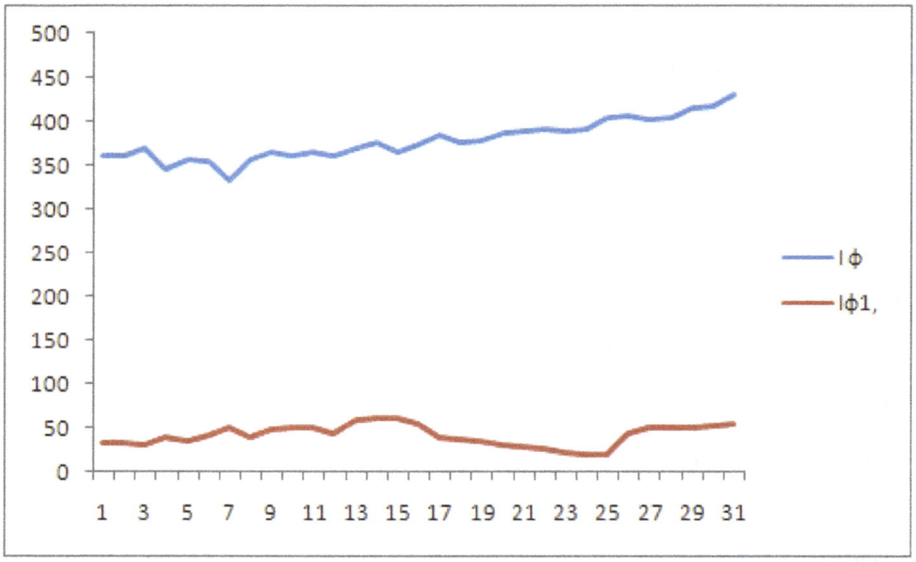

Figure 17. Change in photocurrent for the reflected (ϕ2) and passed through the BO (ϕ1) luminous flux over time.

explained by the individual characteristics of bioresonance formed as the body responds to the impact of a given medical photostimul program. Sharp changes in temperature and optical (traversed and reflected light beam) parameters in the first 10-20 minutes of the medical applications further institutionalize these changes in the future. This is due to the introduction of the body into a state of atony 10 minutes after the start of the medical program, which was recorded in the experiments. The results obtained are consistent with the literature described in the physiology processes of the human body during the stimulation of photoreceptors.

The results regarding changes in the temperature of the coefficient of reflection and light transmission scheme of the BO confirm the possibility of using this method and being a recommended optoelectronic element for implementation.

3.2. Problem of wide using light stimulation

The main problem of wide using light stimulation and the development of a controlled relaxation effect based on light stimulation consists of the absence of the principles of developing light stimulation tools with controlling light stimulation processes and providing their interactive conducting.

One promising solution to this problem is to formulate the basic medical and technical requirements and principles of development and the application in medical practice photos medicines, the basis of which would assign new advances in the field of research in optoelectronics and the design of specialized OES for specific FT. Such OES based on the thermal, plasma-discharge, and semiconductor sources of directional incoherent radiation can be created on the BO stimulation exposure with programmably controlled dynamics (according to the space-frequency characteristics of the irradiated BO) and continuous non-destructive control of a session for the analysis of change from the optical parameters of the BO.

Special attention should be paid to light stimulation through visual analyzers. They perceive a given light stimulation program of the communication channels of the cerebral cortex' excited centers, which is introduced because of the state of relaxation. Particularly, providing guidelines of photo-pulse to the very retina causes induced electrical potentials. It is clear from these reactions that retinal neurons convert photo-stimuli benched by force to electrical signals benched by amplitude. This is a relatively low-frequency modulation of the resting potential leading to relaxation.

This effect is accompanied by the increased thermogenesis caused by the increase of the blood flow in small vessels and capillaries. As a result, the optical properties of tissues, namely the coefficients of transmission and reflection of the light beam in the interaction with certain connective tissues, change. This confirmed the results of medical research conducted (Figs. 18, 19) [19-21].

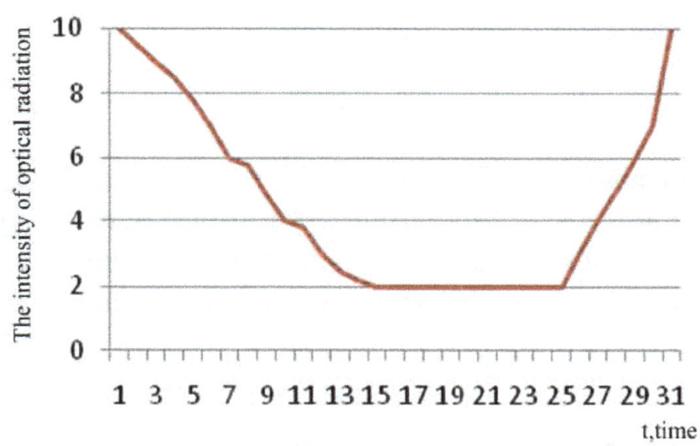

Figure 18. Change in the intensity of the optical radiation device during the treatment session.

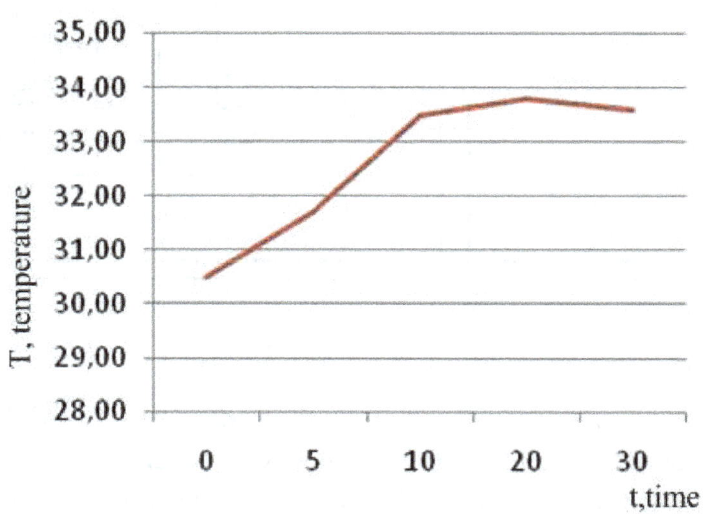

Figure 19. The temperature variation of tissue (skin) of the patient during the treatment session.

Taking the studies of the impact of stimulating light on the human body into account, the opto-electronic system (IPS) is proposed based on incoherent optical emitters, which holds a stimulating light through visual analyzers. The system consists of a generator, a generator's frequency control unit, a power supply, an output stage amplification, a LED matrix, a LED matrix control unit, and a light guide. A change of modes is provided by the generator, controlling the frequency and output power that can be set to choose medical program exposure modes (Fig. 20).

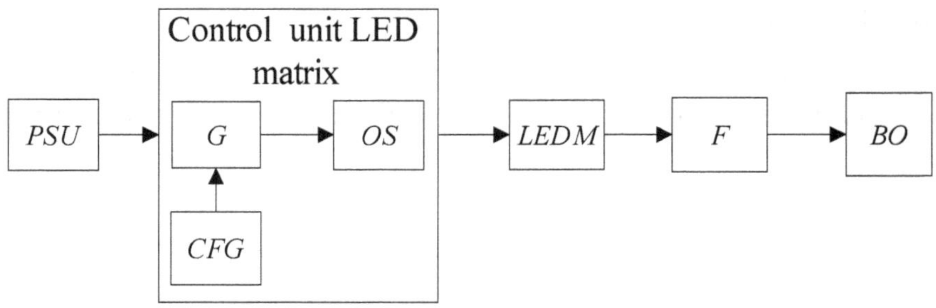

Figure 20. Block diagram of the opto-electronic system for radiation.

In order to provide information density and interactive conduction of photo-stimulation, we propose an OES, being noninvasive and harmless, for recording psychophysical state changes in the BO based on three converters using the method of simultaneous comparison of changes in optical properties and heat flows caused by changes in the temperature of peripheral BO's (Fig. 21).

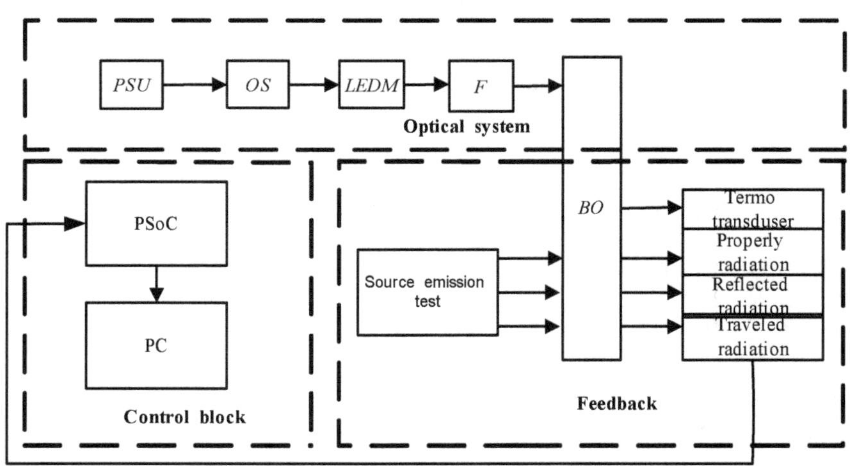

Figure 21. Block diagram of stimulating light.

The system consists of a block of optical radiation, the control block, and the control unit. Block optical radiation consists of the following components: a power supply (PSU), the output stage (OS), and the radiation source (LEDM) with a light guide (F) and creates the required for the irradiation luminous flux in the BO in the process of the treatment session. Under the influence of this radiation, changes in the optical and thermal parameters of BO, which is fixed by the circle feedback control unit, is implemented. Controlled optical signals in the state of the BO registered respective transducers and processed microcontroller PSoC. Based on the results of the analysis, a control output stage radiation source was carried out. Additionally, obtained results are transmitted via a serial interface to a PC for software processing and displayed on the screen in interactive mode.

Upon experimental investigations, it was detected that the BO's measured signals change during 30–40 min of a treatment session. The obtained dependencies underlie the algorithm of the BO's state monitoring system function. The limits of the controlled intensity parameters of the BO's own radiation, the radiation reflected, and radiation that passed through a BO were determined. On the basis of these three parameters, one can evaluate changes in the physiological state of the total organism.

In the process of the monitoring session, the dominant parameter of the most expressed changes of the dynamics was selected. The remaining parameters are additionally used. One of the three software options were selected by means of determining the value of maximum change (Fig.22).

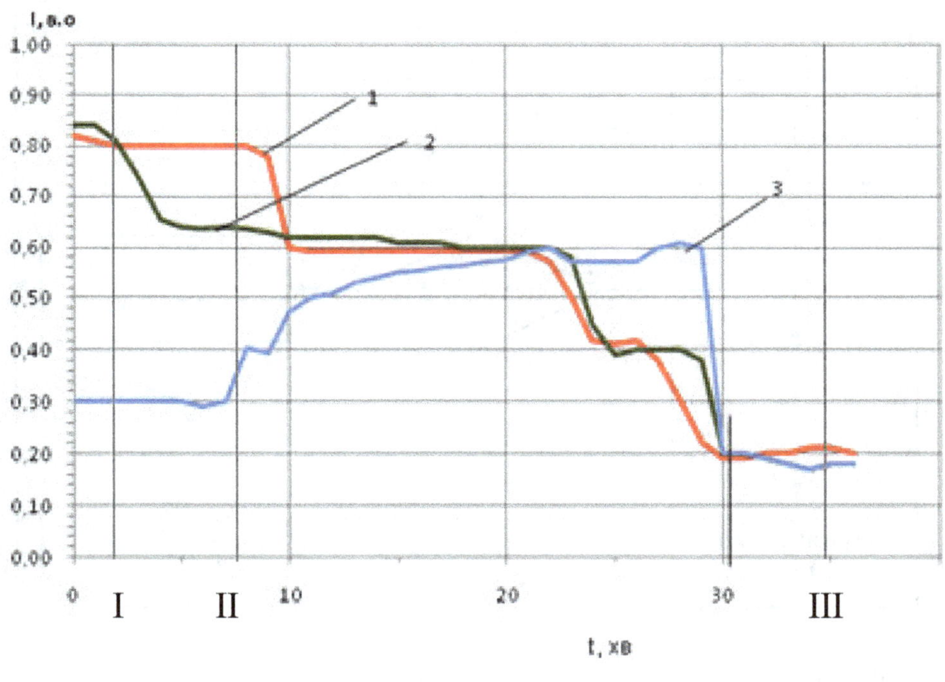

Figure 22. Experimental results dependencies of the intensity controlled light streams: 1 - passed through BO luminous flux test, 2 - test the reflected light beam, 3 - own stream of optical radiation.

As seen from the experimental results at phase I, which is the beginning of the treatment session, the dynamics of change between the absorbed and reflected radiation tissue does not. After exposing to stimulating light for 5-7 min (phase II), there is a change in the dynamics maximum values at $\Delta I = 0.15$ r.u., therefore, the patient is in a state of relaxation. In the next 15 minutes, time stabilization occurs, achieving the desired excitement centers of the brain that shows an effect of relaxation. In the last 2 min of the treatment session, the dynamics observed at ΔI = 0.1 r.u. This shows the state of the vigil centers of the brain, therefore, the patient is in a state of relaxation. As seen from the graph, the intensity of the absorbed and reflected radiation at the third stage, after a 30 min session is almost identical and very low at ≈ 0,2 r.u. and is in a condition for the termination of the treatment session.

In order to implement the interactive operation mode, the following algorithm for analysis of the parameters of the controlled optical was proposed (Fig. 23).

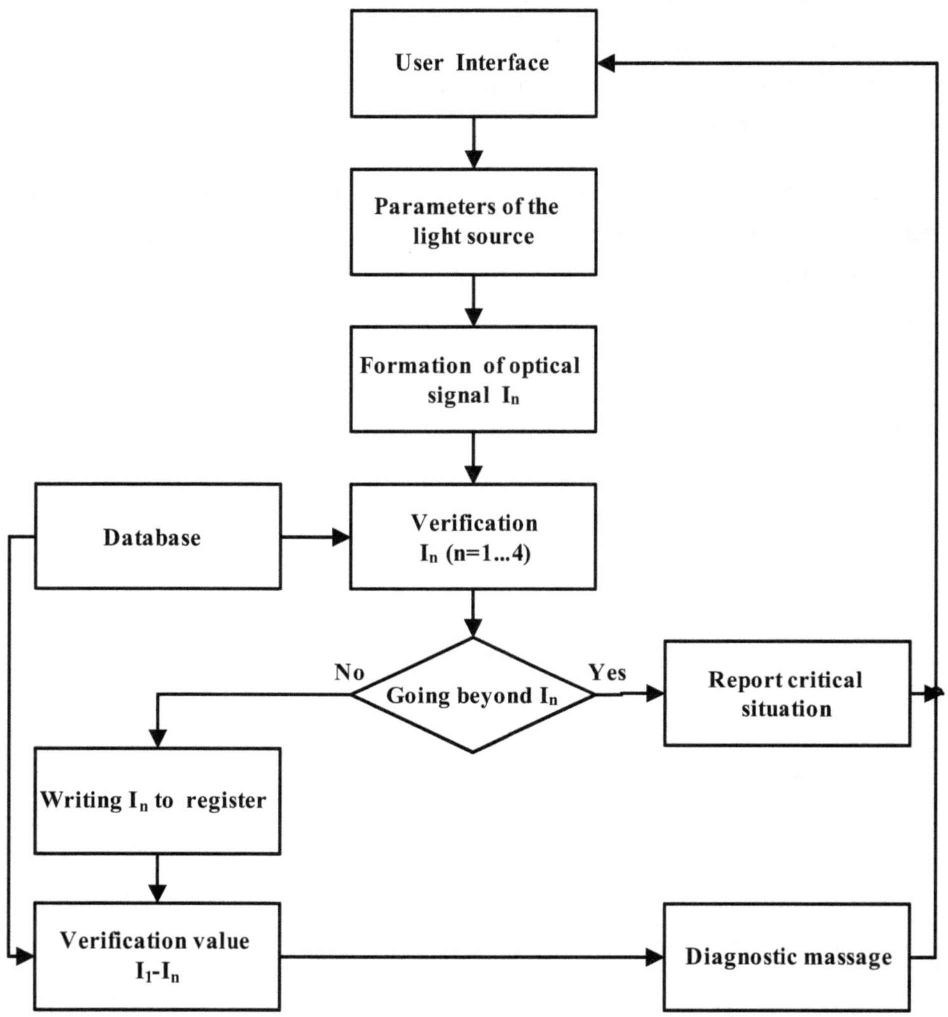

Figure 23. The algorithm interactive mode of a light stimulated session.

Process management is carried out using a PC through an informative user interface. Before starting the work in an automatic or manual mode, one has to set the parameters of the LED matrix: duration of light signals, amplitude, and limits of the spectrum of test signals. After starting the process, each of the input parameters I1 - I4 is controlled continuously. At the same time, each parameter is checked for compliance with the value ranges defined in the database. In the case when the controlled parameter In exceeds the allowable limits, a message about the achievement of the relaxation effect is formed, and the light stimulation session is automatically terminated. Checking the values of all the parameters In remained continuous throughout the session. After the session, on the basis of the results, a diagnostic message is formed according to a combination of optical parameters In set in the database of the system.

The view of the test sample of the controlled light stimulation device is shown in Fig. 24. Channel 1 provides programmable light pulses to a visual analyzer via a light guide cable to dimming glasses. Channel 2 provides a programmable photo-stimulation through other receptor areas [22, 23].

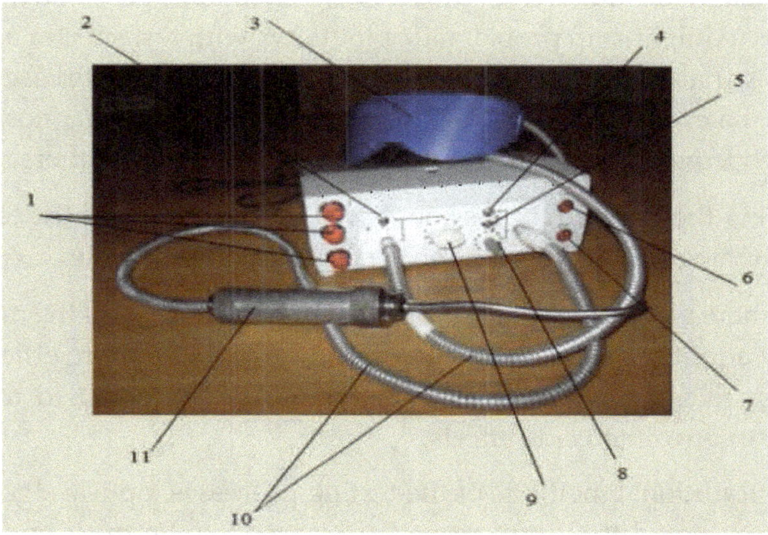

Figure 24. on / off RGB 1st – channel, 2 – mode switching manual / automatic 1-st channel, 3 – glasses for a patient, 4 – mode switching manual / automatic 2-channel, 5 – choice programs in incorporating LEDs 2nd channel, 6 –RESET reset switch program of the 2nd channel, 7 – STOP program stop switch of the 2nd channel, 8 – switching frequency control of the 2nd channel in manual mode, 9 – frequency control of the 1st channel, 10 – fiber optics cables, 11 fiber-tip – instrument.

In the developed OES, a hardware and software system was implemented that allows contactless obtaining of effective results of the current changes in the the psychophysical state of the organism during the session of the light-stimulation radiation program.

The software and hardware of the developed OES with the current values of change in indicators reflect the results of the analysis and provide a notification of the preschedule termination of the current session or unnecessary further sessions.

3.3. Application of the principle of comparing the importance of signals for the multiparameter diagnostic technology

The doctor who conducts the medical session observes the changes in the optical parameters of the BFO on display. The display shown on a computer monitor are the results of the analysis of at least three optical parameters of the BFO. Having to work interactively, the doctor decides on the adequacy of medical procedures and the need for its repetition or appropriate further treatment of the chosen scheme. Subject to the attainment of atony, the doctor may use the second channel with a dynamic radiation exposure for the chosen color therapeutic program as the treatment for other diseases.

The results given were of the clinical research in the Lviv National Medical Institute. Daniel Galician indicates the functional ability of the proposed device and the prospects of its application in otorhinolaryngology.

Recently, in the treatment of skin diseases, there is an urgent need for express diagnosis. This necessitates the development of new methods and creating comfortable, lightweight, compact, simple, and inexpensive electronic means. This means that for the detection and identification of skin diseases by survey, the skin surface of the patient were developed on the principles of contactless electromagnetic radiation of the OES. The method is based on a comparison of the spectra in the reflected light skin problematic plot flows with test spectra accumulated in the database for the external manifestations of common dermatological diseases. This enables the physician, during the examination of the patient, to make a quick diagnosis of skin diseases such as eczema, seborrhea, surface manifestations of psoriasis, and others.

At the end or during the technology session, based on the results of the tested information, messages on a technology action result to an accurate medical decision.

What is important here is that the information content of the results of the medical procedures ensures the timely adoption of the medical decision. It could be more efficient to reduce the technology process in general, making possible cases which result to the redundancy of selective exposure to unwanted side effects.

If we consider the traditional methods of the testing processes, optical signals, and decision-making, it should be noted that until recently, licenses for the most widely-called tabular algorithm is based on the calculation with using tables. Due to the lack of sensitivity to the situation, some values (sometimes the most informative) is not taken into consideration for whatever reason. These drawbacks can be eliminated when using flexible algorithms based on a probabilistic approach. A Bayesian method or the method of sequential statistical analysis method (Wald method) is used [24].

The decision to this problem must be based on a probabilistic approach using a sequential statistical analysis of the results of the continuous automatic testing of the patient and peer reviews for developing the variants of solution by the doctor. The use of logic functions that form the basis of artificial intelligence elements has a huge importance.

At the first approach, logic functions can be viewed as a process of manipulating information and could present certain information signals, such as X1,..., XK. In this approach, the input signals are primary and logical reasoning is secondary. According to Bayes' formula, the measure of the reliability of conclusions about the effectiveness of either efficiency or inefficiency of medical procedures is the probability P (Yj / Xi) in a recurrent form:

$$P\left(Y_j / X_1,\ldots,X_K\right) = P\left(Y_j / X_1,\ldots,X_{K-1}\right)\frac{P\left(X_K / Y_j\right)}{P\left(X_K\right)} \quad (1)$$

If the analysis of one of the two available options, and provided that P (Y1) = P (Y2), the following are valid for statistically independent features:

$$\frac{P(Y_1 / X_1,...,X_K)}{P(Y_2 / X_1,...,X_K)} = \prod_{i=1}^{K} \frac{P(X_i / Y_1)}{P(X_i / Y_2)} \tag{2}$$

Or after the logarithm in a recurrent form:

$$u_K = u_{K-1} + \ln z_K \tag{3}$$

The decision rules in this case as follows:

$$\begin{aligned} u_K &\geq 0 \rightarrow X \in Y_1, \\ u_K &< 0 \rightarrow X \in Y_2. \end{aligned} \tag{4}$$

Together with equation (3) and isolating rule (4), the formula is interpreted as follows: if, after the consideration of another parameter the signal of sign magnitude U_K has not changed, there are grounds for the termination of the analysis.

On the basis of the implementation of a flexible algorithm (Fig. 25).

For each time interval, the test values and the dynamics of change exceeded the critical standards of the emerging reports critical condition - Y_K on the input signal X_I. In addition, for each time interval, the entire procedure is checked. In the case of exceeding the present time and the absence of the dynamics parameters of the dominant signals, the message YNE forms the effectiveness of the treatment procedure in the case of emergencies, which formed a disturbing message about the need for an immediate cessation in the treatments. The resulting data set is stored in a database that is used in the design of similar devices.

According to the proposed analysis algorithm for the test signal, the designed structural diagram of a medical tested device is based on the microcontroller PSoC CY8C24x94 family with built-in full-speed USB [25-27].

The basis of the microcontroller is a powerful microprocessor M8C with Harvard architecture, which controls the operation of the digital and analog blocks and surrounding devices. The digital part consists of four 8-bit internal blocks PSoC and interface modules of the user, which are used for connecting external devices: keyboard, display, power audio signal, and PC. The analog part consists of six indoor units, which allows the use of the following features: analog-to-digital and digital-to-analog conversion, amplification, programmable gain, pulse width modulation, detection, comparison, multiplexing, and correlation. The processing of input information carried out under the control of the embedded software, which is used to store energy-volatile memory (EEPROM and Flash). To store intermediate results during processing, an internal random access memory (RAM) is used.

According to the proposed analysis algorithm for the test signal, a structural diagram of a medical tested device based on the microcontroller PSoC CY8C24x94 family with built-in full-speed USB (Fig.26) was designed.

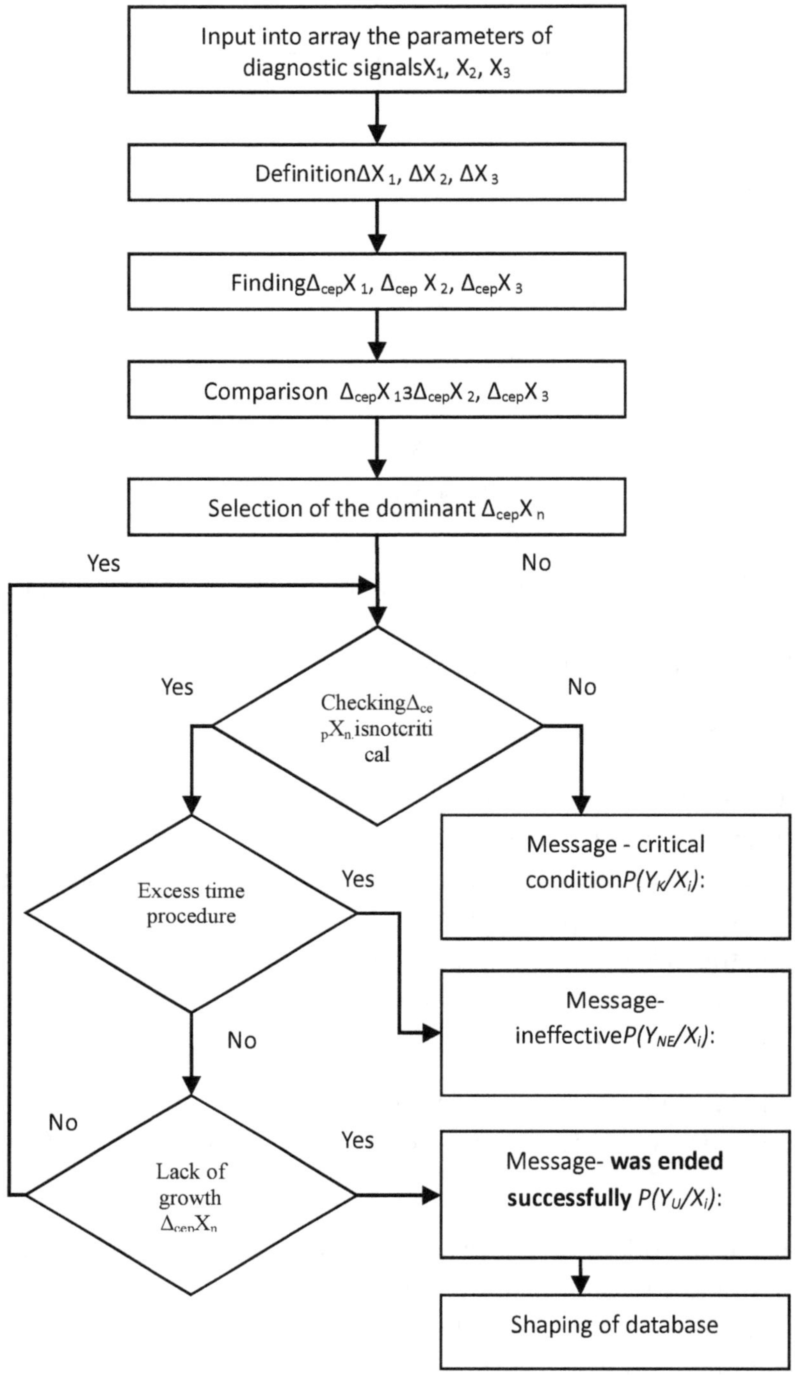

Figure 25. The flexible processing of the algorithm.

For the practical implementation of the device, which uses a PSoC Designer software by which the internal configuration of the microcontroller PSoC CY8C24x94 is debugged according to the functional diagram. Input signals X1-X3 after amplifiers AMR1-AMR3 filtered SF1-SF3 and through the multiplexer MUX to the ADC. Information obtained during the conversion process code that corresponds to the value of one parameter in a fixed point in time is stored in the RAM, the main memory uses for analysis, according to the developed algorithm. For the

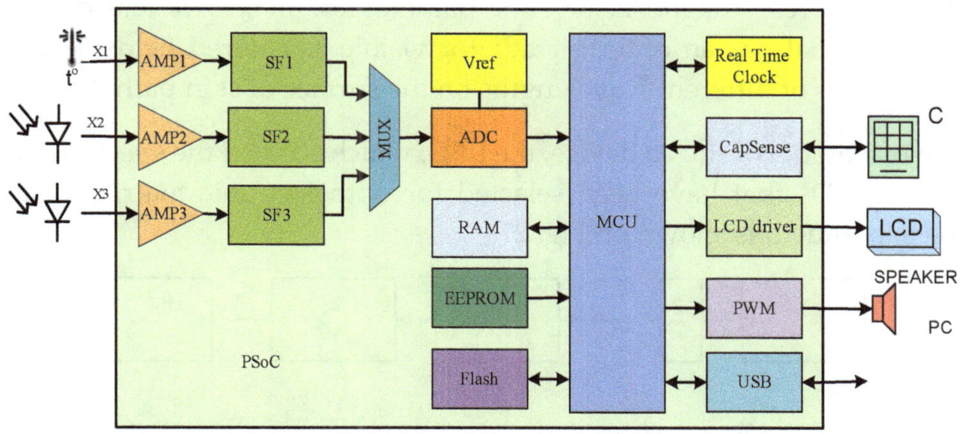

Figure 26. Functional diagram of implementation in the medical device.

formation of the timeframe during which the registration of input signals is carried, a Real-Time Clock is used. The managing works of the device is performed by using the keyboard via the corresponding interface module CapSense. The modes shown on the displayed operation is determined by the internal driver of the LCD module. In the event of a critical situation, additionally formed sound signals, its frequency and duration is given by module PWM. For a detailed analysis and the formation of the database of input parameters and corresponding test states, the results are transferred via USB to a PC.

It is proposed to improve the interactive physician format of the flexible new algorithm based on the probabilistic approach using sequential statistical analysis of the results of the continuous automatic testing of the patient for medical technologies efficacy of the procedure with the adoption of the basis of user solution.

The proposed method uses the elements of artificial intelligence and algorithm analysis with sufficient accuracy.

The results allow the creation of a new class of devices for medical technologies with continuous automatic evaluation of the effectiveness of a technological action results for an accurate medical decision-making and the creation of reports on previous technological strategies.

4. Research on an optoelectronic system for dermatology

Modern medical diagnosis requires the solution of problems related to the study and comparison of healthy (normal) biological tissues with corresponding areas of certain systemic diseases (e.g., the diagnosis and stage presence of diseases such as psoriasis, eczema, seborrhea, dermatitis, etc.). One of the achievements of recent years can be called transillumination method that is widely used, particularly in dermatology for the noninvasive diagnosis and identification of disease [19].

The necessary noninvasive diagnosis can be done by measuring and comparing the spectrophotometric parameters of normal and pathological areas. For a simplified version, transillumination, in the express diagnosis of skin pathologies, makes it feasible to create an inexpensive device that has an optical irradiator visible as colored fields and many diode matrix for light fluxes of different wavelengths on the surface of skin pathologies.

In the creation of a model of such a device, the LED, which studied the spectral characteristics of multi-colored LEDs that have been selected for optoelectronic matrix, is selected. The scheme of the experiment is shown on Fig. 27.

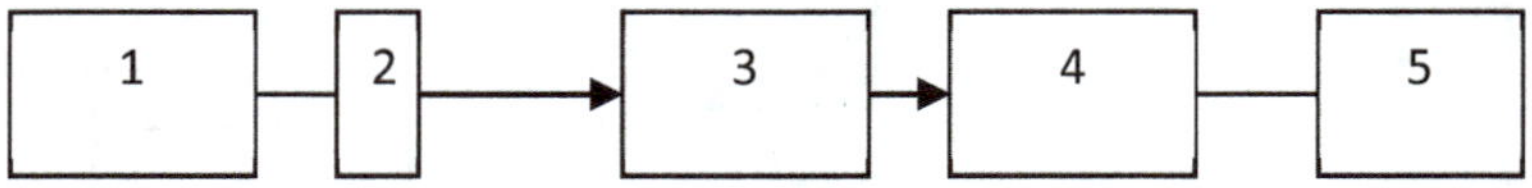

Figure 27. Block diagram investigation of the spectral characteristics of LED: 1 - LED power supply, 2 – LED, 3 – monochromator, 4 – photodetector, 5 - power meter radiation.

The research results presented in Table 4 are based on the spectral characteristics of present values of intensity (Figs. 28, 29) selected for the model of LEDs.

red		orange		yellow		green		azure		blue		violet	
λ, nm	I, a.u.	λ, nm	I, a.u.	λ, nm	I, a.u.	λ, nm	I, a.u.	λ, nm	I, a.u.	λ, nm	I, a.u.	λ, nm	I, a.u.
610	0	633	0	568	0	508	0	400	0	416	0	420	0
633	0.57	639	1	586	0.35	530	0.26	420	1	429	0.8	426	0.16
639	0.69	666	0.77	590	1	541	1	442	0.5	435	1	435	0.25
666	1	680	0.35	599	0.31	565	0.52	469	0	442	0.6	443	0.12
680	0.35	710	0	610	0	592	0			444	0	452	0
709	0												

Table 4. The dependence of the radiation intensity of LED I on the wavelength λ.

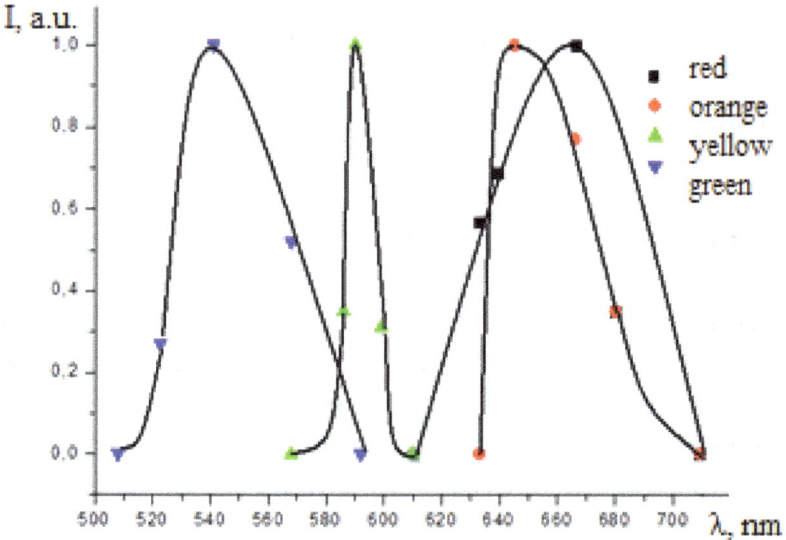

Figure 28. Spectral characteristics of longwave LEDs chosen for the developed model.

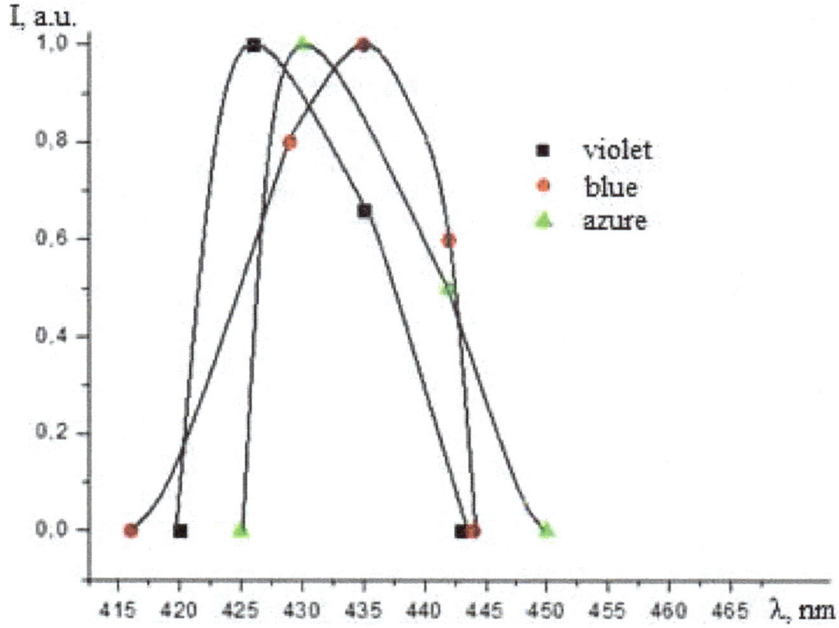

Figure 29. Spectral characteristics of shortwave LEDs chosen for the developed model.

From the graphs (Figs. 28, 29), the full range of radiation selected of LEDs completely covers the need for a radiation area. Thus, the radiation that is created by the LED strip covers the following wavelengths: blue - 508 to 592 nm, yellow - 568 to 610 nm, red - 610 to 709 nm, orange - 633 to 709 nm, azure - 429 to 450 nm, blue - 416 to 444 nm, and purple - 420 to 443 nm. The maximum radiation intensity of LED corresponds to these wavelengths: green - $\lambda = 541$ nm, yellow - $\lambda = 590$ nm, red - $\lambda = 666$ nm, orange - $\lambda = 639$ nm, blue - $\lambda = 430$ nm, blue - $\lambda = 435$ nm, and violet - $\lambda = 426$ nm.

Based on the selected LEDs set, the LED matrix optoelectronic element mock (Fig. 30).

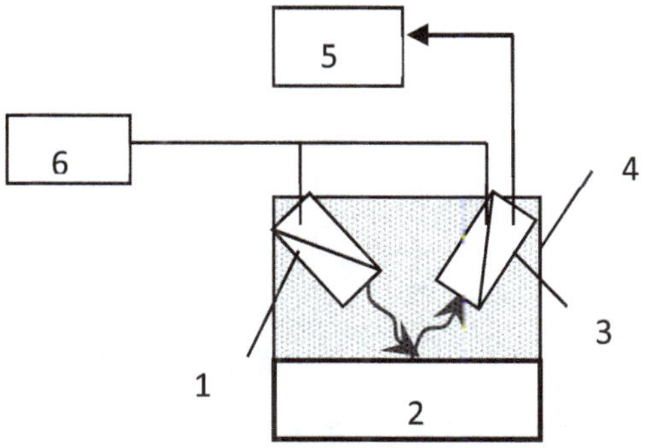

Figure 30. Scheme of researches: 1 – photodetector, 2 - color test sample, 3 - LED 4, - dimming building, 5 - radiation power meter, 6 - power supply.

The LED and the photodetector are located on the reflecting surface (sample) at angles close to 45º so that the radiation coming from the LED is reflected and reaches the photodetector.

The LED and photodetector were in the dimming housing for protection against visible light when they were measured.

The obtained results are reduced to a common matrix (Table 5).

Color sample	red	orange	yellow	green	azure	blue	violet
LED							
red	8.26	8.25	8.26	4.92	7.47	4.05	6.58
orange	5.6	7.04	7.64	3.76	5.22	2.63	4.38
yellow	1.47	5.12	5.48	1.47	2.74	1.44	1.47
green	3.75	4.72	8.23	6.13	8.24	4.03	4.12
azure	6.48	5.14	7.59	7.08	3.36	8.25	7.84
blue	5.44	6.17	7.85	3.87	8.17	3.78	3.84
violet	6.22	5.63	6.21	3.02	4.78	2.29	3.64

Table 5. Matrix intensity of the reflection surface of the skin for different parts of the visible spectrum.

Based on the data obtained, the field developed a policolor probe as part of the dermatological OES, allowing the identification of the individual colors of the reflected radiation intensity for the matrices in seven major areas of the visible spectrum. It is aimed at the examined surface radiation source that constitutes LEDs that turn on their radiation spectra to cover the entire visible wavelength range, located at an angle to the optical axis and coaxially around the sensor mounted on. With the possibility to be away from the examined surface, the reflected light beams from each of the LEDs are enable alternate or simultaneous switching.

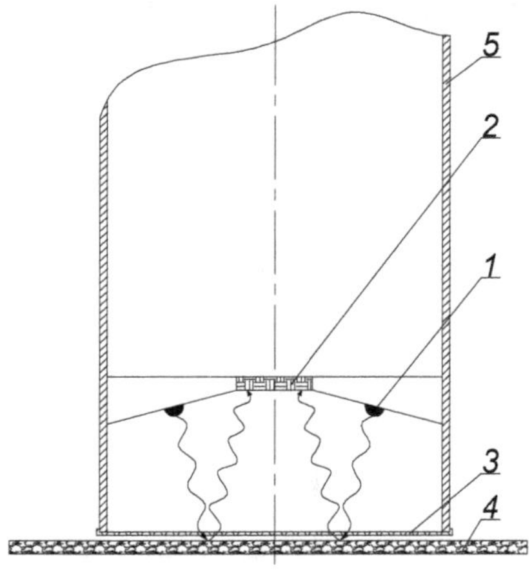

Figure 31. Opticoelectronic probe: 1 – LED matrixs, 2 – photodetector, 3 - Protective optically transparent element, 4 – biomedical object, 5 – hull.

The optoelectronic probe system consists of LED matrix 1 consisting of LED type MUV9C, photodetector 2, which serves as a photodiode PD-24, and an optically transparent protective element 3, which is a translucent polystyrene plate (Fig. 31). The falling light output of the LED matrix 1 after passing through the optically transparent protective element 3 creates a reflection from the surface of the investigated object. The biomedical luminous flux again passes through an optically transparent protective element 3 and reaches the photodetector 2. The signal from the sensor enters the amplifier and then measures the photocurrent contained in the measuring-controlling unit.

On the basis of experimental and computational studies created a working model (Fig. 32).

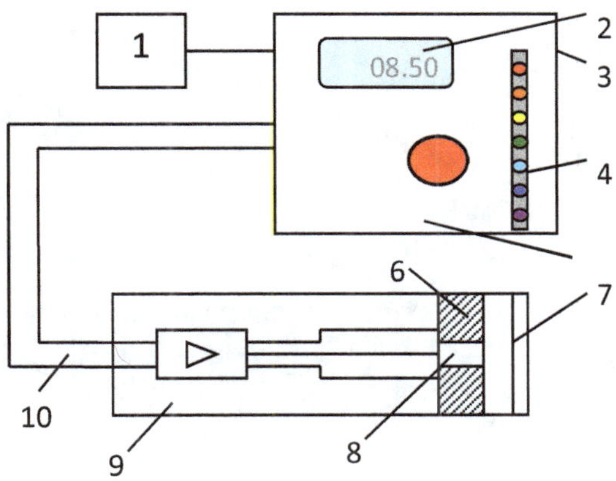

Figure 32. The structure of the model: 1 - power supply, 2 - value of photocurrent, 3 - hull, 4 – indicator, 5 - measurement and a control unit, 6 - LED matrix, 7 - protective cover, 8 – photodetector, 9 - optoelectronic unit, 10 - cable.

The dimensional control block is designed for the control of optoelectronic measuring unit and its signal. By pressing the "Start", the LEDs are alternately turned on in the LED matrix composed of red, orange, yellow, green, blue, violet, and blue LEDs. The indicators reflect a powerful monitor LED. The luminous flux of the LED matrix reflected by the research object reaches the photodetector. The signal from the photodetector reaches the amplifier and gets the value of the photocurrent.

In medical practice, for the detection and identification of disease by the spectral characteristics of skin manifestations and with appropriate spectral characteristics, it is convenient to use the tabulated values of the intensity of the reflected light from surfaces of such flows. For each of these tables, there are LED test matrices to identify the disease in the course of its treatment (Table 6).

LED		**red**	**orange**	**yellow**	**green**	**azure**	**blue**	**violet**
Healthy skin	**I, в.о.**	8.14	5.5	3.07	5.7	8	5.86	4.55
Psoriasis		7.82	5.72	2.66	5.17	7.26	5.24	4.53
Seborrhea		6.74	4.6	2.09	5.23	6.78	4.67	3.78
Chronic dermatitis		8.16	4.96	2.3	4.98	6.98	5.32	4.58

Table 6. The test matrix of skin surface.

the reflective optical imaging device.

On the basis of what has been detected during the mathematical and physical modeling regularities in the interaction of light streams with pathological skin areas, optimized selection and relative positions photodetectors, and irradiative components of the proposed device to hospitals on the basis of a specialized OES (optical-electronic probe), a display and access to a personal computer is included in each visit (Fig. 33) [28].

Figure 33. OEC components and probe express diagnostics of skin diseases (from left to right - probe measurement unit, PC). 1 - LED matrix, 2 – photodetector, 3 - protective optically transparent element, 4 – BO, 5 – enclosure.

The methodology for light diagnostic research by means of the developed device is an external surface diagnosing skin with a suspected pathology of supply optoelectronic probe to the problem area. A new application of the specialized OES allows the doctor to make long-term monitoring of the changes in cell behavior, unaccompanied by heated skin, and the different effects on biochemical, histological, and functional levels and replenishment on this bases of biological tissues in diseases including digital pictures of the cutaneous manifestations of disease. Clinical studies have shown that the application of the proposed specialized OES significantly reduced the process of diagnosing the level with a rapid process and expanded the diagnostic capabilities for the identification of detected pathologies.

Author details

Barylo Hryhoriy[1], Gotra Zenon[1], Ivakh Mariya[1], Kozhukhar Oleksandr[1*], Makara Ivanna[2] and Virt Volodymyr[1]

*Address all correspondence to: akozhukha@ukr.net

1 The ED Department, Lviv Polytechnic National University, Ukraine

2 The IM Department, Lutsk National Technical University, Ukraine

References

[1] Tuchin V.V. "Handbook of Optical Biomedical Diagnostics", SPIE Press, 2002 - 1093 p.

[2] Rodney Cotterill "Biophysics: an introduction", 2002, 408 p.

[3] Ivakh Mariya, Kozhukhar Alexander, Zazuljak Andriy, Kosyi Evgen "Expansion of therapeutic possibilities based on the developed optronic systems monitoring progress of medical treatment", Biomedical Engineering and Electronics, №1, 2012, http://www.es.rae.ru:8888/biofbe/.

[4] Kozhukhar O.T., Kurgan M.G., Skira M.S., Kurgan D.M. "Problem device for illumination in treatment by photopheresis"in 4th International Forum on Innovative Technologies for Medicine (ITMED 2010), Bialystok, 2010, P. 32.

[5] Oleksandr Kozhukhar, Mariya Skira, Yurij Kuzio "Descriptions of optoelectronic elements are for control of changes of optical properties of blood after the photopheresis" in Xth International Conference "Modern problems of radio engineering, telecommunications and computer science", 23–27 Feb. 2010, P. 122.

[6] Alternative approaches to improving ultraviolet actuators photopheresis (in Ukrainian) / Z. Gotra, M Ivakh, A. Kozhukhar, K. Voytsihovska / Systems of information processing, № 2 (92), 2011, pp. 239-243.

[7] Gotra Zenon, Mariya Ivakh, Kozhukhar Alexander "System of testing the patient's blood advisability photopheresis treatment" in 4th International Forum Science and Technology Days Poland –East, 12-14 April, 2011,Bialowieza, 2011, P. 31.

[8] Z.Y. Hotra, A.T. Kozhukhar, A.M. Zazuljak, M.S. Ivakh "Development of noncoherent optical-electronic systems in modern photo medical technology" Proc. SPIE 8698, Optical Fibers and Their Applications 2012, 86980J (January 11, 2013); doi:10.1117/12.2019268

[9] Zenon Gotra, Alexander Kozhukhar, Igor Didych, Andrij Zazuljak, Marija Skira, Olena Tkachenko. The universal sensory optronic system for estimation of blood technology quality, The conference catalogue chronicles theproceedings of the 3rd International Forum on Innovative Technologies for Medicine, Białystok, 2009, p. 39.

[10] Device for photostimulation therapy in hematology (in Ukrainian): patent Ukraine 62114 / Hotra Zenon; Kozhukhar Oleksandr; Ivakh Mariia (2011)

[11] Optical electronic control of activity of treatment procedures / Kozhukhar Alexander, Andriy Zazuljak, Ivakh Mariya // 4th International Forum Science and Technology Days Poland –East: Forum procedeengs. – Bialowieza: Innovative Eastern Poland Asociation, 2012, pp. 47-52.

[12] Grishchenko G. Bioresonance stimulation of the human organism as a substitute for application of chemical medicines (in Ukrainian), Proceedings [electronic resource], 2010, Mode of access:http://bibl.kma.mk.ua/index.php?m=10&s=3&t=6.

[13] *S.Tonnies "Entspannung fur Tinnitusbetrofenne durch Photostimylation" Springer Medizin Verlag,* № *54, 2006, pp.481-486.*

[14] Kozhukhar O.T., Zazuliak A.M., Kuchak E.V. "The influence fotostymuliv with frequencies on human biorhythms body" Electronics" Bulletin ou "LP", № 680, 2010, pp. 225-230.

[15] I Didych, A. Zazulyak, O. Kozhukhar "The programable system is for photomedic technologies". In XIII Miedzynarod. Szkoly Komputerowego wspomaganja projektovanja. Wytwarzania I eksploatacji. Jurata. 11-15 05.2009.

[16] A. Kozhukhar, M. Kostiv "Photomedic light sources with controlled dynamics radiation" (in Ukrainian) in First All-Ukrainian Congress of "Medical and Biological Informatics and Cybernetics" with international participation. - Kyiv. 23-26 June 2010, p. 254.

[17] Hotra Z., Zazuliak A., Kozhukhar A. "Otorhinolaryngological photo therapeutics programmatically controlled complex with a continuous the assessment of efficiency procedure" Perspective technologies and devices, № 2, 2012, p. 30-36.

[18] *Zazuljak, A. Kozhuкhar, O.Tkachenko "Therapeutic and diagnostic devices based on encephalographic frequencies programmed low intensity photostimulation on visual Receptors" in 2nd Forum Science&Nechnology Days Poland., 2009, p. 42.*

[19] Tkachuk R., Makara I., Kozhukhar O., Tkachenko O. "Exspress diagnostics and authentication of illnesse is with application of the new optical-electronic systems" in 6th International Forum on Innovative Technologies for Medicine (ITMED 2012), 21–23 November 2012, P. 24-25.

[20] Z. Hotra, A. Zazuljak, M. Ivakh "New approaches to continuos information support and evaluation effectiveness of physiotherapeutic procedures" in 12-th International conference "The experience of desigining and application of cad systems in microelectronics", 19-23 Feb. 2013, Polyana-Svalyava (Zakarpattya), 2013, pp. 152-153.

[21] Zenon Gotra, Hryhoriy Barylo, Aleksandr Kozhukhar, Mariya Ivakh, "Implementation of flexible algorithms for continuous monitoring and operative correction of medical technologies Innovative eastern poland association" in 7th International forum on innovative technologies for medicine, 5-7 decem. 2013, Bialystok, 2013, p. 10.

[22] Device for phototherapy via visual receptor (in Ukrainian): patent Ukraine 42525: Kozhukhar Oleksandr; Skunts Nazar; Zazuliak Andrii (2009)

[23] Noninvasive sensor device for registration of changes in psycho-physical state of human person (in Ukrainian): patent Ukraine 60600, Hotra Zenovii; Kozhukhar Oleksandr; Zazuliak Andrii; Kuchak Yevhen (2010)

[24] Guosheng Yin Bayesian Generalized Method of Moments, Bayesian Analysis (2009) Number 2, pp. 191-208

[25] Robert Ashby. Designer's Guide to the Cypress PSoC (Embedded Technology).

[26] D. Kilochek "Design on programmable systems-on-chip PSoC Cypress" Components and technology, № 4, 2006.

[27] G.I. Barylo, M.S. Ivakh, A.M. Zazulyak, O.T. Kozhukhar "Information support of medical procedures based on automated analysis perephericperformance" in the twentieth international scientific-practical. conf. "Information Technology: Science, Technology, education, health: CH.III (May 15-17, 2012, Kharkiv), 2012. - P. 96

[28] Light-diagnostic survey device: patent Ukraine 53154 / Melnyk Ivanna; Kozhukhar Oleksandr.

6

InP-Based Antimony-Free MQW Lasers in 2-3 μm Band

Yi Gu and Yong-Gang Zhang

Abstract

Mid-infrared semiconductor lasers in the wavelength range of 2-3 μm have aroused increasing interests as they are highly desired for a wide range of applications ranging from medical diagnostics to environmental sensing. Access to this wavelength range was mainly achieved by antimony-containing compound semiconductor structures on GaSb substrates. Besides, InP-based $In_xGa_{1-x}As$ (x>0.53) type-I multiple quantum well laser is a promising antimony-free approach in this band. The emission wavelength can be tailored to the 2-3 μm band by increasing the indium composition in the quantum wells. During the demonstration of this kind of lasers, controlling the strain and keeping fair structural quality is the main obstacle.

In this chapter, the route for developing this kind of lasers is reviewed. The schemes of pseudomorphic and metamorphic structures are discussed for the 2-2.5 μm and 2.5-3 μm range, respectively. In the pseudomorphic scheme, triangular quantum wells grown by digital alloy technology are applied to restrict the strain and increase the lasing wavelength. Lasers at 2.43 μm were demonstrated under continuous wave operation at room temperature. To extend the emission wavelength longer, an InP-based metamorphic template with larger lattice constant was produced and InAs quantum wells were then grown. The lasing wavelength was further increased up to 2.71 μm. The details on the gas source molecular beam epitaxial growth, device processing as well as performance characterization are presented.

Keywords: Lasers, Quantum well, InP, 2-3 μm, Antimony-free

1. Introduction

The 2-3 μm wavelength range covers both the short side of the mid-infrared (2.5-25 μm) band and the long side of the near-infrared (0.7-2.5 μm) band. Semiconductor laser diodes emitting in this spectral range are very attractive for free space communications, ultra-low loss fluoride fiber communication, solid or fiber lasers (amplifiers) pumping and seeding, light detection and ranging (LIDAR), tunable diode laser absorption spectroscopy (TDLAS), etc., due to the abundant spectroscopic features in this wavelength range. For example, the 2.1-2.4 μm band is free of water and carbon dioxide absorption as shown in Figure 1, which forms a clear atmosphere window; on the other hand, the strong and wide absorption band of water (mainly around 2.7 μm) and the moderate absorption band of carbon dioxide (around 2.0-2.1μm and 2.65-2.85 μm) forms a high contrast area, which is usable for humidity and carbon dioxide monitoring, active spectroscopic imaging of water containing subjects, etc.

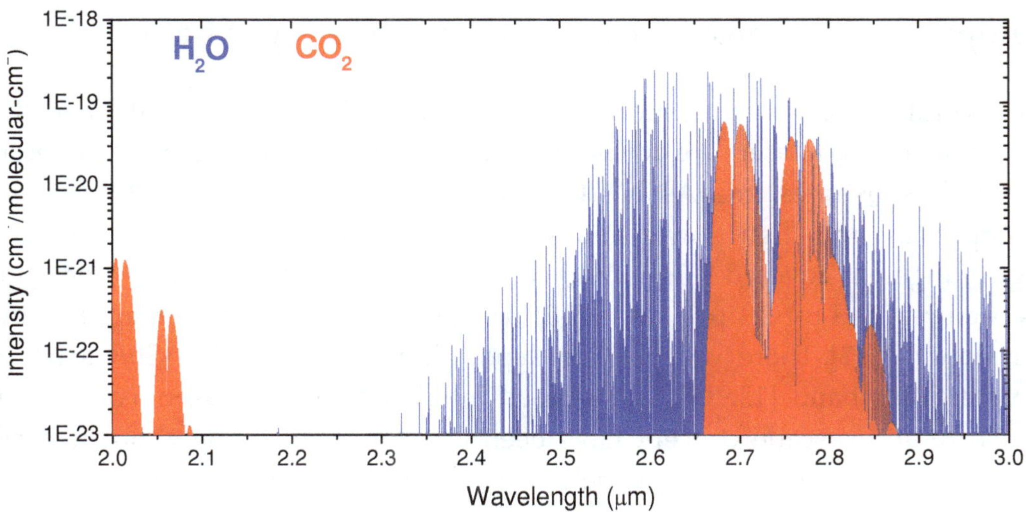

Figure 1. Absorption intensity of H_2O and CO_2 in the 2-3 μm band.

Furthermore, as shown in Figure 2, the fingerprint absorption lines of several molecules also exist in this band, and therefore, semiconductor lasers in this band are of great interest in atmospheric pollution monitoring and medical diagnostics by using TDLAS [1, 2]. As a monochromatic light source, the laser diode in this wavelength band is also desired for the evaluation of optoelectronic materials and devices such as photodetectors and focal plane arrays operated in this wavelength range, because in the measurement schemes wavelength match is very important to acquire their actual features [3, 4].

The choices and developments of semiconductor lasers in the 2-3 μm band will be reviewed at first. Then we will introduce the recent progress for the design, growth and demonstration of InP-based antimony-free multiple quantum well (MQW) lasers in this band. Schemes of pseudomorphic and metamorphic structures will be discussed for the 2-2.5 μm and 2.5-3 μm ranges, respectively.

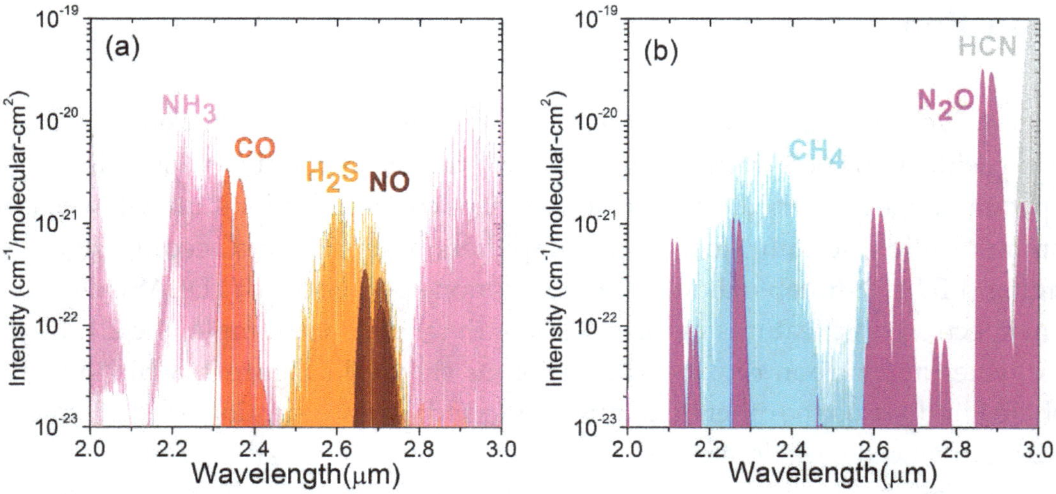

Figure 2. Absorption intensity of (a) NH_3, CO, H_2S, NO molecules and (b) CH_4, N_2O, HCN molecules in the 2-3 μm band.

2. Development of semiconductor lasers in the 2-3 μm band

There are several options of semiconductor lasers in the 2-3 μm wavelength band. Traditionally, such semiconductor lasers are developed in an antimony (Sb)-containing material system on GaSb or InAs substrates. InGaAsSb/AlGa(In)AsSb type-I quantum well (QW) lasers are the primary approach, and high power and single-mode lasers have been achieved and the wavelength range has covered 2-3.8 μm [5-10], and continuous wave (CW) operation above 80°C has been reached for 2.1 μm lasers [10]. The laser diodes have been applied in TDLAS demonstrations [11, 12]. There were also some sporadic reports on InGaAsSb/GaSb type-II QW lasers covering this band [13, 14]. GaSb-based interband cascade lasers combining type-II heterostructure and cascade concept have been developed adequately, but it is still a big challenge to shift the lasing wavelength below 3 μm, which requires new material systems [15].

Compared to GaSb, InP substrates display superior quality and are easier to acquire. Sb-containing InGaAs/GaAsSb type-II structures on InP substrates have been researched while room temperature (RT) photoluminescence (PL) and sporadic device results have been reported [16, 17]. InAsSb quantum dots (QDs) on InP are another potential scheme to demonstrate light sources in 2-3 μm, but the improvements of material quality are still needed to achieve more experimental progress [18]. On the other hand, the growth and processing technologies of the Sb-containing materials are still less mature than those of InP-based Sb-free materials. Besides, the thermal characteristics of antimonide are much inferior to the phosphides and arsenide [19, 20]. Many impressive results on mid-infrared intersubband emission of InP-based Sb-free InGaAs/InAlAs quantum cascade lasers have been reported in the recent two decades, whereas the emission wavelengths are mainly longer than 3 μm [21].

Besides the aforementioned options, InP-based Sb-free type-I $In_xGa_{1-x}As$ ($x>0.53$) QW laser is a promising approach in this band. As the bandgap wavelength of InAs at RT is as long as 3.5 μm, the emission wavelength of In(Ga)As QW can be tailored to the 2-3 μm range [22]. For this InP-based Sb-free In(Ga)As system, mature growth and processing technologies can be relied on, and a simple structure can be applied for the type-I QW lasers using interband emission, so the laser diodes with better performance could be expected. The QW lasers applying InP-lattice-matched InGaAsP as QW layers have been well developed with the aim of telecom lasers [23]. However, the lasing wavelength is shorter than 1.7 μm. To extend the wavelength beyond 2 μm, a higher InAs composition should be applied and thus strain would be introduced into the QW layers.

The InP-based InGaAs strained QW lasers were first demonstrated by Forouhar et al. in the early 1990s [24]. The lasing wavelengths are around 2 μm at RT, and $In_{0.75}Ga_{0.25}As$ layers grown by metal-organic vapor phase epitaxy (MOVPE) were applied as the QW layers and other layers were all lattice matched to InP [25-27]. As shown in Figure 3(a), in these pseudomorphic structures, the InGaAs (x>0.53) QW layers are compressively strained whereas the other layers, such as cladding, waveguide and barrier layers, are all nearly lattice-matched to InP. The Fabry-Perot (FP) and distributed feedback buried (DFB) devices around 2.07 μm were then demonstrated by NTT's group [28-31]. After the year 2000, FP and vertical-cavity surface-emitting lasers (VCSELs) at 2.3 μm were demonstrated by Amann et al. using molecular beam epitaxy (MBE) grown InAs-containing triangular QW active region [32-34]. Using MOVPE grown 5 nm pure InAs as the QW layer, Mitsuhara et al. in NTT's group reported FP and DFB lasers with lasing wavelength at 2.33 μm [35-37]. Lasers up to 2.4 μm were recently demonstrated by using triangular QWs grown by MBE digital alloy technology [38].

As mentioned earlier, further increase of well width is restrained by the strain between the QW part and InP substrate, and thus the longest emission wavelength is limited. To extend the emission wavelength to longer wavelengths, a metamorphic "virtual substrate" can be produced with a larger lattice constant than InP, e.g. metamorphic $In_{0.8}Ga_{0.2}As$ layer as shown in Figure 3(b). In this case, the critical thicknesses can be larger than those of pseudomorphic structures, therefore the QW width can be increased, which would increase the emission wavelength. Recently, step-graded $InAs_xP_{1-x}$ buffers on InP have been applied to achieve 2.8-3 μm PL at RT [39-41], where 16.5 nm InAs or 22 nm $InAs_{0.94}P_{0.06}$ were used as the well layers. By using 15 nm InAs QWs on continuously-graded metamorphic $In_xAl_{1-x}As$ buffers, RT PL of QWs at 3.05 μm [42] and low temperature lasing of laser diodes at 2.7 μm have been achieved [43]. This scheme can even be explored to mid-infrared metamorphic InAs QWs on GaAs substrate, which is even more attractive than on InP substrate [44].

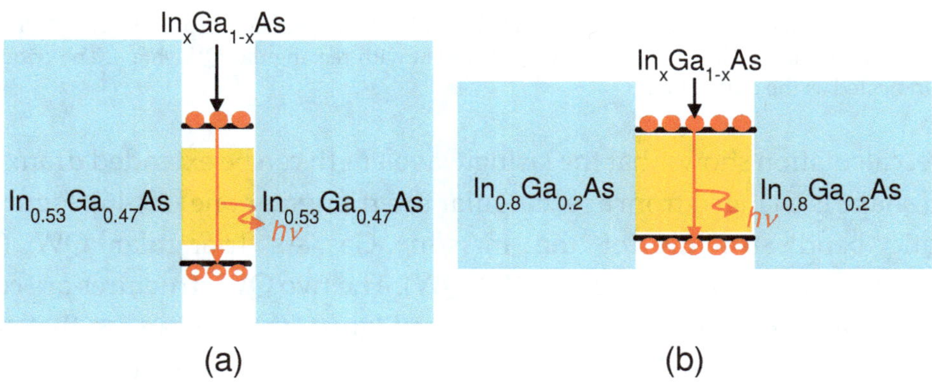

Figure 3. Schematic diagrams of the transition for the InP-based (a) pseudomorphic and (b) metamorphic $In_xGa_{1-x}As$ QWs.

3. Theoretical evaluation of InP-based Sb-free type-I QW lasers

At first, we theoretically evaluate the type-I transition energy and emission wavelength of rectangular pseudomorphic $In_xGa_{1-x}As/In_{0.53}Ga_{0.47}As$ (x>0.53) and metamorphic $In_xGa_{1-x}As/In_{0.8}Ga_{0.2}As$ (x>0.8) QWs. The subband energies of the QWs were calculated by numerically solving the Schrödinger equation with strain-contained 4×4 Luttinger-Kohn Hamiltonian, and the transition energy from the first electron energy level to the first heavy hole energy level was calculated. The emission wavelength versus the well width of $In_xGa_{1-x}As/In_{0.53}Ga_{0.47}As$ (x>0.53) and $In_xGa_{1-x}As/In_{0.8}Ga_{0.2}As$ (x>0.8) QWs are shown in Figure 4(a) and Figure 4(b), respectively. The well width is limited by the critical thickness calculated from the force balance model as shown in Figure 4. It is seen that the critical thickness can be extended by using the metamorphic scheme, and therefore, a thicker well width can be used, leading to a longer wavelength.

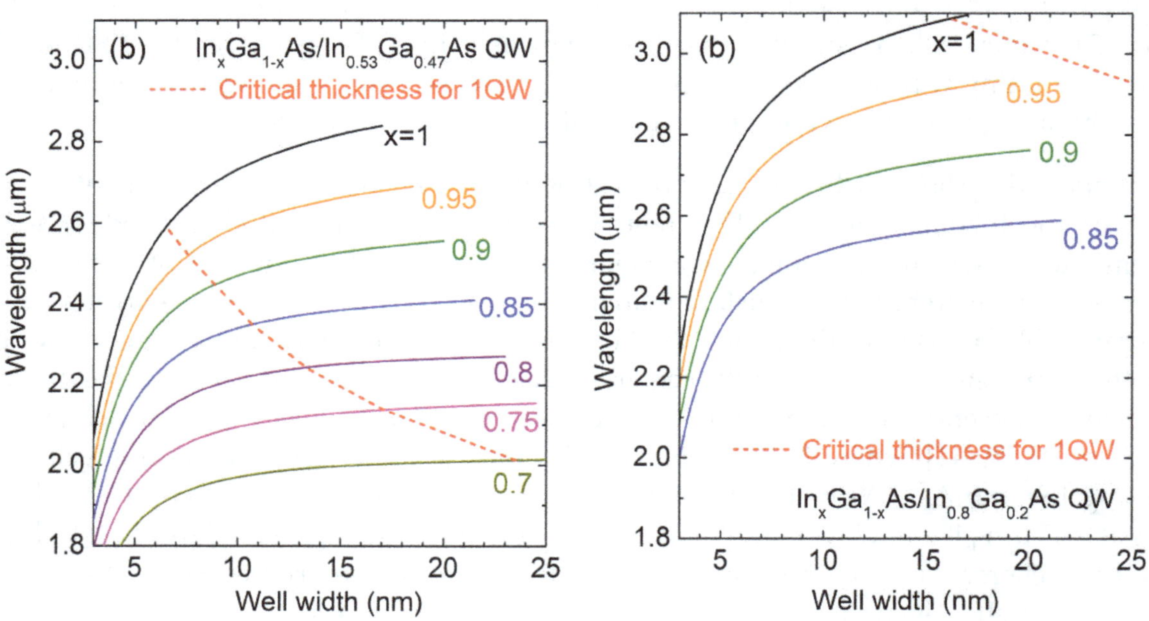

Figure 4. Calculated transition wavelength versus well width of (a) pseudomorphic $In_xGa_{1-x}As/In_{0.53}Ga_{0.47}As$ QW (x>0.53) and (b) metamorphic $In_xGa_{1-x}As/In_{0.8}Ga_{0.2}As$ QW (x>0.8) with rectangular QW shape. The critical thickness for a single QW is indicated as the dotted line.

Furthermore, calculation shows that the lasing wavelength can be extended dramatically if the energy band shape is changed from a rectangular to a triangular one [45, 46]. Figure 5(a) shows the schematic band structure of the $InAs/In_{0.53}Ga_{0.47}As$ triangular QW (TQW) and $In_{0.765}Ga_{0.235}As/In_{0.53}Ga_{0.47}As$ rectangular QW (RQW). The two QW structures are considered to have the same strain extent in the QW part. From Figure 5(b), we can see that changing the energy band from the rectangular shape to a triangular one redshifts the lasing wavelength markedly. For the well width of 16 nm, the calculated wavelength of RQW is only about 2.1 μm, but it is beyond 2.5 μm for TQW. The theoretical estimations show the promising potential of the TQW to extend the wavelength and improve the performances.

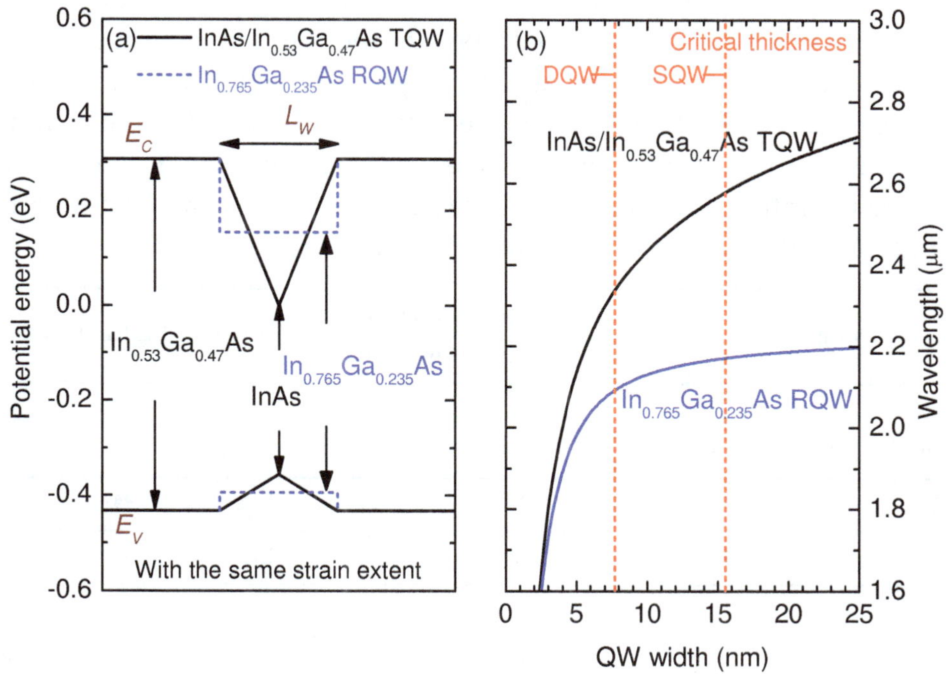

Figure 5. (a) Schematic band structures and (b) calculated transition wavelength as a function of well width of InAs/$In_{0.53}Ga_{0.47}As$ TQW and $In_{0.765}Ga_{0.235}As/In_{0.53}Ga_{0.47}As$ RQW with the same strain extent. The critical thicknesses for single QW and double QWs are indicated.

4. Pseudomorphic scheme for lasing wavelength below 2.5 μm

The pseudomorphic scheme is based on the lattice-mismatched material system except the active QW region. Therefore, the main concern of such lasers is the control of strain in the QW and to keep fair structural quality. The gas source molecular beam epitaxy (GSMBE) growth parameters and the structures of the QW parts need to be optimized at first [47, 48]. Then the whole laser structures were grown and the lasers were demonstrated and analyzed.

4.1. GSMBE growth and optimization of pseudomorphic triangular QWs

To construct the QWs with triangular shape in practice, usually two methods can be applied. The first one uses analogue alloy where the emission source flux changes gradually with an extremely low rate [49], and the other one uses the so-called digital alloy, where short period superlattice composed of two materials are grown at digitally setting thicknesses [50]. The growth rate is normal for the latter one, and this method is especially much more convenient for molecular beam epitaxy (MBE). In MBE, the time for opening and closing a shutter is far below 1 s, so the control precision of the short period superlattice can be high enough. It is also found that digital alloy technology can extend the critical thickness and restrain the three-dimensional material growth [51, 52].

A VG Semicon V80H GSMBE system was applied for the growth. The best background vacuum achieved in this system was about 1×10^{-11} Torr. The elemental indium and gallium cells with two heaters as well as aluminum standard cell were used as group III sources. The fluxes of these group III emission sources were controlled by the cell temperatures. Group V sources were As_2 and P_2 cracked from Arsine (AsH_3) and phosphine (PH_3) cracking cells at around 1000 °C. The fluxes of As_2 and P_2 were controlled by pressure. Standard beryllium and silicon cells were used as *p*- and *n*-type doping sources, respectively. The doping levels were controlled by the temperatures of the emission cells. The growth rates of InP-lattice-matched InGaAs(P) and InP layers were all controlled to be around 1 μm/h.

Sample	Designed			T_g (°C)	Parameters from HRXRD simulation		PL wavelengths (μm)		PL FWHM (meV)	
	λ_{300} (μm)	d_W (nm)	d_B (nm)		d_W (nm)	d_B (nm)	300 K	12 K	300 K	12 K
1	2.1	9	15	500	9.1	16.2	2.14	1.94	35	24
2	2.1	9	15	530	9.1	16.4	2.11	1.93	35	19
3	2.3	13	15	530	14.0	16.4	2.25	2.07	31	20
4	2.5	19	15	530	19.2	16.4	2.38	2.16	36	24
5	2.5	19	20	530	19.2	20.6	2.38	2.16	33	17
6	2.6	25	20	530	N/A	N/A	N/A	2.26 & 2.31	N/A	24 & 31

Table 1. The structure, growth parameters, HRXRD and PL results of the samples. λ_{300} is the wavelength at 300 K, T_g is the growth temperature, d_W and d_B is well width and barrier width, respectively.

The grown QW samples consisted of a 200-nm-thick InP buffer layer, two triangular QWs and a 100-nm-thick InP cap layer. The triangular QWs were constructed by growing InGaAs/InAs digital alloy with very short periods. In each period the thicknesses of $In_{0.53}Ga_{0.47}As$ (d_1) and InAs (d_2) is designed from the following equation:

$$\begin{cases} (d_1 \times 0.53 + d_2)/d = \alpha \\ d_1 + d_2 = d \end{cases} \tag{1}$$

where α is the expected average indium composition in each short period, and d is the period of the short period setting to be 1 nm here. The barrier layer is AlInGaAs grown using the growth parameters of InP-lattice-matched $In_{0.52}Al_{0.48}As$ and $In_{0.53}Ga_{0.47}As$, thus the composition is around $Al_{0.33}In_{0.36}Ga_{0.31}As$.

Table 1 lists the structure parameters, growth conditions, high resolution X-ray diffraction (HRXRD) and PL results of the grown QW samples. The HRXRD (004) scanning curves of all samples are shown in Figure 6. For all the samples, the upper and lower curves show the measured and simulated results, respectively. The good agreement between the measurement and simulation in samples 1-5 indicates the pseudomorphic growth. By simulation, the average indium composition in the well and barrier layers were about 0.765 and 0.355 in samples 1-5, the deduced well widths d_w and barrier widths d_B for all the samples are also listed in Table 1, agreeing well with the designed values. Figure 7 presents the PL spectra of samples 1-6 at RT and 12 K, respectively. The peak wavelengths and full-width at half-maximum (FWHM) of the PL spectra are listed in Table 1.

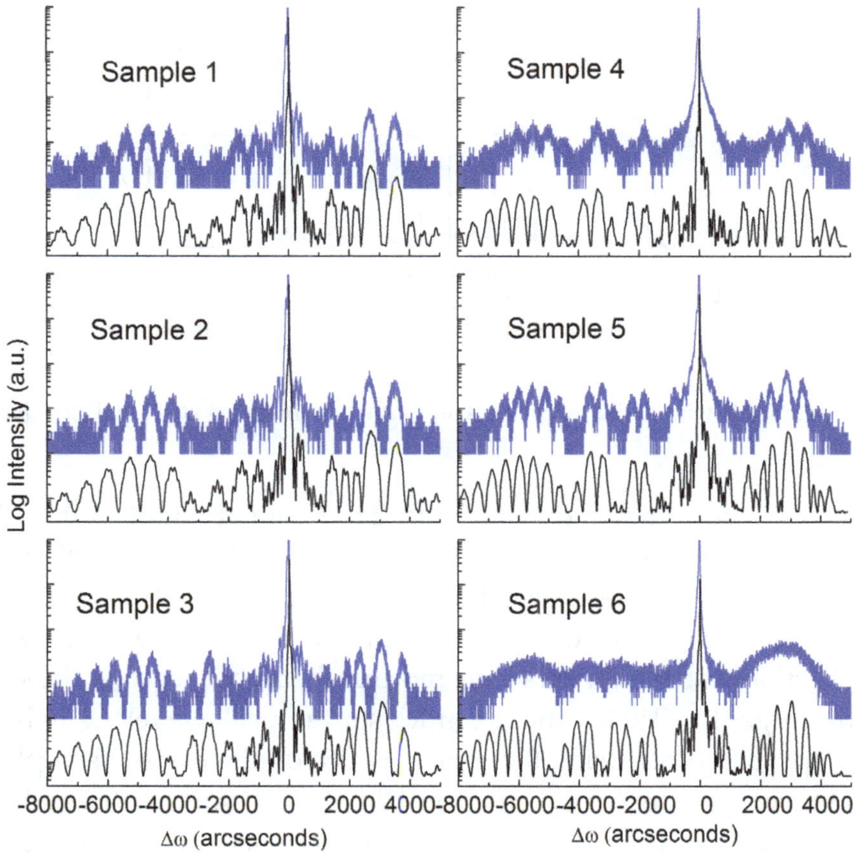

Figure 6. HRXRD (004) scanning curves of the TQWs with various structures and growth parameters [48]. (Reprinted with permission from Elsevier)

The growth temperature T_g is a crucial growth parameter for high crystalline and optical quality during the MBE growth. For the growth of high indium materials, generally T_g needs to be lowered. However, for the $In_{0.53}Ga_{0.47}As$ barriers and the Al-containing AlInGaAs barriers, a relatively higher growth temperature is preferred to improve the optical properties. It is really unpractical to increase and decrease T_g in the interface of well and barrier layers, because the interface quality would be significantly deteriorated by the growth interruption. Therefore, T_g is needed to be optimized, especially for the QW growth. Samples 1 and 2 are with different growth temperatures at 500 °C and 530 °C, respectively. Figure 6 shows the sharp satellite peaks in the HRXRD scanning curves of samples 1 and 2, denoting the nice crystalline quality of both samples. Similarly, PL spectra of the two samples have been observed at 300 K as shown in Figure 7, whereas at 12 K the PL intensity of sample 2 is stronger than that of sample 1. The FWHM at 12 K is 19 meV for sample 2, and 24 meV for sample 1. It is shown that the sample grown at a relatively higher temperature shows better optical quality.

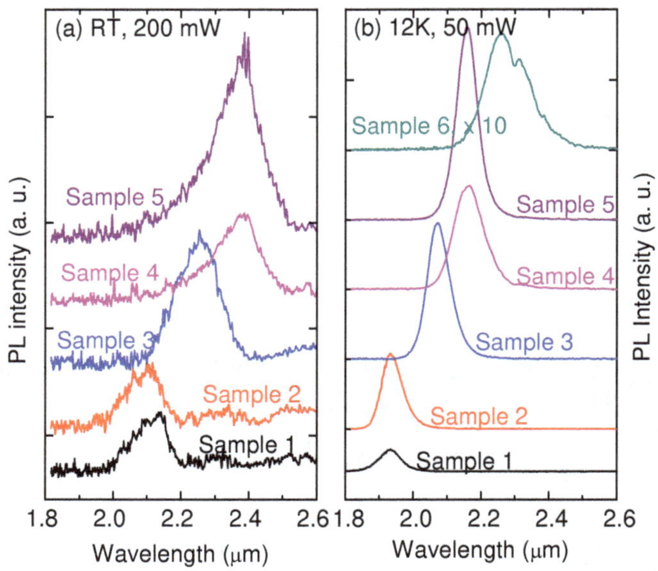

Figure 7. PL of triangular QWs with various structures and growth parameters at RT and 12 K [48]. (Reprinted with permission from Elsevier)

By increasing the well width from 9 nm of sample 2 to 13 nm of sample 3 and 19 nm of sample 4 while keeping the same barrier width, the HRXRD curve becomes indistinct as shown in Figure 6, indicating the deteriorated crystalline quality of sample 4. As the QW width increases, more defects may be generated due to the larger lattice mismatch. The wider QW also weakens the quantum restriction, and the carriers have more probability to occupy the second subband, which induces the larger PL FWHM and the much shorter peak wavelength than designed for sample 4. For sample 5, by increasing the barrier width on the basis of sample 4, the satellite peaks of HRXRD curves recover to be distinct and the PL intensity is enhanced. The PL FWHM is reduced to 17 meV at 12 K and 33 meV at 300 K for sample 5, comparing to 24 meV at 12 K and 36 meV at 300 K for sample 4. The well width of sample 6 is further enlarged to 25 nm in order to extend the wavelength further. It can be seen from Figure 6 that the satellite peaks of the HRXRD curve become indistinct, and two PL peaks at 12 K exist as shown in Figure 6(b), whereas no PL signal is observed at 300 K. It suggests that 2.4 μm is around the ceiling wavelength for the In(Ga)As strained QW.

4.2. Demonstration of CW-operated lasers above room temperature

In our preliminary work, InP layers were applied as waveguides for the convenience of the growth, but the lasing could only be achieved below 150 K [53, 54]. Subsequently, InGaAsP quaternary alloys with bandgap around 1.1 eV and lattice-matched to InP were used as the waveguide layers, and the laser performances were dramatically improved. In this section, the performances of RQW and TQW lasers at 2.2 μm will be presented. And the demonstration of TQW lasers beyond 2.4 μm with increased QW width will also be introduced.

For the lasers with designed lasing wavelength at 2.2 μm, two structures with RQW and TQW were grown [52]. The active QW regions of both samples were sandwiched between 120-nm-thick InGaAsP waveguide layers, as well as 1000-nm-thick *n*-type bottom and 1700-nm-thick *p*-type upper InP cladding layers, and 300-nm-thick p-type $In_{0.53}Ga_{0.47}As$ layers were grown as a top contact layer. The active QW region was formed by four QW layers sandwiched between 20-nm-thick $In_{0.53}Ga_{0.47}As$ barriers for both samples. The QW layers were 3-nm-thick compressive InAs layers in the RQW structure, and 10-nm-thick triangular $In_{0.53}Ga_{0.47}As$/InAs/$In_{0.53}Ga_{0.47}As$ layers in the TQW structure. The total strain content of the TQW was still larger than that of the RQW due to the much thicker well width.

Ridge waveguide lasers with a strip width of 6 μm were fabricated by standard lithography and wet chemical etching process. Then, a 300-nm-thick Si_3N_4 layer was deposited by using plasma enhanced chemical vapor deposition (PECVD). On the top of the ridge a 4-μm-wide window was opened. The top and bottom metallic contacts were sputtered Ti/Pt/Au and evaporated Ge/Au/Ni/Au, respectively. The chips were cleaved into 0.8-mm-cavity-length bars with uncoated facet, soldered on copper heat sinks and wire bonded.

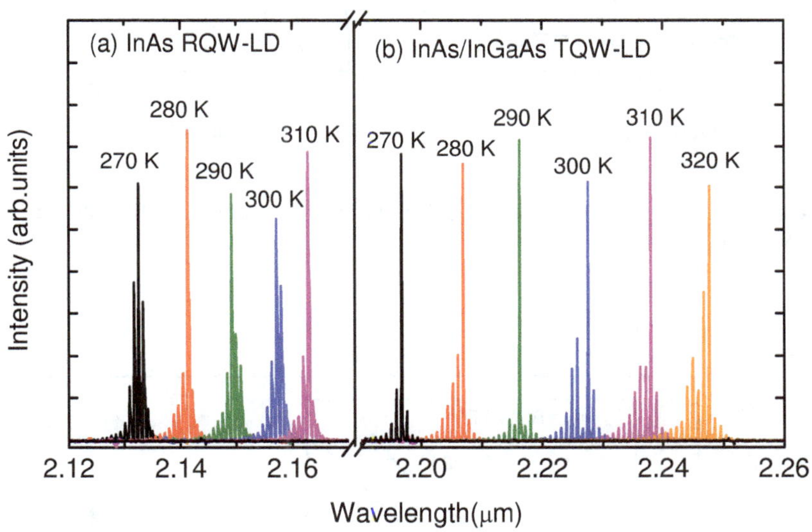

Figure 8. Lasing spectra of (a) rectangular QW laser and (b) triangular QW laser around 2.2 μm at various temperatures under CW driving condition.

Figure 8 shows the CW lasing spectra at a driving current of 1.2 times the threshold current in a temperature step of 10 K. The maximum operation temperature of the RQW laser is about 310 K and the lasing peak is about 2.16 μm at 310 K, whereas the TQW laser shows an increased maximum operation temperature up to 330 K. Over the whole temperature range, the lasing peaks of the TQW laser are about 60 nm longer than those of the RQW laser, probably due to the model error in theoretical design as well as due to the growth error. These results show that the TQW laser can operate at higher temperatures even with a longer lasing wavelength, proving the benefit of TQW. The average temperature coefficient of the lasing wavelength $\Delta\lambda/\Delta T$ is about 0.8 nm/K for the RQW laser and 1 nm/K for the TQW laser.

The temperature-dependent I-P characteristics and the I-V curve at 300 K of the lasers are shown in Figure 9. The threshold current of the RQW laser is 124 mA at 300 K, corresponding to a threshold current density of 2.58 kA/cm², and the output power is 3.6 mW/facet at an injection current of 400 mA. In contrast, the TQW laser shows a much lower threshold current of 68 mA (1.42 kA/cm²) at 300 K. Also, a much higher output power of 10.4 mW/facet is obtained at 300 K, almost three times that of the RQW laser. The turn-on voltage and the differential resistance for both lasers are almost the same at about 1.25 V and 1.87 Ω at 300 K.

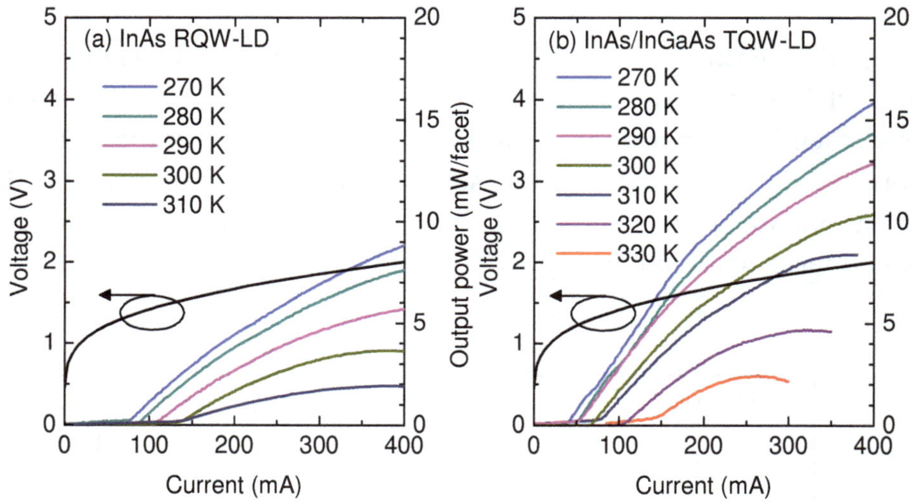

Figure 9. I-P and I-V characteristics of (a) rectangular QW laser and (b) triangular QW laser around 2.2 μm at various temperatures under CW driving condition.

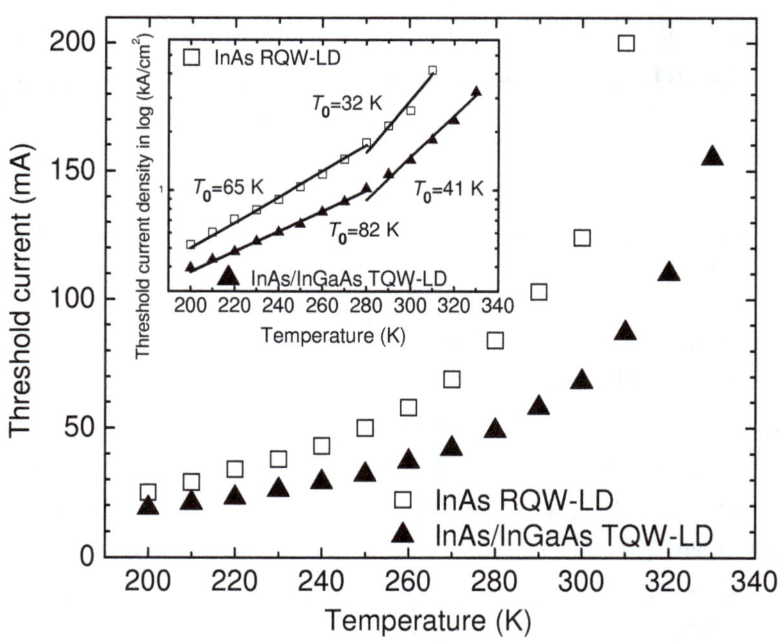

Figure 10. Temperature-dependent threshold current of rectangular QW laser and triangular QW laser around 2.2 μm at various temperatures under CW driving condition. The inset shows the temperature-dependent threshold current densities and characteristics temperatures.

The threshold current as a function of temperature is shown in Figure 10, covering the temperature range from 200 K to 310 K and 330 K for the RQW and TQW lasers, respectively. The characteristic temperatures were achieved from the temperature-dependent threshold current density as shown in the inset of Figure 10. In the low temperature range 200-280 K, a characteristic temperature of T_0=82.1 K was obtained for the TQW laser, higher than the value of T_0=65.2 K for the RQW laser. In the high temperature range beyond 290 K, T_0 is decreased for both lasers, but the TQW laser also has a higher value of T_0=40.5 K than the T_0=32.1 K for the RQW laser.

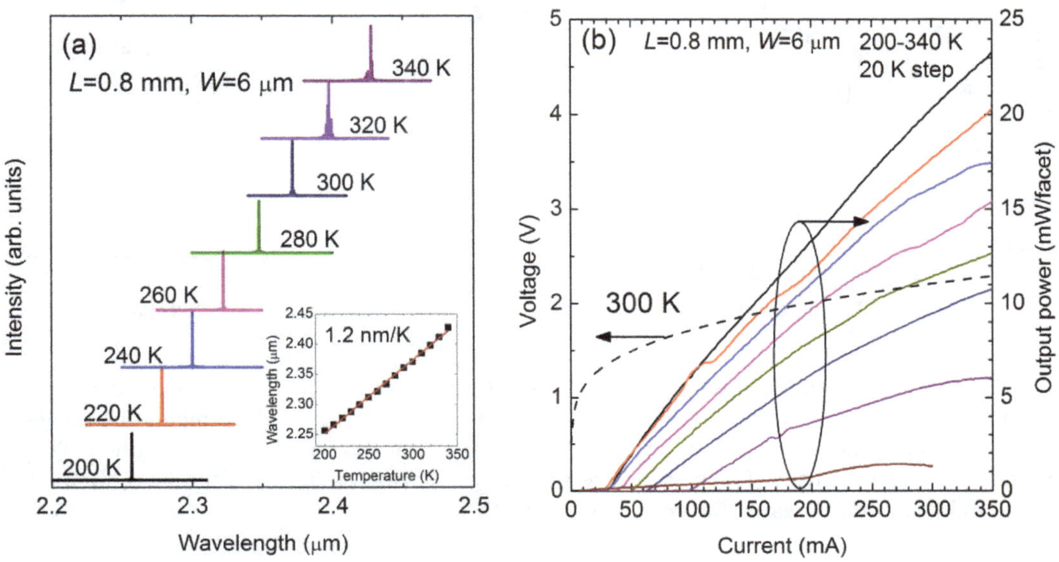

Figure 11. (a) The CW lasing spectra of the around 2.4 μm TQW laser at various temperatures. The inset shows the temperature-dependent lasing wavelength. (b) Output power at different temperatures and voltage at 300 K as a function of injected current of the around 2.4 μm TQW laser [38].

By further increasing the well width of TQW laser to 16 nm, the lasing wavelength of the device has been extended. Figure 11(a) shows the CW lasing spectra in the temperature range of 200 K to 340 K. The lasing wavelength is 2.37 μm at 300 K and red-shifts to 2.43 μm at 340 K. The average temperature coefficient of the wavelength is 1.2 nm/K in 200-340 K as shown in the inset of Figure 10(a). As shown in Figure 11(b), the output power at 200 K in CW mode is 23 mW/facet injected by a current of 350 mA and at 300 K the power dropped to around 11 mW/facet. At 200 K the threshold current is 24 mA and at 300 K the threshold current is increased to 62 mA. The corresponding threshold current density is 1.3 kA/cm^2 for the laser with four QWs and 325 mA/cm^2 for each QW, respectively. At 340 K the maximum output power is decreased to 1.4 mW/facet with a threshold current of 186 mA.

The laser threshold current density and external differential quantum efficiency η_d as a function of temperature are shown in Figure 12(a). The characteristic temperature T_0 is derived to be about 99 K in the temperature range of 200-300 K, and decreases to 35 K in 300-340 K range. The external differential quantum efficiency at 200 K is 41% and is decreased to 23% at 300 K and is further dropped to 7% at 340 K. The quantum efficiency characteristic temperature of T_1=142 K is obtained by fitting the external differential quantum efficiency in 200-320 K temperature range.

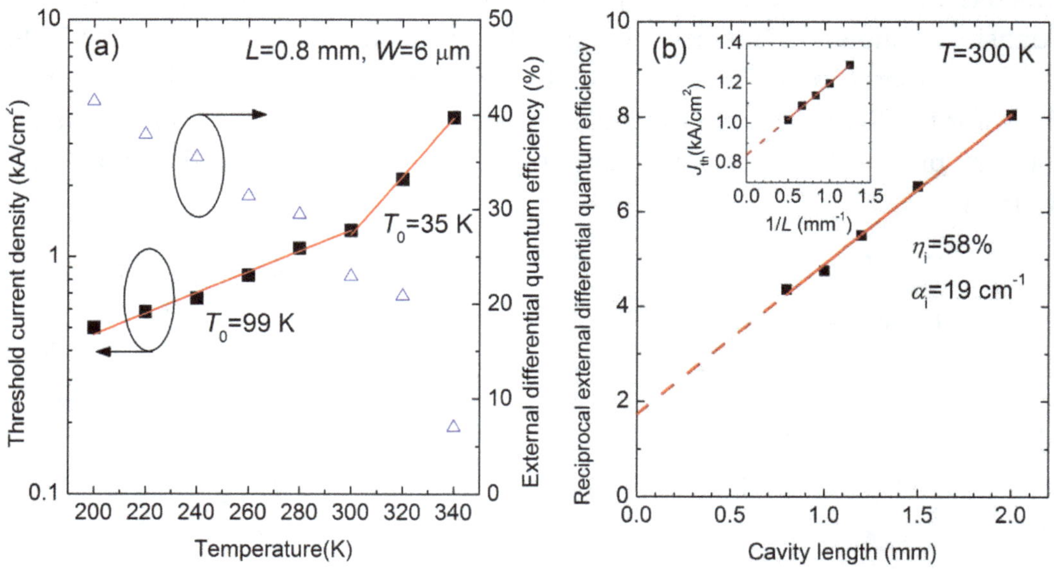

Figure 12. (a) Temperature dependence of the threshold current density and external differential quantum efficiency of the 2.4 μm TQW laser. The lines are the fitting lines of threshold current density in two temperature ranges. (b) Reciprocal external differential quantum efficiency of the TQW lasers as a function of the cavity length at 300 K. The inset shows the threshold current density (J_{th}) versus the reciprocal cavity length (1/L). The solid lines are the fitting of the measured data, and the dashed lines are the extrapolated data [38].

The reciprocal external differential quantum efficiency η_d as a function of the cavity length is shown in figure 12(b). The internal quantum efficiency η_i and internal loss α_i are calculated by fitting the data following $1/\eta_d=(1/\eta_i)[1+\alpha_i L/\ln(1+R)]$, where L is the cavity length and R is the facet reflectivity. Taking R as 0.35 for the as-cleaved facet in the calculation, the internal quantum efficiency and internal loss is calculated to be 58% and 19 cm^{-1}, respectively. Note that the internal quantum efficiency value is favorable but the loss is relatively high, which limits the laser performances. The intervalence band absorption in the *p*-type InP up cladding is one main cause of the large internal loss. The inset of Figure 11(b) illustrates the threshold current density versus the reciprocal cavity length. By extrapolation, the threshold current density at infinite cavity length is 841 A/cm^2 (210 A/cm^2 per QW).

5. Metamorphic scheme for lasing wavelength beyond 2.5 μm

For the metamorphic scheme, InAlAs graded buffers were grown on InP substrate to form the metamorphic template with larger lattice constant than InP. The InAs QW structures were then grown on the template with thicker QW width, and the wavelength could be beyond 2.5 μm. For this kind of metamorphic laser structures, the material quality is very crucial for the device performance, therefore both of the buffer layers and growth parameters were optimized to improve the buffer quality. On the basis of the buffer optimization, the lasers were grown, demonstrated and characterized.

5.1. GSMBE growth and optimization of metamorphic buffers and QWs

As a preliminary experiment, the metamorphic buffer composed of a 2.5-μm-thick $In_xAl_{1-x}As$ continuously graded buffer and a 1-μm-thick $In_{0.8}Ga_{0.2}As$ virtual substrate layer were grown on InP substrate after the growth of a 100-nm-thick InP buffer layer. In the graded buffer, the indium composition x was graded from 0.52 to 0.8 through the simultaneously linear increase of indium source temperature and decrease of aluminum source temperature. After that, two periods of $InAs/In_{0.53}Ga_{0.47}As$ strained QWs were grown. The $In_{0.53}Ga_{0.47}As$ barrier is used to compensate the compressive strain of InAs well comparing to the virtual substrate layer. The lattice mismatch of InAs well and $In_{0.53}Ga_{0.47}As$ barrier with respect to $In_{0.8}Ga_{0.2}As$ virtual substrate layer is +1.3% and -1.8%, respectively. Two samples were grown and the widths of InAs well and $In_{0.53}Ga_{0.47}As$ barrier were 10 nm/7 nm and 15 nm/10 nm, respectively to form a strain compensated QW structure with minimal residual strain. The $In_{0.53}Ga_{0.47}As$ barrier can also stop the diffusion of exited carriers between the QW region and the rest part of the sample.

The grown samples show mirror-like surface morphology without haziness under optical microscopy. Regular cross-hatch patterns are observed on the surfaces with ridges and troughs along the [110] and [1-10] crystal directions. The root mean square (RMS) roughness measured by atomic force microscope (AFM) over 40×40 μm^2 is 3.9 nm and 6.1 nm for the 10 nm and 15 nm QW lasers, respectively. From the cross-sectional transmission electron microscope (XTEM) images of 15 nm QW laser measured at 160 kV as shown in Figure 13, the dislocations are mainly localized within the $In_xAl_{1-x}As$ graded buffer and do not propagate into the upper structures. The $In_{0.8}Ga_{0.2}As$ virtual substrate layer is free of dislocations in the XTEM measurements, which means the threading dislocation density is under the detection limit (about 10^7 cm^{-2}). In the amplificatory image of the $InAs/In_{0.53}Ga_{0.47}As$ QWs region, distinct interfaces between the well and barrier layers are observed. The thickness fluctuations of wells and barriers are revealed around ±1 nm by a careful inspection.

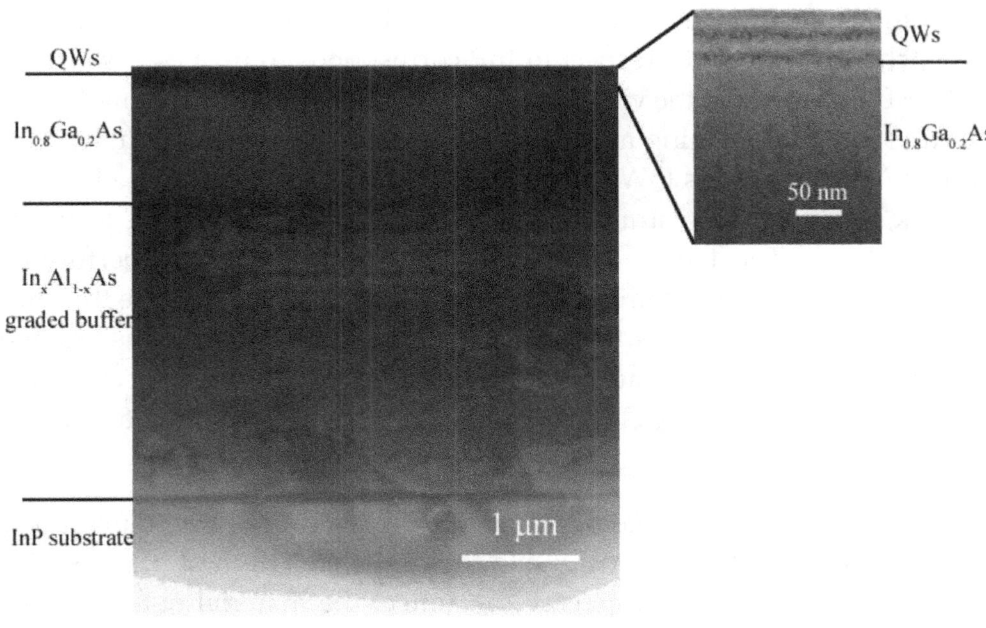

Figure 13. Bright-field XTEM images of the InP-based metamorphic QW epitaxy structure [42]. (Reprinted with permission from AIP)

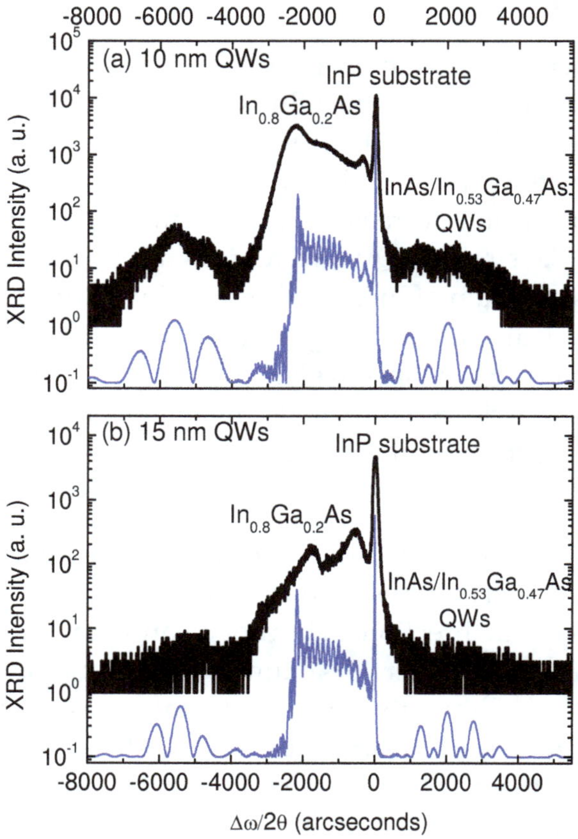

Figure 14. HRXRD (004) scanning curves of InP-based metamorphic (a) 10 nm QWs and (b) 15 nm QWs. The thick and thin lines show the measured and simulated results, respectively [42]. (Reprinted with permission from AIP)

The HRXRD ω/2θ (004) scanning curves were measured and shown in Figure 14 (thick lines). The peak with the largest intensity corresponds to InP substrate, and the wide peak with a relaxed mismatch of 1.8×10^{-2} with respect to InP corresponds to $In_{0.8}Ga_{0.2}As$ virtual substrate layer. The HRXRD signals from the virtual substrate of the two samples show some difference, but become the same after etching away the QW regions at surface. Rough satellite peaks, related to the $InAs/In_{0.53}Ga_{0.47}As$ QWs, are observed symmetrically on the both sides of the $In_{0.8}Ga_{0.2}As$ peak, indicating the strain compensation effect with respect to the $In_{0.8}Ga_{0.2}As$ virtual substrate layer. The lattice dynamical simulations were also performed and the simulated curves are shown in Figure 14 (thin lines). In the simulation, the thicknesses of the wells/barriers were set the same as the design. The $In_xAl_{1-x}As$ buffer was assumed graded relaxed and the $In_{0.8}Ga_{0.2}As$ virtual substrate layer fully relaxed with respect to InP substrate, meanwhile InAs wells and $In_{0.53}Ga_{0.47}As$ barriers were fully strained with respect to $In_{0.8}Ga_{0.2}As$ virtual substrate layer.

The PL measurements were performed as shown in Figure 15. The photo-excited carriers are very sensitive to threading dislocations due to the diffusion length of photo-excited carriers in micrometers. Therefore, the PL intensity can reflect the material quality. The strong PL intensity at 300 K in Figure 15(a) indicates few threading dislocations in QW structures. For each sample, two emission peaks can be observed at 300 K, where the peak at about 2.4 μm corresponds to $In_{0.8}Ga_{0.2}As$ virtual substrate layer, the peaks at 2.90 μm and 3.05 μm for the

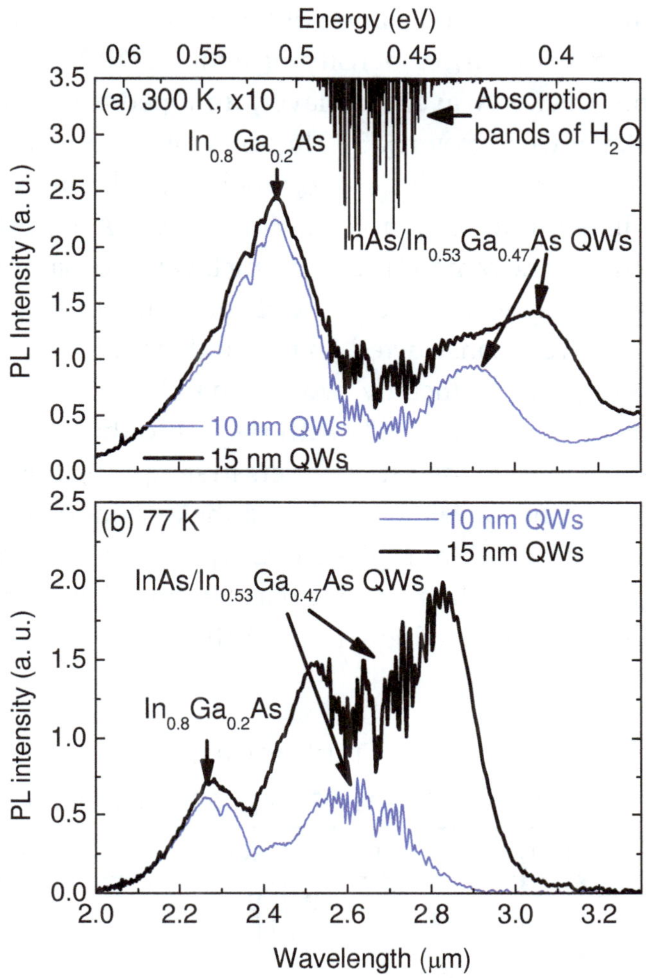

Figure 15. PL results of the metamorphic QW samples at (a) 300 K and (b) 77 K. The thin and thick lines show the results of 10 nm and 15 nm QWs, respectively. The absorption bands of water are also indicated [42]. (Reprinted with permission from AIP)

two samples correspond to the active QWs. In Figure 15(b) the peak corresponds to $In_{0.8}Ga_{0.2}As$ buffer blue-shifts to about 2.3 μm at 77 K. The PL signals of active QWs blue-shift to about 2.6 μm and 2.7 μm at 77 K for the 10 nm and 15 nm QW lasers, respectively, and the peak shapes are affected significantly by the water absorption. The relatively wide PL peaks even at low temperatures are possibly due to the presents of lattice defects and the poor carrier confinement on the top side of the QW. The metamorphic QWs with 15 nm thick InAs layer still show moderate optical quality. This structure is promising for the lasers around 3 μm.

5.2. Demonstration of metamorphic CW-operated lasers at low temperatures

To demonstrate the metamorphic lasers, the effects of strain-compensated QWs and separate confinement heterostructure (SCH) were studied. The growth of the samples started with a 200-nm-thick n^+ InP buffer layer grown, followed by a 1800-nm n^+ $In_xAl_{1-x}As$ continuously graded buffer layer, both highly doped with Si to about 2×10^{18} cm^{-3}. In the graded buffer the indium composition was graded from 0.52 to 0.84 introducing an indium composition overshoot of 0.04 to achieve the full relaxation of the metamorphic layer. Then an 800-nm n^+ $In_{0.8}Al_{0.2}As$ cladding layer with carrier density of about 2×10^{18} cm^{-3} was grown to form a template. Then a 150-nm-thick n^- waveguide layer, undoped active double QWs and a 150-nm-thick p^- waveguide layer were grown [55]. The schematic band energy of the active regions for samples A, B and C can be found in Figure 16, where the solid lines indicate samples A and B whereas the dotted lines indicate sample C. For sample A $In_{0.8}Al_{0.2}As$ was used as the waveguide layers, which was the same as the cladding layers. For samples B and C, $In_{0.8}Ga_{0.2}As$ was used as the waveguide layers so that the SCH consisting of $In_{0.8}Ga_{0.2}As$ waveguide and $In_{0.8}Al_{0.2}As$ cladding layers were constructed. In the double QW region, a strain-compensated structure was used for sample A, formed by two 12-nm InAs wells sandwiched by $In_{0.6}Ga_{0.4}As$ barriers with the strains of +1.4% and -1.4% with respect to the $In_{0.8}Al_{0.2}As$ template. On the other hand, $In_{0.8}Ga_{0.2}As$ was used as the barrier layers in sample B, thus no strain compensation was applied in the QW region. Sample C combined the SCH and strain-compensated QWs, where 15-nm-thick InAs well and 15-nm-thick $In_{0.6}Ga_{0.4}As$ barrier layers were applied. At last, a 1700-nm-thick p^+ $In_{0.8}Al_{0.2}As$ cladding layer and a 300-nm-thick p^+ $In_{0.8}Ga_{0.2}As$ contact layer were grown, heavily doped with Be to higher than 5×10^{18} cm^{-3}.

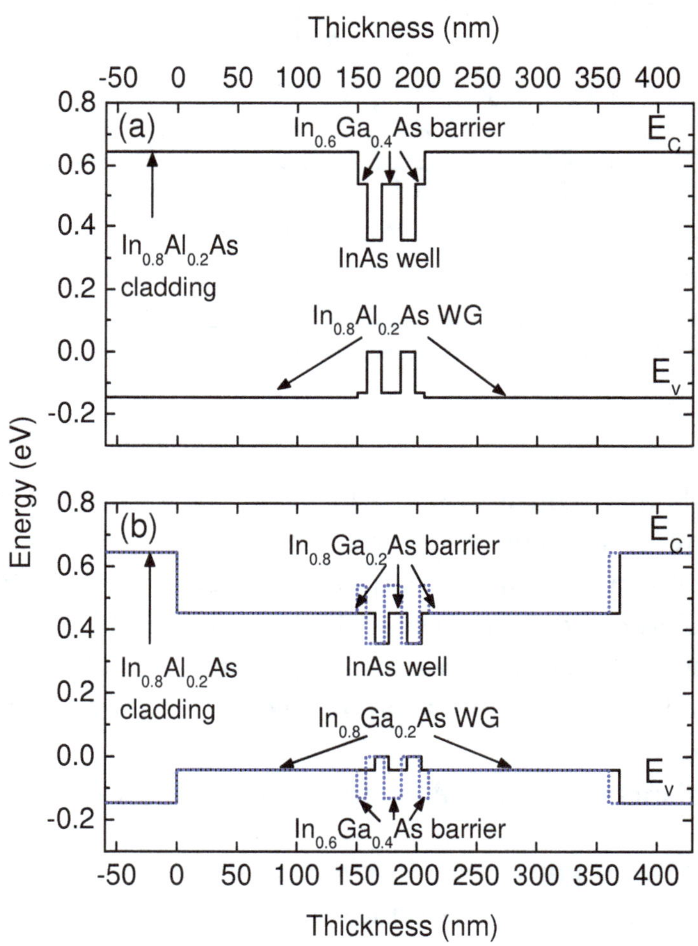

Figure 16. Schematic energy-band diagrams of the active regions of (a) sample A, (b) sample B (solid line) and sample C (dotted line) [55]. (Reprinted with permission from IOP)

Figure 17 shows the HRXRD scanning curves and PL spectra of the samples after etching away the p-type contact layer and p$^+$ $In_{0.8}Al_{0.2}As$ cladding layer. Rough satellite peaks, related to the QWs, can be observed on the left side of the $In_{0.8}Al_{0.2}As/In_{0.8}Ga_{0.2}As$ peaks for the samples, although they are not so distinct. Comparing the intensities of $In_{0.8}Al_{0.2}As/In_{0.8}Ga_{0.2}As$ and QW peaks for the three samples, it is observed that the layer peak intensities of samples A and C with strain-compensated QWs are stronger than those of sample B without strain compensation. The lattice dynamical simulations were also performed and the HRXRD intensities of layer peaks on the left side of InP substrate were very close for all the samples in the simulations. Therefore, the measured stronger layer peak intensities of samples A and C indicate their better material quality than sample B due to the strain compensation of QW regions.

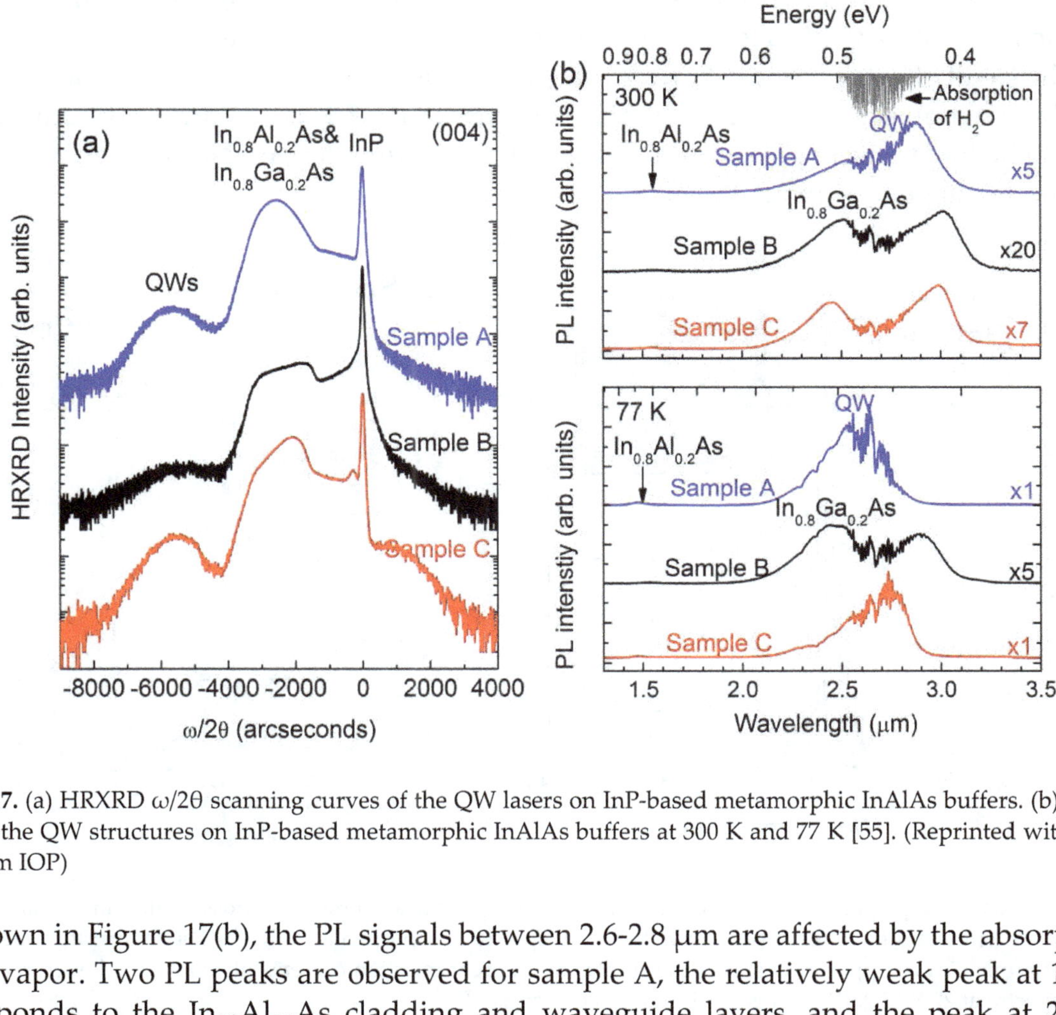

Figure 17. (a) HRXRD ω/2θ scanning curves of the QW lasers on InP-based metamorphic InAlAs buffers. (b) PL emissions of the QW structures on InP-based metamorphic InAlAs buffers at 300 K and 77 K [55]. (Reprinted with permission from IOP)

As shown in Figure 17(b), the PL signals between 2.6-2.8 μm are affected by the absorption of water vapor. Two PL peaks are observed for sample A, the relatively weak peak at 1.55 μm corresponds to the $In_{0.8}Al_{0.2}As$ cladding and waveguide layers, and the peak at 2.88 μm corresponds to the QWs. For sample B, there are also two PL peaks, and the PL signal of $In_{0.8}Al_{0.2}As$ is not clear in linear scale. The two PL peaks at 2.51 μm and 3.02 μm correspond to $In_{0.8}Ga_{0.2}As$ waveguide layers and the QWs, respectively. The PL intensities of the QWs for samples A and C are several times of that of sample B at both 300 K and 77 K, confirming the better material quality of samples A and C. The phenomenon of PL measurements is consistent with the results of HRXRD.

The ridge waveguide lasers were then demonstrated and the cross-section of the ridge waveguide structure was shown in the upper inset of Figure 18(a), where the upside of the ridge was narrowed due to the lateral etching. For the devices fabricated by sample A, no lasing was observed even at 77 K and injected by current up to 1 A. For sample B, around 2.3 μm lasing was observed at low temperatures. This implies that the SCH is very crucial for the device performances, even more important than the crystalline quality of the QWs. For sample B the optical confinement factor is calculated about 0.22, but for sample A the confinement factor is only 0.12 because only the QWs provide a very limited optical confinement in sample A. Figure 18 shows the typical 77 K CW spectrum of the device fabricated by sample B injected by the current of 195 mA. The lasing was declined markedly as the temperature increases. The maximum operation temperature was about 100 K in CW mode and 170 K in pulsed mode with a 5% duty cycle. The lasing transition energy is close to the transition from $In_{0.8}Ga_{0.2}As$ waveguide layers, indicating that excessive band filling exists and higher gain from the QW transition is required. Quite a few carriers may escape from the QWs to the $In_{0.8}Ga_{0.2}As$ layers, due to the small offsets of conductive and valence bands between InAs and $In_{0.8}Ga_{0.2}As$ layers, especially when injected by relatively high currents. The 77 K I-V-P curves of sample B are shown in the lower inset of Figure 18(a). The output power was smaller than 1 mW/facet and the turn-on voltage was about 0.5 V.

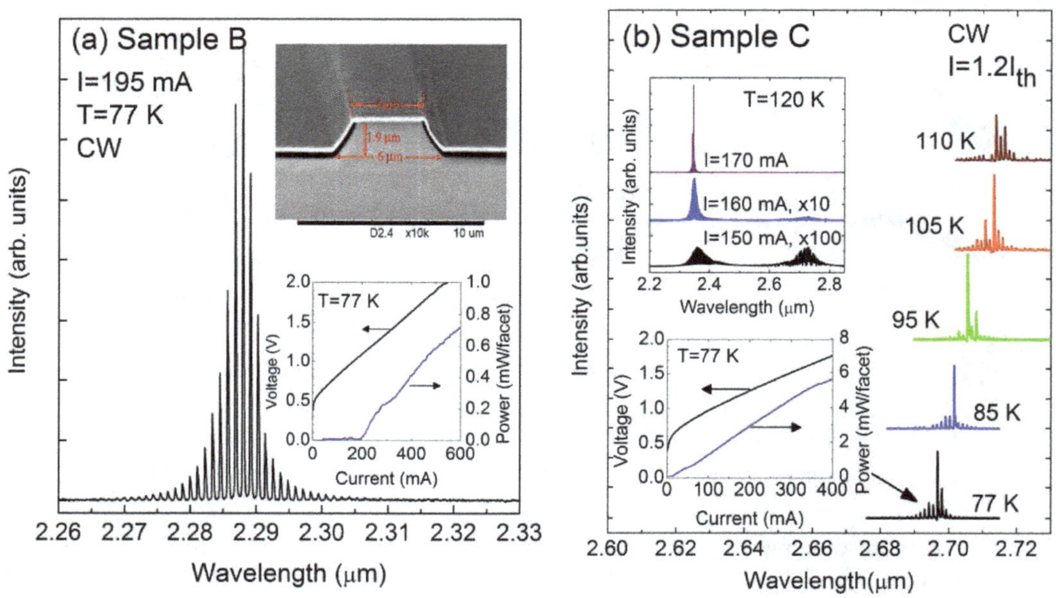

Figure 18. (a) CW lasing spectrum of sample B at 77 K. The upper inset shows the cross-section image of the ridge waveguide structure and the lower inset shows the I-V-P spectra of sample B at 77 K. (b) The temperature-dependent spectra of sample C. The upper inset shows the spectra at different injection currents at 120 K. The lower inset shows the I-V-P spectra at 77 K [55]. (Reprinted with permission from IOP)

For sample C, the CW lasing as long as 2.7 μm was observed, and the temperature-dependent lasing spectra are shown in Figure 18(b). The laser was operated in CW mode and injected by the current of 1.2 times the threshold current. As shown in the lower inset of Figure 18(b), the output power at 77 K was 5.7 mW/facet injected by the current of 400 mA and the turn-on voltage was about 0.65 V. The lasing wavelength moved from 2.70 μm at 77 K to above 2.71 μm at 110 K. When the temperature increased to 120 K, the lasing bounded to around 2.34 μm. The luminescence spectra at different injection currents at 120 K are shown in the upper inset of Figure 18(b). Two luminescence envelops at about 2.34 μm and 2.7 μm can be observed under the injection current of 150 mA. As the injection current increased the 2.34 μm luminescence was enhanced whereas the 2.7 μm luminescence was declined, and only the lasing at 2.34 μm could be observed when the injection current was increased to 170 mA.

Sample C combines with SCH and strain-compensated QWs and shows the best performances in these three metamorphic QW laser structures. The calculated optical confinement factor of sample C is about 0.21, close to that of sample B with SCH. The larger bandgap of $In_{0.6}Ga_{0.4}As$ barrier layers compared to that of $In_{0.8}Ga_{0.2}As$ barriers in sample B can also enhance the carrier confinement and gain of QWs. On the other hand, the strain-compensated QWs ensure the feasible material quality. Nevertheless, the confinements in this laser structure are still unfavorable, as the bandgap of $In_{0.8}Ga_{0.2}As$ waveguide layers is smaller than that of $In_{0.6}Ga_{0.4}As$ barriers, and the refraction index is larger in $In_{0.8}Ga_{0.2}As$ waveguide. The excessive band filling carrier overflow still exists and the lasing at 2.34 μm is possibly due to the transition from the waveguide layers. Generally, increasing the QW number is an effective approach to improve the optical gain and confinement, but more QWs may introduce more strain in the QW region. The growth process needs to be optimized and a tradeoff needs to be considered between the material quality and the QW number.

6. Conclusion

In conclusion, we have demonstrated InP-based Sb-free QW lasers in 2-3 μm wavelength band by optimizing the structural design, GSMBE growth and device processing. To control the strain in the pseudomorphic QW region, the QWs with triangular shape were grown by using digital alloy technology. RT-CW lasers up to 2.43 μm have been successfully achieved by this approach and the performances are competitive to those well-developed GaSb-based QW lasers. Metamorphic InAs QWs were grown on InP-based $In_{0.8}Al_{0.2}As$ template to extend the lasing wavelength even longer. The lasers with CW lasing up to 2.71 μm have been demonstrated by applying 15 nm thick InAs QWs, although they still can only work at a low temperature of 110 K. Our efforts make clear that the extending of the lasing wavelength of InP-based Sb-free QW lasers into 2-3 μm band is not only attractive, but also practical. For CW operation of the laser at RT, the lasing wavelength up to about 2.5 μm should be possible through the fine design of the epitaxial structure, as well as optimization of the growth and processing technologies, adopting pseudomorphic or metamorphic schemes. For longer wavelengths, limited by the strain in pseudomorphic scheme or poor material quality and confinements of both photons and carriers in metamorphic scheme, the reachable performances of the lasers are degraded remarkably.

Acknowledgements

This work is supported by the National Basic Research Program of China under grant No. 2012CB619200 and the National Natural Science Foundation of China under grant Nos. 61275113, 61204133 and 61405232.

Author details

Yi Gu[*] and Yong-Gang Zhang

*Address all correspondence to: ygu@mail.sim.ac.cn

State Key Laboratory of Functional Materials for Informatics, Shanghai Institute of Microsystem and Information Technology, Chinese Academy of Sciences, China

References

[1] Lei W, Jagadish C. Lasers and photodetectors for mid-infrared 2-3 μm applications. J. Appl. Phys. 2008; 104(9), 091101.

[2] Tittel F K, Richter D, Fried A. Mid-infrared laser applications in spectroscopy. Topics Appl. Phys. 2003; 89, 445–516.

[3] Zhang Y G, Gu Y. Gas source MBE grown wavelength extending InGaAs photodetectors. In: Gian-Franco Dalla Betta (ed.) Advances in photodiodes. Rijeka: InTech; 2011. pp. 349–376.

[4] Dhar N K, Dat R, Sood A K. Advances in infrared detector array technology. In: Pyshkin S L, Ballato J M (ed.) Optoelectronics – Advanced materials and devices. Rijeka: InTech; 2013. pp. 149–190.

[5] Bleuel T, Müller M, Forchel A. 2-μm GaInSb-AlGaAsSb distributed-feedback lasers. IEEE Photon. Technol. Letts. 2001; 13(6), 553.

[6] Belenky G, Shterengas L, Kipshidze G, Hosoda T. Type-I diode lasers for spectral region above 3 μm. IEEE J. Select Top. Qutuantum Electron. 2011; 17, 1426.

[7] Sanchez D, Cerutti L, Tournie E. Single-mode monolithic GaSb vertical-cavity surface-emitting laser. Optics Express 2012; 20, 15540.

[8] Vizbaras K, Amann M C. Room-temperature 3.73 μm GaSb-based type-I quantum-well lasers with quinternary barriers. Semicond. Sci. Technol. 2012; 27, 032001.

[9] Zhang Y G, Li A Z, Zheng Y L, Lin C, Jian G Z. MBE grown 2.0 μm InGaAsSb/AlGaAsSb MQW ridge waveguide laser diodes. J. Cryst. Growth 2001; 227–228, 582–585.

[10] Zhang Y G, Zheng Y L, Lin C, Li A Z, Liu S. Continuous wave performance and tunability of MBE grown 2.1 μm InGaAsSb/AlGaAsSb MQW lasers. Chin. Phys. Lett. 2006; 23(8), 2262.

[11] Zhang Y G, Zhang X J, Zhu X R, Li A Z, Liu S. Tunable diode laser absorption spectroscopy detection of N_2O at 2.1 μm using antimonide laser and InGaAs photodiode. Chin. Phys. Lett. 2007; 24(8), 230.

[12] Zhang Y G, Gu Y, Zhang X J, Li A Z, Tian Z B. Gas sensor using a robust approach under time multiplexing scheme with a twin laser chip for absorption and reference. Chin. Phys. Lett. 2008; 25(9), 3246–3249.

[13] Baranov A N, Cuminal Y, Boissier G, Alibert C, Joullié A. Low-threshold laser diodes based on type-II GaInAsSb/GaSb quantum-wells operating at 2.36 μm at room temperature. Electron. Lett. 1996; 32, 2279.

[14] Rossner K, Hummer M, Lehnhardt T, Muller M, Forchel A, Fischer M, Koeth J. Continuous-wave operation of GaInAsSb/GaSb type-II ridge waveguide lasers emitting at 2.8 μm. IEEE Photon. Technol. Lett. 2006; 18(13), 1424.

[15] Yang R Q. Infrared laser based on intersubband transitions in quantum wells. Superlattices Microstruct. 1995; 17(1), 77.

[16] Sprengel S, Andrejew A, Vizbaras K, Gruendl T, Geiger K, Boehm G, Grasse C, Amann M C. Type-II InP-based lasers emitting at 2.55 μm. Appl. Phys. Lett. 2012; 100(4), 041109.

[17] Chang C H, Li Z L, Pan C H, Lu H T, Lee C P, Lin S D. Room-temperature mid-infrared "M"-type GaAsSb/InGaAs quantum well lasers on InP substrate. J. Appl. Phys. 2014; 115(6), 063104.

[18] Cornet C, Doré F, Ballestar A, Even J, Bertru N, Corre A, Loualiche S. InAsSb/InP quantum dots for midwave infrared emitters: A theoretical study. J. Appl. Phys. 2005; 98, 126105.

[19] Zhu C, Zhang Y G, Li A Z, Zheng Y L. Comparison of thermal characteristics of antimonide and phosphide MQW lasers. Semicond. Sci. Technol. 2005; 20, 563.

[20] Zhu C, Zhang Y G, Li A Z, Zheng Y L, Tang T. Heat management of MBE-grown antimonide lasers. J. Cryst. Growth 2005; 278, 173.

[21] Razeghi M, Bandyopadhyay N, Bai Y B, Lu Q Y, Slivken S. Recent advances in mid infrared (3–5 μm) quantum cascade lasers. Opt. Mater. Express 2013; 3(11), 1872.

[22] Vurgaftman I, Meyer J R, Ram-Mohan L R. Band parameters for III–V compound semiconductors and their alloys. J. Appl. Phys. 2001; 89, 5815.

[23] Zhang Y G, Chen J X, Chen Y Q, Qi M, Li A Z, Frojdh K, Stoltz B. Characteristics of strain compensated 1.3 μm InAsP/InGaAsP ridge waveguide laser diodes grown by gas source MBE. J. Cryst. Growth 2001; 227–228, 329–333.

[24] Forouhar S, Ksendzov A, Larsson A, Temkin H. InGaAs/InGaAsP/InP strained layer quantum well lasers at ~2 μm. Electron. Lett. 1992; 28(15), 1431.

[25] Forouhar S, Keo S, Larsson A, Ksendzov A, Temkin H. Low threshold continuous operation of InGaAs/InGaAsP quantum well lasers at ~2.0 μm. Electron. Lett. 1993; 29(7), 574.

[26] Major J S, Nam D W, Osinski J S, Welch D F. High-power 2.0 μm InGaAsP laser diodes. IEEE Photon. Technol. Lett. 1993; 5(6), 594.

[27] Martinelli R U, Menna R J, Triano A, Harvey M G, Olsen G H. Temperature dependence of 2 μm strained-quantum-well InGaAs/InGaAsP/InP diode lasers. Electron. Lett. 1994; 30(4), 324.

[28] Ochiai M, Temkin H, Forouhar S, Logan R A. InGaAs-InGaAsP buried heterostructure lasers operating at 2.0 μm. IEEE Photon. Technol. Lett. 1995; 7(8), 825.

[29] Oishi M, Yamamoto M, Kasaya K. 2.0-μm single-mode operation of InGaAs-InGaAsP distributed-feedback buried-heterostructure quantum-well lasers. IEEE Photon. Technol. Lett. 1997; 9(4), 431.

[30] Mitsuhara M, Ogasawara M, Oishi M, Sugiura H. 2.05-μm wavelength InGaAs-InGaAs distributed-feedback multiquantum-well lasers with 10-mW output power. IEEE Photon. Technol. Lett. 1999; 11(1), 33.

[31] Mitsuhara M, Ogasawara M, Oishi M, Sugiura H. Metalorganic molecular-beam-epitaxy-grown $In_{0.77}Ga_{0.23}As$/InGaAs multiple quantum well lasers emitting at 2.07 μm wavelength. Appl. Phys. Lett. 1998; 72(24), 3106.

[32] Kuang G K, Böhm G, Grau M, Rösel G, Meyer R, Amann M C. 2.12 μm InGaAs-InGaAlAs-InP diode lasers grown in solid-source-molecular-beam epitaxy. Appl. Phys. Lett. 2000; 77(8), 1091.

[33] Ortsiefer M, Böhm G, Grau M, Windhorn K, Rönneberg E, Rosskopf J, Shau R, Dier O, Amann M C. Electrically pumped room temperature CW VCSELs with 2.3 μm emission wavelength. Electron. Lett. 2006; 42(11), 640.

[34] Boehm G, Grau M, Dier O, Windhorn K, Roenneberg E, Rosskopf J, Shau R, Meyer R, Ortsiefer M, Amann M C, Growth of InAs-containing quantum wells for InP-based VCSELs emitting at 2.3 μm. J. Cryst. Growth. 2007; 301–302, 941–944.

[35] Sato T, Mitsuhara M, Kondo Y. 2.33 μm-wavelength InAs/InGaAs multiple-quantum-well lasers grown by MOVPE. Electron. Lett. 2007; 43(21), 1143.

[36] Sato T, Mitsuhara M, Nunoya N, Kasaya K, Kano F, Kondo Y. 2.33-μm-wavelength distributed feedback lasers with InAs-$In_{0.53}Ga_{0.47}As$ multiple-quantum wells on InP substrates. IEEE Photon. Technol. Lett. 2008; 20(12), 1045.

[37] Sato T, Mitsuhara M, Kakitsuka T, Fujisawa T, Kondo Y. Metalorganic vapor phase epitaxial growth of InAs/InGaAs multiple quantum well structures on InP substrates. IEEE J. Sel. Topics Quan. Electron. 2008; 14(4), 992.

[38] Gu Y, Zhang Y G, Cao Y Y, Zhou L, Chen X Y, Li Hsby, Xi S P. 2.4-μm InP-based antimony-free triangular quantum well lasers in continuous-wave operation above room temperature. Appl. Phys. Express 2014; 7, 032701.

[39] Krier A, Chubb D, Krier S E, Hopkinson M, Hill G. Light sources for wavelengths >2 μm grown by MBE on InP using a strain relaxed buffer. IEE Proc. 1998; 145(5), 292.

[40] Kirch J, Garrod T, Kim S, Park J H, Shin J C, Mawst L J, Kuech T F, Song X, Babcock S E, Vurgaftman I, Meyer J R, Kuan T S. $InAs_yP_{1-y}$ metamorphic buffer layers on InP substrates for mid-IR diode lasers. J. Cryst. Growth 2010; 312, 1165–1169.

[41] Jung D, Song Y, Yu L, Wasserman D, Lee M L. 2.8 μm emission from type-I quantum wells grown on InAsP/InP metamorphic graded buffer. Appl. Phys. Lett. 2012; 101, 251107.

[42] Gu Y, Zhang Y G, Wang K, Fang X, Li C, Cao Y Y, Li A Z, Li Y Y. InP-based InAs/ InGaAs quantum wells with type-I emission beyond 3 μm. Appl. Phys. Lett. 2011; 99, 081914.

[43] Cao Y Y, Zhang Y G, Gu Y, Chen X Y, Zhou L, Li Hsby. 2.7 μm InAs quantum well lasers on InP-based InAlAs metamorphic buffer layers. Appl. Phys. Lett. 2013; 102, 201111.

[44] Gu Y, Zhang Y G, Chen X Y, Cao Y Y, Fang X, Ding G Q, Zhou L. $InAs/In_{0.83}Al_{0.17}As$ quantum wells on GaAs substrate with type-I emission at 2.9 μm. Appl. Phys. Lett. 2013; 102, 121110.

[45] Gu Y, Zhang Y G, Liu S. Strain compensated AlInGaAs/InGaAs/InAs triangular quantum wells for lasing wavelength beyond 2 μm. Chin. Phys. Lett. 2007; 24, 3237–3240.

[46] Gu Y, Zhang Y G. Properties of strain compensated symmetrical triangular quantum wells composed of InGaAs/InAs chirped superlattice grown using gas source molecular beam epitaxy. Chin. Phys. Lett. 2008; 25, 726–729.

[47] Chen J X, Li A Z, Chen Y Q, Guo F M, Lin C, Zhang Y G, Qi M. Quasi RT-CW operation of InGaAs/InGaAsP strained quantum well lasers. J. Cryst. Growth 2001; 227–228, 338–342.

[48] Gu Y, Zhang Y G, Wang K, Li A Z, Li Y Y. Optimization of AlInGaAs/InGaAs/InAs strain compensated triangular quantum wells grown by gas source molecular beam epitaxy for laser applications in 2.1–2.4 μm range. J. Cryst. Growth 2009; 311, 1935–1938.

[49] Giugni S, Tansley T L, Green F, Shwe C, Gal M. Optical transitions in symmetric, compositionally graded triangular AlGaAs quantum wells grown by molecular beam epitaxy. J. Appl. Phys. 1992; 71, 3486–3491.

[50] Lin D Y, Lin F C, Huang Y S, Qiang H, Pollak Fred H, Mathine D L, Maracas G N. Piezoreflectance and photoreflectance study of GaAs/AlGaAs digital alloy compositional graded structures. J. Appl. Phys. 1996; 79, 460–466.

[51] Jourba S, Gendry M, Marty O, Pitaval M, Hollinger G. High-quality highly strained InGaAs quantum wells grown on InP using $(InAs)_n(GaAs)_{0.25}$ fractional monolayer superlattices. Appl. Phys. Lett. 1999; 75, 220.

[52] Cao Y Y, Zhang Y G, Gu Y, Chen X Y, Zhou L, Li Hsby. Improved performance of 2.2 μm InAs/InGaAs QW lasers on InP by using triangular wells. IEEE Photon. Technol. Lett. 2014; 26(6), 571–574.

[53] Gu Y, Wang K, Li Y Y, Li C, Zhang Y G. InP-based InGaAs/InAlGaAs digital alloy quantum well laser structure at 2 μm. Chin. Phys. B 2010; 19, 077304.

[54] Cao Y Y, Gu Y, Zhang Y G, Li Y Y, Fang X, Li A Z, Zhou L, Li Hsby. InAs/InGaAs digital alloy strain-compensated quantum well lasers. J. Infrared Millim. Waves 2014; 33(3), 213–217.

[55] Gu Y, Chen X Y, Zhang Y G, Cao Y Y, Fang X, Li Hsby. Type-I mid-infrared InAs/InGaAs quantum well lasers on InP-based metamorphic InAlAs buffers. J. Phys. D: Appl. Phys. 2013; 46, 505103.

7

OFDM and SC-FDMA over Fiber using Directly Modulated VCSELs

Henrique M. Salgado, Rúben E. Neto,
Luís M. Pessoa and Pedro J. Batista

Abstract

Radio-over-fiber technology, used in the transport of radio signals over optical fiber by means of an optical carrier between a remote site and a central node of a cellular network, is an attractive solution for backhauling of a large number of remote antennas, enabling the shifting of the hardware complexity from base stations to a central station.

Integration of both optical and wireless broadband infrastructures into the same backhaul network leads to significant simplification and cost reduction of base stations permitting equipment sharing and dynamic allocation of resources, which in turn leads to simplified system operation and maintenance.

Wireless systems on the other hand are evolving rapidly and new standards are appearing, such as the Long-Term Evolution aiming at satisfying the required need for increasing bandwidth. Radio-over-fiber systems are known to be susceptible to noise and non-linear distortion in particular to the large peak-to-average power ratio of orthogonal frequency division multiplexing signals employed in these standards. In this work we compare, experimentally and through simulation, the performance of orthogonal frequency division multiplexing and single carrier frequency division multiple access signals, in radio-over-fiber applications, using directly modulated VCSELs.

Keywords: RoF, VCSEL, OFDM, SC-FDMA

1. Introduction

Radio-over-fiber (RoF) technology is used in transporting radio signals over optical fiber by means of an optical carrier between a remote site and a central node of a cellular network. RoF techniques are increasingly seen as a promising solution to facilitate the backhauling of a large number of remote antennas, enabling the shifting of the hardware complexity from base stations to the central station [1]. RoF allows for the combination of the advantages of optical systems such as their high bandwidth and low power consumption, with the advantages of wireless systems, namely the flexibility to use multiple standards such as Long-Term Evolution (LTE) or other upcoming standards.

The number of LTE users has been increasing due to advantages such as the ability to reach a peak throughput of 300 Mbps on the downlink and 75 Mbps on the uplink [2]. To achieve high radio spectral efficiency as well as enable efficient scheduling in time and frequency domains, a multicarrier approach for multiple access was chosen by the 3rd Generation Partnership Project. Orthogonal frequency division multiple access (OFDMA) and single carrier frequency division multiple access (SC-FDMA) were selected for the downlink and uplink, respectively, the latter also known as discrete Fourier transform–spread OFDMA.

2. Motivation

When OFDMA or SC-FDMA type of signals are transmitted through directly modulated vertical-cavity surface emitting lasers (VCSELs), they suffer from intermodulation distortion, due to the large number of electrical subcarrier combination in the laser cavity which degrades system performance in addition to relative intensity noise (RIN), clipping noise at the transmitter as well as shot noise and thermal noise at the receiver.

This phenomenon is well known in the literature [3], and is a result of the interaction between the electrons and photons in the active region, which is generally well described by the rate equations [4]. As a motivation for this problem, Figure 1 and Figure 2 show the spectrum of an OFDM electrical signal at the laser output (after conversion by means of a photodiode), for two values of the laser bias current, 3 and 6 mA, respectively. The original OFDM signal is centred at 1 GHz, and directly modulates the VCSEL. The high nonlinear distortion of the signal (with centre at 1 GHz) can be clearly seen, which is caused by intermodulation distortion. Third order intermodulation products (IMPs) of the type, $f_i + f_j - f_k$ or $2f_i - f_j$, where $f_{i,j,k}$, represents the frequency of the OFDM subcarriers, coincide with the transmission band of the signal and severely limit system performance. The number of intermodulation terms falling on channel r, $IM_{111}^{N}(r)$, $IM_{21}^{N}(r)$, of type $f_i + f_j - f_k$ and $2f_i - f_j$, respectively, where N is the number of subcarriers, is given as follows [5]:

$$IM_{21}^{N} = \frac{1}{2}\left\{N - 2 - \frac{1}{2}\left[1 - (-1)^{N}\right](-1)^{r}\right\} \tag{1}$$

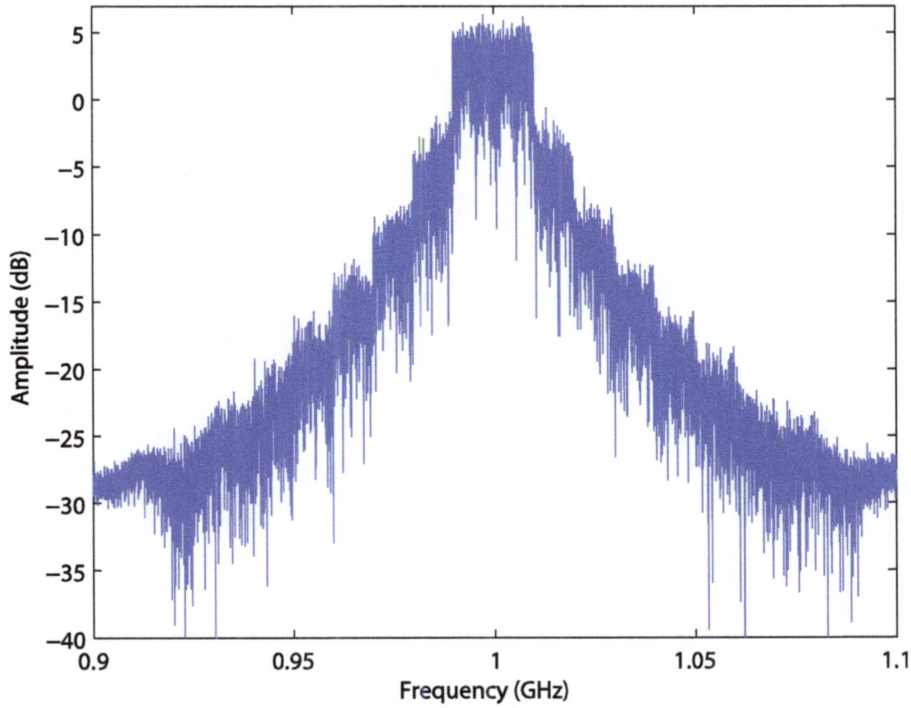

Figure 1. OFDM spectrum at the laser output for 3 mA bias current.

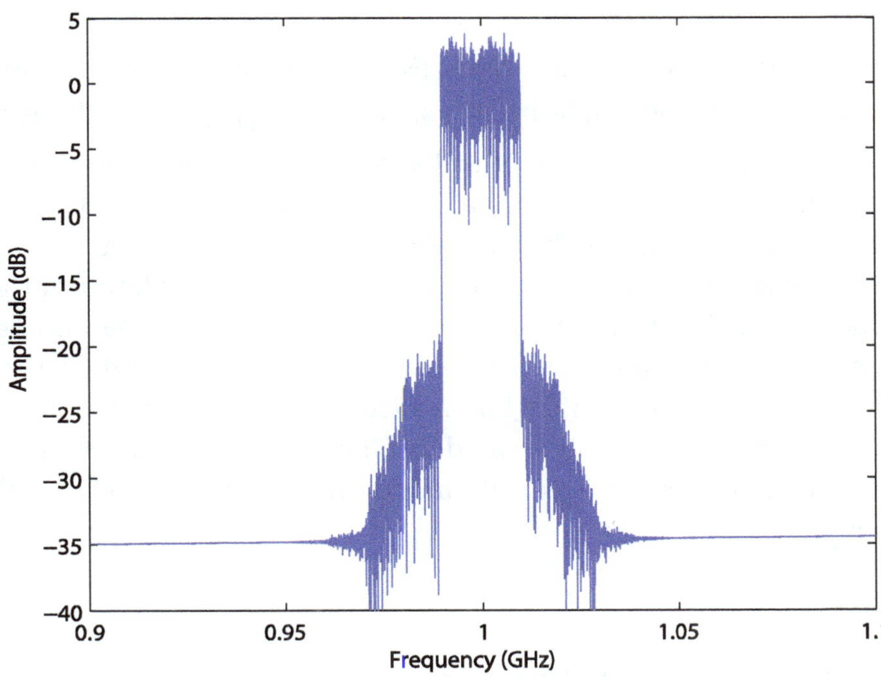

Figure 2. OFDM spectrum at the laser output for 6 mA bias current.

$$IM_{11\bar{1}}^{N} = \frac{r}{2}\left(N - r + 1\right) + \frac{1}{4}\left[\left(N-3\right)^{2} - 5\right] - \frac{1}{8}\left[1 - \left(-1\right)^{N}\right]\left(-1\right)^{N+r} \tag{2}$$

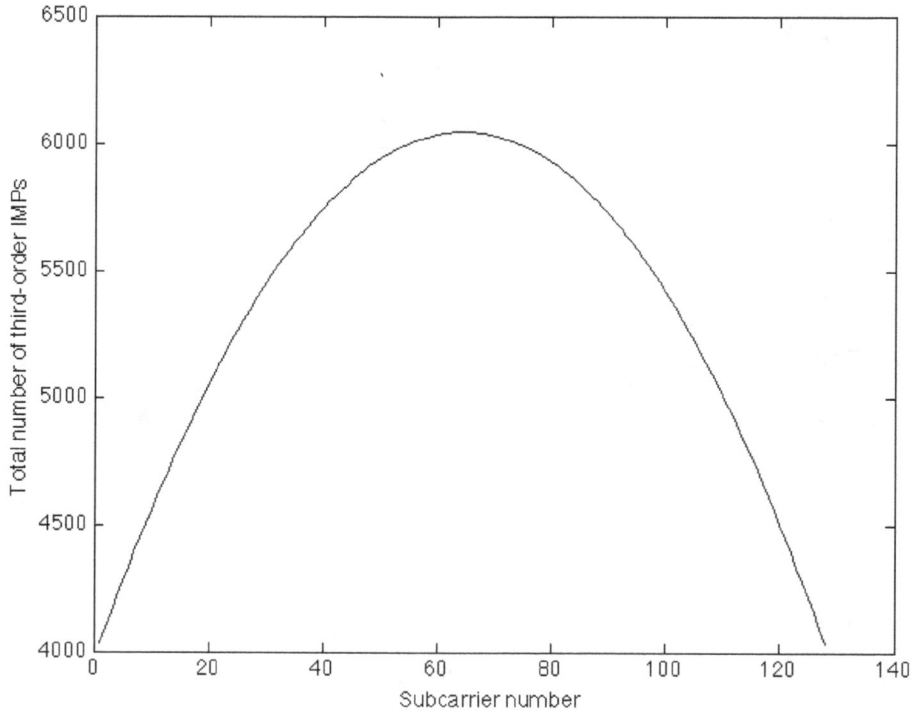

Figure 3. Total number of third-order intermodulation products as a function of subcarrier number for a 128 subcarriers signal.

Figure 3 shows the total number of third-order IMPs as a function of channel number, for 128 subcarriers. For a large number of subcarriers IM_{111}^{N} and IM_{21}^{N}, approach the asymptotic value of $3N^2/8$ and $N/2$, respectively. On the other hand, the output spectrum is highly dependent on the operating point of the laser and on the allocation of the channels relative to the resonance frequency of the laser. On increasing the bias current to 6 mA, the resonance frequency, which previously was centred at 1 GHz, moves away to 2.5 GHz; therefore, the output signal presents a lower distortion as depicted in Figure 2. The interplay between the biasing of the laser, subcarriers frequency operation and noise, be it shot noise or RIN of the laser, is a complex one that needs to be modeled accurately for an adequate assessment of system performance. Hence, it is important to obtain a realistic model of the VCSEL device, including the electrical circuit associated with the parasitic elements and for that effect to extract the relevant parameters from experimental measurements.

3. Vertical-cavity surface-emitting laser

The VCSEL has emerged as an important class of semiconductor lasers in recent years. Its main characteristics, associated with the vertical-cavity geometry, are light emission perpendicular to the surface of the wafer and single longitudinal mode due to its short cavity length. The VCSEL is a microcavity laser consisting of a thin active region (< 1 μm) sandwiched between epitaxially grown distributed Bragg reflectors (DBRs). Since the first demonstration of the

VCSEL laser in Tokyo Institute of Technology in 1979 and after three decades of research, many potential applications have emerged. As a result several VCSEL manufacturers turned up in the market.

Due to their geometry, VCSELs offer a number of significant advantages over edge-emitting lasers listed below:

- Low threshold currents: the volume of the active region is relatively small, hence they possess low threshold currents and therefore consume less power than edge-emitting lasers.
- Circularly shaped beam: the symmetry in the wafer plane means the laser output is a narrow, low divergence circular beam, permitting high coupling efficiency to optical fibers with relaxed alignment tolerances.
- Single-mode operation: due to the microscopic cavity length, VCSELs inherently operate in a single-longitudinal mode which makes them suitable for high-bit-rate fiber optic communications.
- Low-cost-wafer fabrication: VCSELs allow for a high packing density in the form of two-dimensional arrays and cost-effective fabrication and testing at the wafer level.
- High modulation speed: experimental studies [3] indicate that the VCSELs have very fast intrinsic dynamic properties with relaxation oscillation frequencies as high as 71 GHz.

These advantages make the VCSEL device a suitable candidate for the applications targeted in ROF scenarios. However, the VCSEL as any semiconductor laser exhibits a nonlinear dynamic behavior that needs to be modeled accurately for the assessment of the impact of distortion on system performance. In the next section, the model of the VCSEL based on a set of rate equations for the carrier and photon density is presented. This model will be the basis for the investigation of the laser nonlinear distortion.

3.1. VCSEL modeling

The basic structure of the VCSEL is shown in Figure 4. The laser output is taken vertically through one of the mirrors, in contrast to conventional edge-emitting laser which emits light in the plane of the wafer surface. The conventional structure employs an active region consisting of multiple quantum wells between a n-type and a p-type DBR mirrors. Because of the short cavity length ($\approx 1\ \mu m$) and thickness of the active region, the mirror reflectivities (R) of the DBRs must be greater than 99%. To achieve these high values of reflectivity, the DBR mirrors are made up of 20 to 40 alternating quarter-wavelength thick layers of high and low refractive indices made of semiconductors with different compositions, typically $Al_xGa_{(1-x)}As$.

The operation of a VCSEL can be understood by accounting for the rate of recombination of carriers in the active region and the rate of generation and loss of photons. For laser emission to occur, stimulated emission should be the dominant recombination mechanism. The threshold gain is defined as the gain required to sustain the optical field after travelling one round trip in the cavity. Assuming the optical gain is constant over the whole length of the laser, this leads to the condition

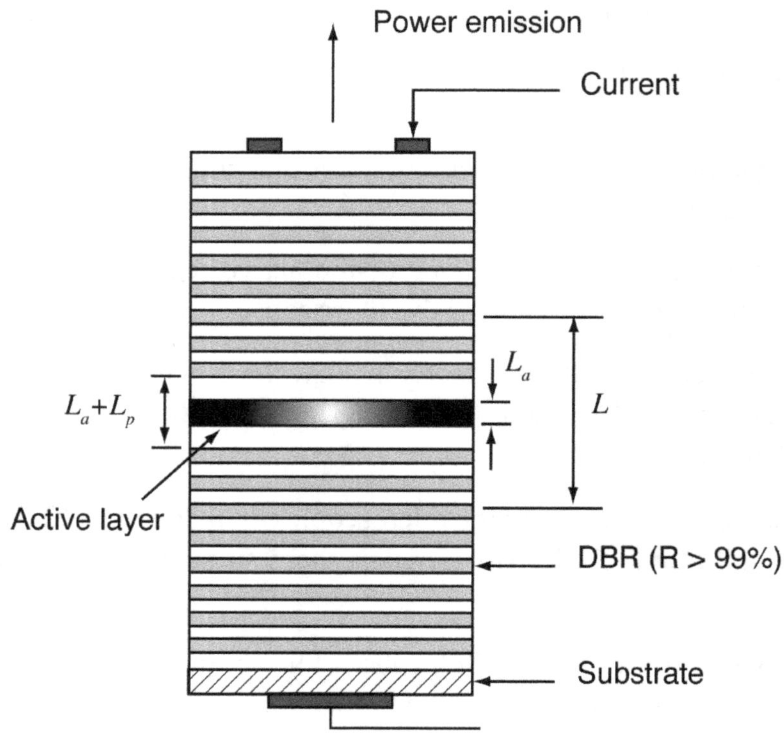

Figure 4. Schematic diagram of the laser structure, indicating active region thickness *La*, and effective cavity length *L*.

$$\Gamma g_{th} = \alpha_i + \frac{1}{L}\ln\left(\frac{1}{R}\right) \tag{3}$$

where Γ is the optical confinement factor, α_i is the average internal loss and the second term is defined as the mirror loss α_m. This equation shows that the gain per unit length must be sufficient to cancel out the optical losses and the losses due to light emission. Since the contribution of spontaneous emission in this simple analysis has not been considered, the actual gain will be slightly lower than the threshold gain. The description of laser operation is complete once the carrier density, N, is related to the injected current, I. This is accomplished through a rate equation that incorporates all the mechanisms by which the carriers are generated or lost inside the active region. The continuity equation which describes the rate of change of carriers in its general form is

$$\frac{dN}{dt} = \frac{\eta_i I}{qV} - \gamma_e(N)N - R_{st}P \tag{4}$$

The first term governs the rate at which the carriers are injected into the active layer due to external pumping; q is the value of the electron charge, η_i is the injection efficiency and V is the volume of the active region. The second term takes into account the carrier loss owing to various recombination processes: spontaneous emission and non-radiation. The last term of equation (4) is due to stimulated emission recombination that leads to coherent emission of light.

A suitable form for the carrier recombination rate $\gamma_e(N)$ and the corresponding carrier lifetime τ, for lightly doped material is

$$\gamma_e(N) = \frac{1}{\tau} = A_{nr} + BN + CN^2 \tag{5}$$

where the terms with the coefficients A_{nr}, B, and C represent defect, bimolecular recombination and Auger recombination, respectively.

A corresponding rate equation for the photon density can be obtained from the Maxwell's equations using a classical approach [4, 6]. By a simple bookkeeping of the supply, annihilation, and creation of carriers and photons inside the laser cavity, we get

$$\frac{dP}{dt} = P\left(\Gamma R_{st} - \frac{1}{\tau_p}\right) + \beta\Gamma R_{sp} \tag{6}$$

In which the photon lifetime is defined by

$$\tau_p^{-1} = v_g\left[\alpha_i + \frac{1}{L}\ln\left(\frac{1}{R}\right)\right] \tag{7}$$

Equation (6) states that the rate of increase in photon density is equal to the photon generation by stimulated emission $\Gamma R_{st}P$ less the loss rate of photons $-P/\tau_p$ (as characterized by the photon lifetime τ_p), plus the rate of spontaneous emission into the photon mode βR_{sp}, where β is the fraction of the total spontaneous emission coupled into the laser mode.

The net stimulated rate which tells us how many photons are generated per unit of time per existing photon, yields a generation rate of new photons dP/dt according to the following equation

$$\frac{dP}{dt} = R_{st}P \tag{8}$$

The resulting stimulated gain coefficient relates to the stimulated emission coefficient, R_{st}, by

$$g = \frac{\text{power emitted per unit volume}}{\text{power crossing a unit area}} = \frac{R_{st}}{v_g} = \frac{R_{st}\bar{\mu}}{c} \tag{9}$$

where v_g is the group velocity, c is the free-space velocity of light, and $\bar{\mu}$ is the group refractive index of the material taking dispersion into account: $[\bar{\mu}=(\mu+\nu d\mu/d\nu)]$. For a multiple quantum well laser, a logarithmic function of the carrier density fits the gain well over a wide range of N, (see [7]),

$$g(N)=g_c\log\left(\frac{N}{N_{0m}}\right) \tag{10}$$

where N_{0m} is the carrier density for transparency (zero gain) and g_c is the gain coefficient. The gain function may be linearized about the carrier density at transparency yielding [8],

$$g(N)\approx a\left(N-N_{0m}\right)\left(1-\varepsilon P\right) \tag{11}$$

where a is the differential gain, $\partial g/\partial N=g_c\cdot m/N_{0m}$ and m is a linearization parameter obtained so that $g_{log}(N)=g_{lin}(N)$ at threshold, that is for $N=N_{th}$. Gain compression is also accounted phenomenologically through the term $(1-\varepsilon P)$, ε being the gain compression factor expressed in cubic meters.

Thus, we can rewrite the carrier and photon rate equations as

$$\frac{dN}{dt}=\frac{\eta_i I}{qV}-\frac{N}{\tau}-v_g a\left(N-N_{0m}\right)\left(1-\varepsilon P\right)P \tag{12}$$

$$\frac{dP}{dt}=\Gamma v_g a\left(N-N_{0m}\right)\left(1-\varepsilon P\right)P-\frac{1}{\tau_p}P+\beta\Gamma R_{sp} \tag{13}$$

The evolution of the signal transmitted over the optical fiber requires knowledge of the phase of the electric field. To account for dispersion effects, these equations may be complemented by an additional equation for the phase:

$$\frac{d\phi}{dt}=\frac{\alpha}{2}\left[\Gamma a v_g\left(N-N_{0m}\right)-\frac{1}{\tau_p}\right] \tag{14}$$

where α is the linewidth enhancement factor. Equations (11-13) represent the basic relations for describing the dynamic characteristics of laser diodes, as long as the noise sources may be omitted.

The first-order transfer function of the intrinsic laser can be obtained by linearization of the previous equations and is given by the following equation

$$\frac{H_1(\omega)}{H_1(0)} = \frac{\gamma(1-\varepsilon p_0)p_0}{(j\omega)^2 + j\omega(\gamma\Gamma\beta j_{th} / p_0 + \gamma\varepsilon p_0) + \gamma(1-\varepsilon p_0)p_0} \tag{15}$$

$$\gamma = \frac{\tau}{\tau_p} \tag{16}$$

where for the ease of numerical calculation, the physical quantities have been normalized according to [9] and p_0, n_0 correspond to the steady-state values of the photon and carrier density within the active region, associated with the bias current, j_{th}

$$n_0 = g_0 \tau_p N_0 \tag{17}$$

$$p_0 = \frac{\beta n_0}{1/\Gamma + n_{0m} - n_0} \tag{18}$$

$$j_{th} \cong n_{th} = 1/\Gamma + n_{0m} \tag{19}$$

3.2. Package and chip parasitic elements

When dealing with high-frequency electronics, the frequency limits are usually established by the parasitic elements. It is then required to know whether the laser modulation characteristics are due to the laser alone or due to the parasitic elements. To this aim, one must treat the laser as an electrical element and establish an equivalent circuit that includes the parasitic elements. Characterization of an electrical network at high frequencies is usually done using the scattering parameters.

The elements of the laser-equivalent circuit are derived from the rate equations augmented by the heterojunction voltage-current and space-charge characteristics. The resulting equivalent circuit is a parallel RLC resonant circuit [9, 10]. The carrier density and quasi-Fermi levels are clamped above threshold which manifests in the equivalent circuit as an "ac" short and no voltage can develop. The magnitude of the impedance of the entire circuit $|Z(\omega)|$ is therefore essentially zero at all frequencies except near the relaxation oscillation resonance, where its value does not exceed $\approx 1\ \Omega$. For deriving the relation between the total external current I_s and the current through the active region I_a and in comparison to the relatively large external elements, the intrinsic laser diode can be regarded as a short circuit at all frequencies. Under zero bias, the intrinsic laser can be modeled by the active layer space-charge capacitance [11].

Chip parasitic elements vary widely among different laser structures. In practice, they take the form of a resistance in series with the intrinsic device combined with a shunt capacitance. An equivalent circuit model of the package and chip parasitic elements is shown in Figure 5; it includes a series inductor representing the wirebond, a shunting capacitor representing the contact capacitance, and a series resistor representing the contact resistance and the Bragg mirror stacks.

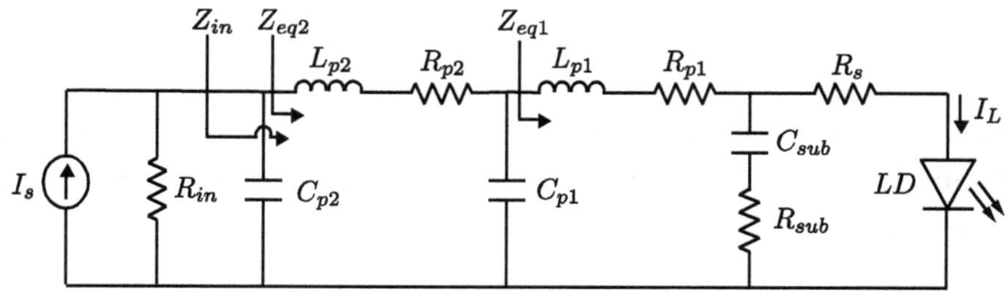

Figure 5. Equivalent circuit model for the laser parasitic elements.

We now define the following impedances:

$$Z_{p2} = R_{p2} + j\omega L_{p2} \tag{20}$$

$$Z_{p1} = R_{p1} + j\omega L_{p1} \tag{21}$$

$$Z_{sub} = R_{sub} + \frac{1}{j\omega C_{sub}} \tag{22}$$

The transfer function of the laser parasitic elements corresponding to ratio of the current flowing through the intrinsic laser, I_L , and the source current, I_S, is given by:

$$Z_{in} = \frac{R_{in}}{1 + j\omega C_{p2} R_{in}} \tag{23}$$

$$Z_{eq1} = \frac{Z_{in} + Z_{p2}}{j\omega C_{p1}\left(Z_{in} + Z_{p2}\right)} \tag{24}$$

$$Z_{eq2} = \frac{Z_{eq1} + Z_{p1}}{Z_{sub} + Z_{eq1} + Z_{p1}} \tag{25}$$

$$\frac{I_L}{I_S} = \frac{Z_{eq1}}{Z_{eq1} + R_s} \frac{Z_{eq2}}{Z_{eq2} + Z_{p1}} \frac{Z_{in}}{Z_{in} + Z_{p2}} \tag{26}$$

The full laser transfer function is then the product of the intrinsic laser [equation (14)] and parasitic transfer functions (equation (26)).

3.3. Laser characterization

In modeling the VCSEL for simulation purposes, it is important to obtain a model as faithful as possible to the real device. This is achieved by extracting the laser parameters from experimental data. The S_{21} and S_{11} parameters are measured using a vectorial network analyzer (Lightwave Component Analyzer which characterizes devices in the electric and optical domains), a current source (Laser Diode Controller LDC-3700B), a bias-T (allows the continuous current injection in the laser), and a test fixture, as shown in Figure 6. The test fixture allows the connection between the laser and the SMA (Subminiature Version A) connector. The latter was designed in Advanced Design System considering Rogers 4000 series as a substrate. It should be noted that it is necessary to subtract the test fixture impact on the measurements using the de-embedding technique [12, 13].

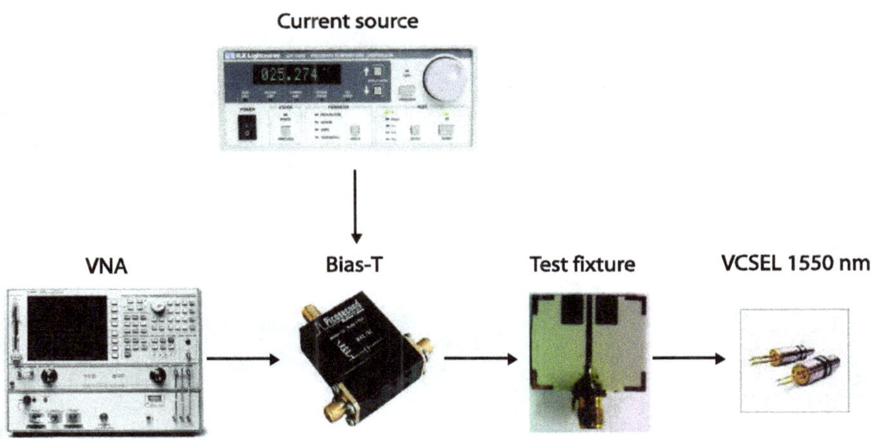

Figure 6. Setup for the measurement of the 1550 nm VCSEL.

3.4. Extraction of laser parasitic elements

It is possible to determine the input impedance Z_{in} using the experimentally measured S_{11} parameter:

$$Z_{in} = 50 \times \frac{1 + S_{11}}{1 - S_{11}} \tag{27}$$

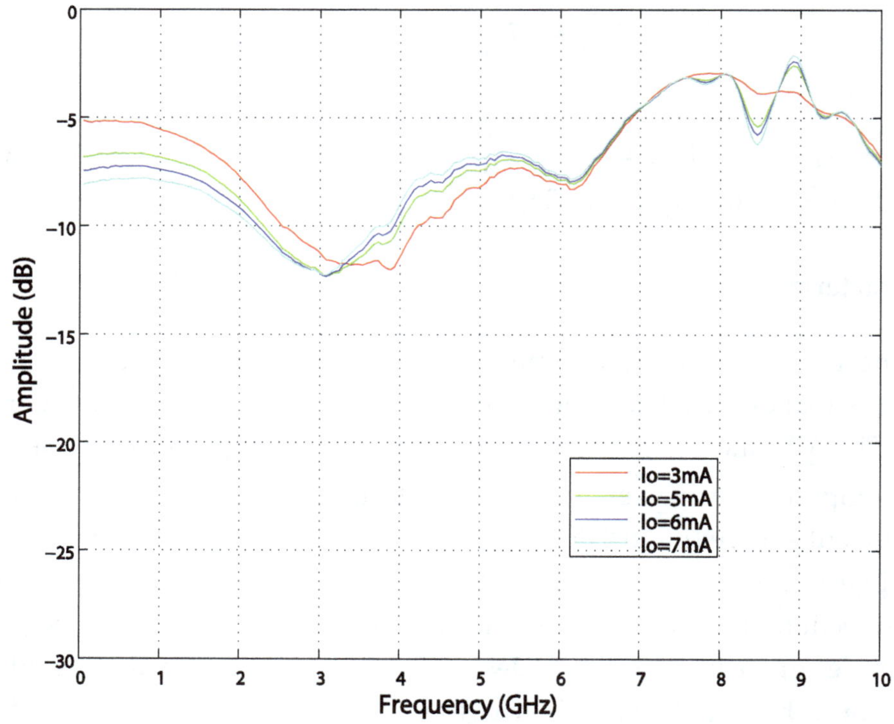

Figure 7. S_{11} parameter for the 1550 nm VCSEL.

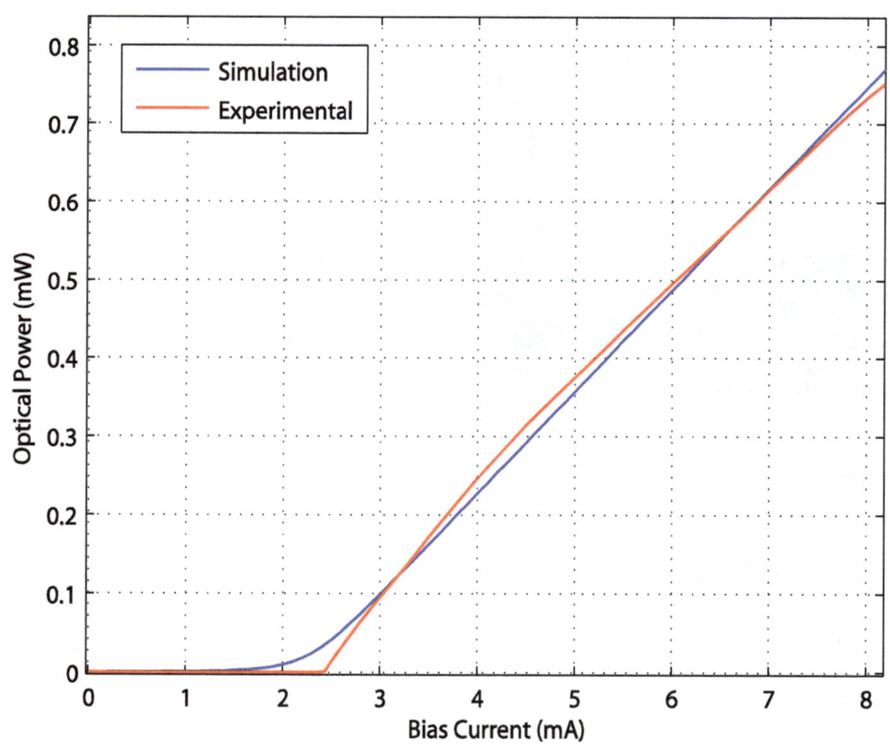

Figure 8. Characteristic curve for the 1550 nm VCSEL.

The measured S_{11} parameter for the 1550 nm VCSEL for different bias currents above threshold is represented in Figure 7. The threshold current (2.14 mA) was obtained by inspection of the optical power versus current (*P-I*) characteristic curve shown in Figure 8.

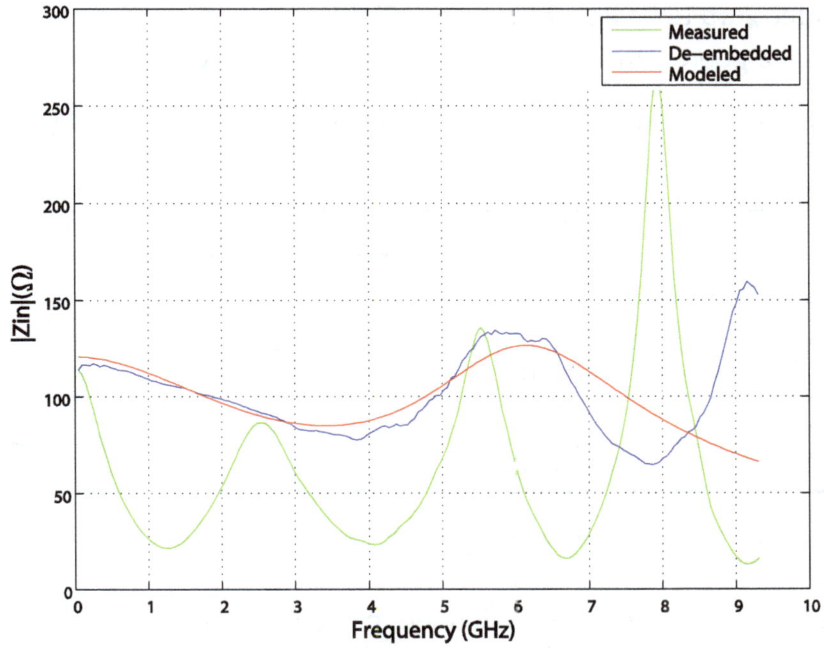

Figure 9. Magnitude of the 1550 nm VCSEL input impedance.

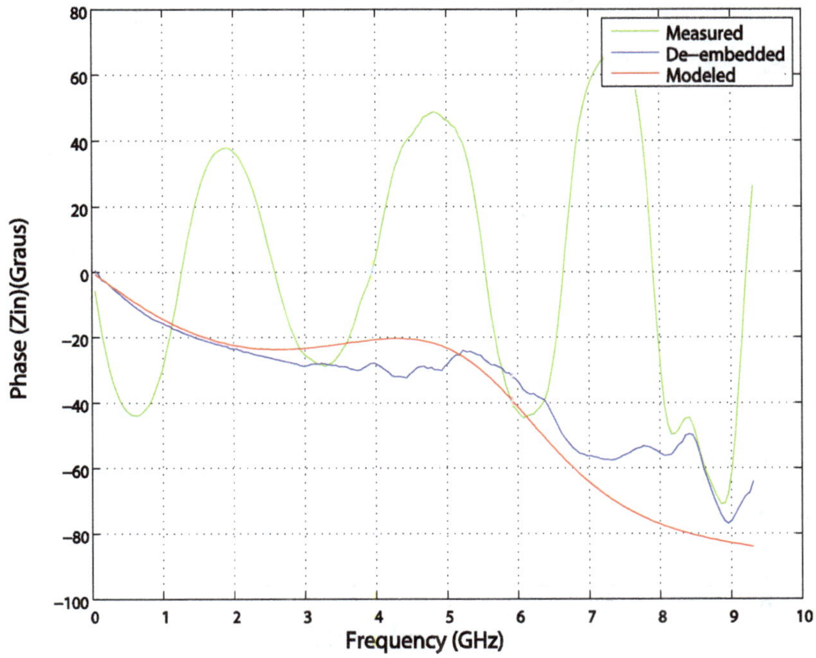

Figure 10. Phase of the 1550 nm VCSEL input impedance.

The laser parasitic elements are then obtained by means of an optimization process that fits the magnitude and phase of the input impedance of the equivalent circuit of the parasitics to the corresponding experimental result after the de-embedding procedure. From the results represented in Figure 9 and Figure 10, it is possible to verify the good approximation between the theoretical model (modeled) obtained using equation (23) and the experimental measurement (de-embedded), up to 7 GHz. Moreover, it is clear the importance of the de-embedding operation to obtain a good estimate of the laser parasitic elements.

The parasitic elements obtained before the optimization process are represented in Table 1:

Element	Value	Element	Value
R_{in}	50 Ω	L_{p1}	0.68 nH
C_{p2}	0.3498 pF	R_{p1}	0.5 Ω
L_{p2}	2.828 nH	C_{sub}	0.04 pF
R_{p2}	52.387 Ω	R_{sub}	0 Ω
C_{p1}	0.6696 pF	R_s	113.02 Ω

Table 1. Parasitic elements of the equivalent circuit

The frequency response of the circuit, as defined in equation (26), is shown in Figure 11, where the lowpass characteristic is observed.

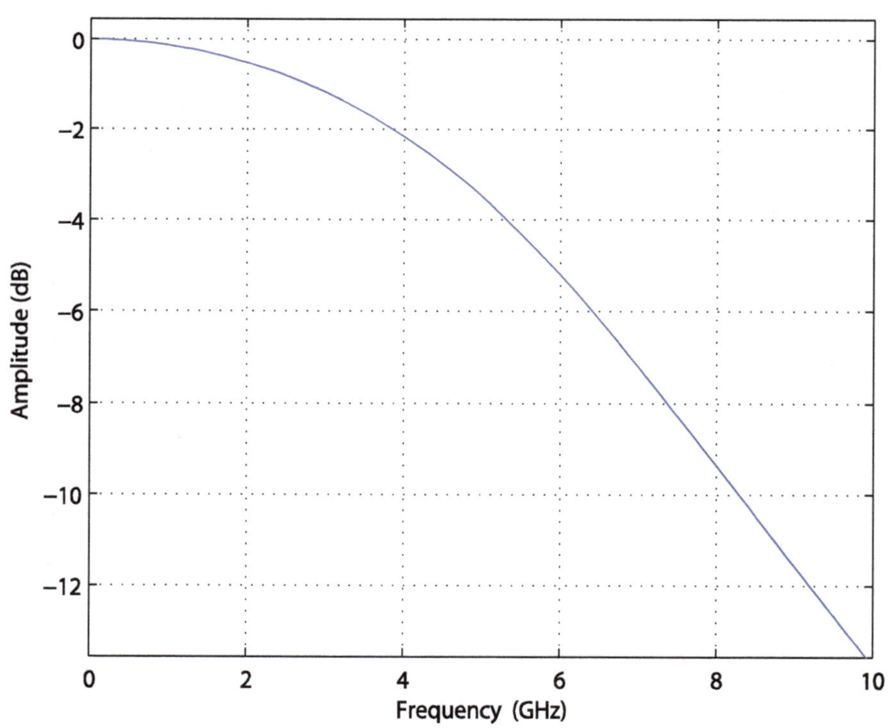

Figure 11. Normalized frequency response of the parasitic elements circuit.

3.5. Extraction of laser intrinsic parameters

The method employed for the extraction of the laser intrinsic parameters follows the frequency subtraction method described in reference [14]. To that purpose the laser transfer functions at different bias currents are obtained, using the S_{21} parameter.

The experimental results, for different bias currents (above the threshold current), are represented as dashed lines in Figure 12. These results were used to extract the laser intrinsic parameters $[H_{ILD}(f)]$, by dividing each curve by the reference transfer function (measured for a bias current of 3 mA). The resulting function does not depend on the parasitic circuit $[H_{PC}(f)]$ or the test fixture $[H_{TC}(f)]$, as shown in equation (28) [15]:

$$\frac{H_{Global}(f, I_{Bias})}{H_{Global}(f, I_{ref})} = \frac{H_{ILD}(f, I_{Bias}) H_{PC}(f) H_{TC}(f)}{H_{ILD}(f, I_{ref}) H_{PC}(f) H_{TC}(f)} = \frac{H_{ILD}(f, I_{Bias})}{H_{ILD}(f, I_{ref})} \tag{28}$$

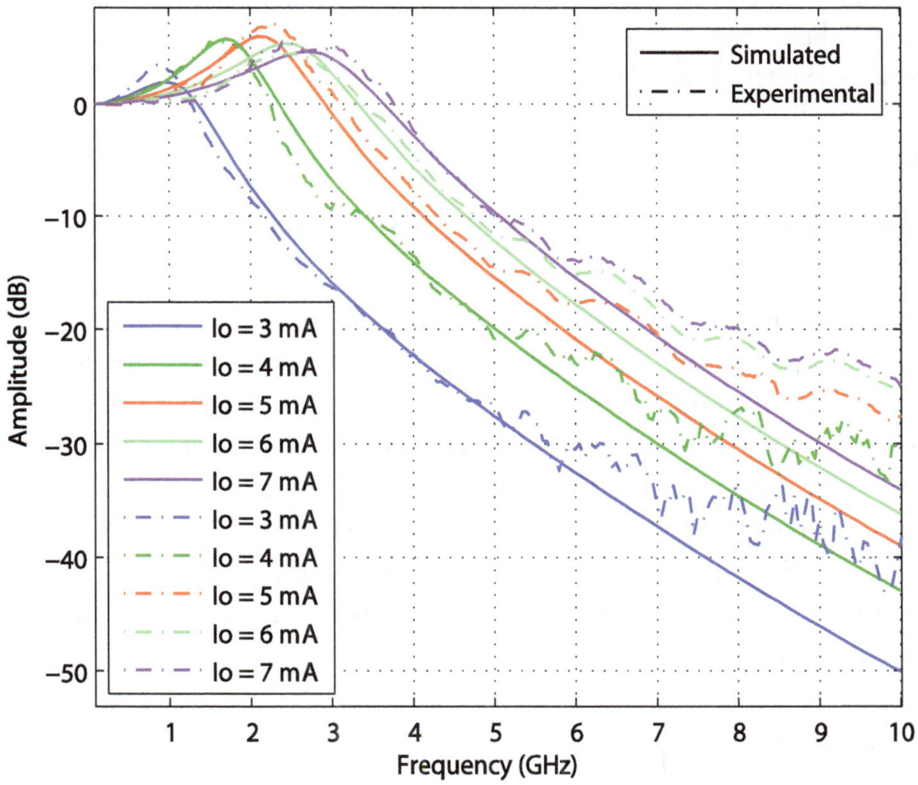

Figure 12. Frequency response of the 1550 nm VCSEL for different bias currents.

It is then possible to fit the corresponding theoretical response to the measured data through an optimization process of the laser parameters. That procedure was applied to the 1550 nm VCSEL, RC33xxx1-F from *RayCan*, using the Optimization Toolbox of MATLAB; the final result is shown in Table 2.

Parameter	Value
V , active region volume	4.93×10^{-18} m^3
$g_{0,}$ gain slope constant	2.50×10^{12} m^{3}s^{-1}
N_{0m} , electron density at transparency	2.71×10^{24} m^{-3}
β , spontaneous emission factor	6.5×10^{-3}
Γ , optical confinement factor	3×10^{-2}
$\tau_{s,}$ electron lifetime	2.6 ns
τ_p , photon lifetime	4.0 ps
ε , gain compression factor	5.0×10^{-23} m^3
η_i , internal quantum efficiency	0.8

Table 2. Extracted intrinsic parameters of the 1550 nm VCSEL

The frequency response of the VCSEL is represented in Figure 12, for different bias currents, including the simulated and experimental results. A good approximation is obtained between the simulated and experimental results.

4. OFDM and SC-FDMA over fiber applications

The VCSEL can be directly modulated using signals such as OFDM or SC-FDMA to encode digital data on multiple subcarrier frequencies. OFDM and SC-FDMA are used in applications such as wireless networks and LTE mobile communications. In order to avoid the high peak-to-average-power ratio (PAPR) inherent to OFDM modulation, the LTE standard employs SC-FDMA [16-18], an alternative modulation technique for the uplink with a similar low-complexity. Additionally, OFDM requires highly linear power amplifiers operating with a large backoff from their peak power, which results in low power efficiency [19]. In this context, it becomes pertinent to study the impact of employing SC-FDMA modulation within a RoF system based on directly modulated VCSELs.

4.1. Method and setup

The OFDM and SC-FDMA signal generation and demodulation is carried out in MATLAB environment. The experimental RoF setup illustrated in Figure 13 includes the Vector Signal Generator for the generation of the radio frequency (RF) signal that directly modulates the VCSEL, the optical fiber, the optical attenuator, and the optical receiver. The received signal is then sampled with a digital sampling oscilloscope at 20 Gsamples/s for offline demodulation in MATLAB. The validation of the transmitter and receiver blocks was performed in back-to-back configuration, by comparison with the theoretical PAPR results from the literature.

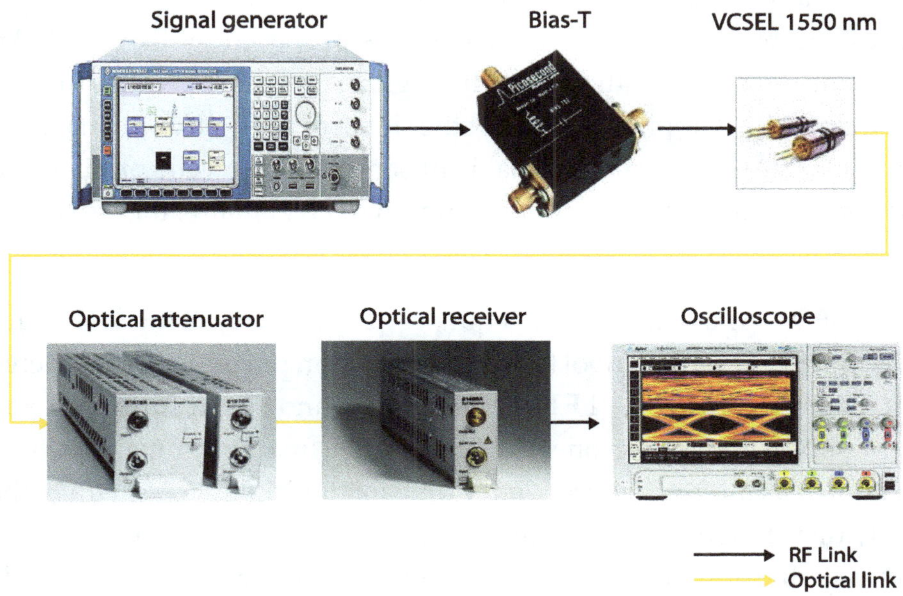

Figure 13. Experimental setup.

4.2. Transmitter

Both OFDM and SC-FDMA signals were modulated with a bit sequence, using three different possible modulation formats: QPSK (Quadrature Phase Shift Keying), 16-QAM (Quadrature Amplitude Modulation) and 64-QAM. Pilot subcarriers were added to the resulting symbols to estimate the effect of channel propagation. A diagram of both transmitters is shown in Figure 14.

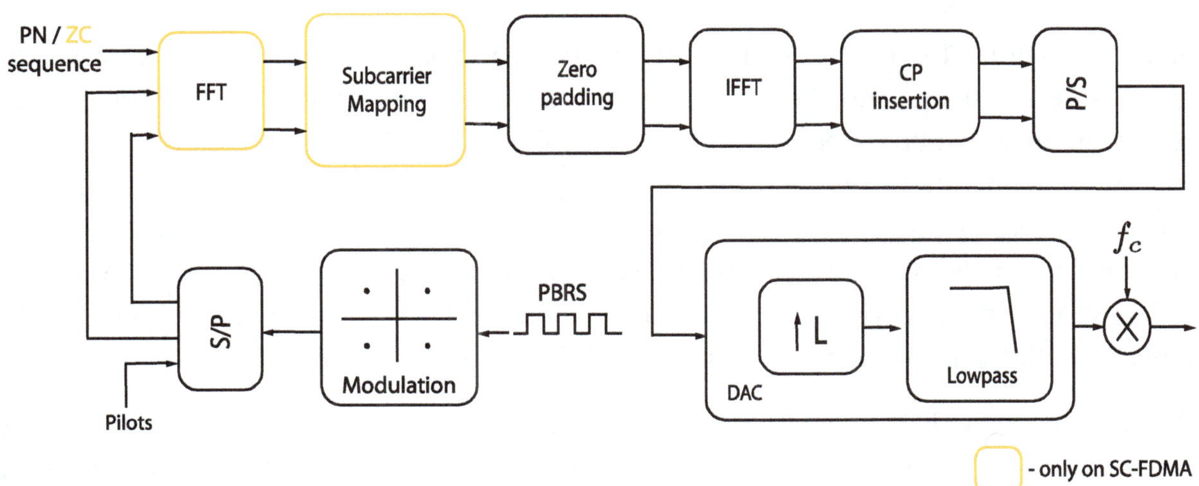

Figure 14. OFDM and SC-FDMA transmitters.

In the case of OFDM, the first symbol is used to facilitate time synchronization (finding the first symbol of the signal) at the receiver. This symbol is obtained using the inverse fast Fourier transform (IFFT) of a sequence formed by the real part of a pseudorandom noise sequence at even frequencies and by zeros at odd frequencies, as proposed by Park [20]. After applying zero padding (adding zeros on both sides of the signal spectrum) in order to ease the filtering operation, a N-point IFFT is applied to convert the signal to the time domain. Before the digital-to-analog converter (DAC) where the signal is upsampled, the cyclic prefix (CP) (copy of the last part of the signal) is added to prevent multipath delay. Finally, the signal is upconverted to a RF carrier.

In the case of SC-FDMA, a Zadoff-Chu sequence [21] is used in the LTE standard, which functions as the first SC-FDMA symbol for synchronization purposes at the receiver. After this, a N-point fast Fourier transform (FFT) is performed and the resulting N subcarriers are mapped into M subcarriers using one of two different mapping methods: the interleaved mapping also known as interleaved frequency division multiple access, where the subcarriers are equidistantly distributed over the entire spectrum; and the localized mapping, also called localized frequency division multiple access, where the subcarriers are confined to a fraction of the spectrum. Thereafter, the zero padding and M-point IFFT are applied. Lastly, the CP is added and, as in the case of OFDM, the upsampling and the upconversion operations are performed.

In order to allow the performance comparison between the two modulation formats, only one user is considered.

The In-phase (I) and Quadrature (Q) components of both modulations formats were loaded to the signal generator, represented in Figure 13, where the frequency and the power of the RF-transmitted signal was specified.

4.3. Receiver

As shown in Figure 15, the received OFDM signal is first baseband filtered, to eliminate the noise outside the band, and then downconverted at the same frequency specified on the generator in order to obtain the baseband signal, followed by the lowpass filtering to remove the harmonics generated. Then the downsampling and quantization operations are applied to the signal followed by a temporal synchronization using Park's method. Then, the inverse of the operations carried out in the OFDM transmitter are performed at the receiver, namely the channel estimation and corresponding equalization in the frequency domain, followed by the removal of both the pilot subcarriers after the FFT and zero padding operations.

In the case of the SC-FDMA receiver, the operations are as follows: first baseband filtering, then temporal synchronization using the cross-correlation between the received signal and the reference signal (Zadoff-Chu sequence), and finally downconversion and downsampling. In order to recover the transmitted symbols, the inverse of the operations carried out at the SC-FDMA transmitter are performed at the receiver, including the channel estimation and the frequency domain equalization, similarly to the case of the OFDM receiver, after the IFFT and the subcarrier demapping.

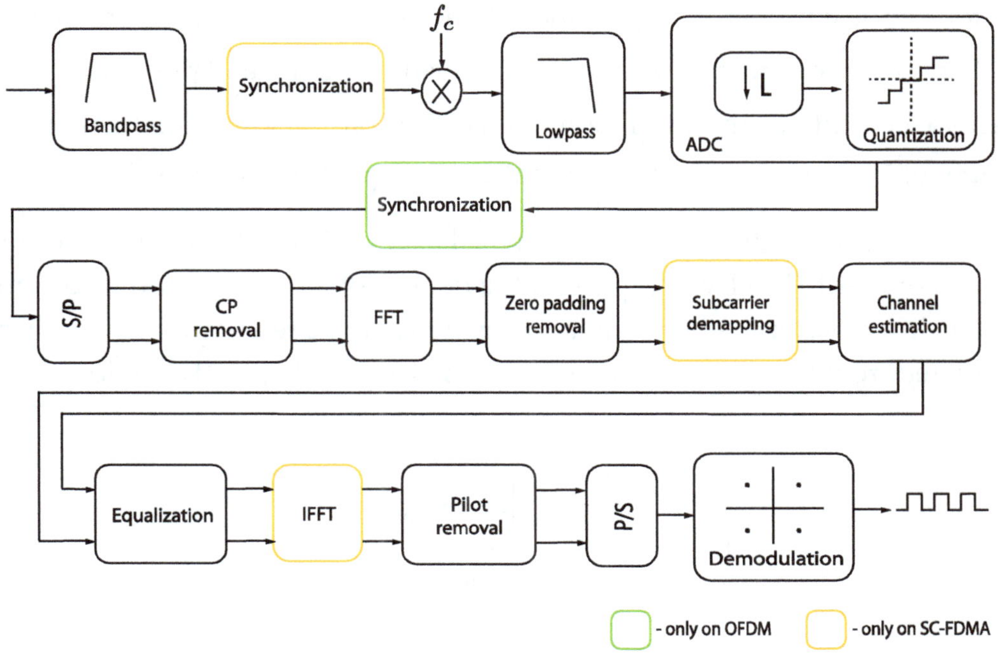

Figure 15. OFDM and SC-FDMA receivers.

5. System performance assessment

In this section, we present and discuss the relevant results using the signal-to-noise ratio (SNR) as a figure of merit to compare the OFDM and SC-FDMA signals with the same signal parameters.

5.1. Signals parameters

The signals generated are characterized by the following parameters: 16-QAM modulation, 10 Mbps, 1 user, 128 subcarriers, 8 pilot subcarriers for channel estimation, $4 \times \text{FFT}_{size}=512$ of zero padding, CP of $\text{FFT}_{size}/4=32$, 11 transmitted symbols (1st symbol for synchronization and the others for data) and RF carrier located at 2.4 GHz. On the receiver side, the zero forcing equalization is used after the least squares channel estimation.

5.2. Analysis

We now assess the performance of a RoF system and compare the performance of OFDM and SC-FDMA wireless signals over fiber. Theoretically, the optical SNR is given by the following equation [22]:

$$\text{SNR} = \frac{I_{RX}^2}{I_{RIN}^2 + I_{SN}^2 + I_{TH}^2 + I_{IMI}^2} \tag{29}$$

where the four noise currents are: I_{RIN}^2 is the RIN noise current, I_{SN}^2 is the shot noise current, I_{TH}^2 is the thermal noise current due to equivalent load resistance and pre-amplifier noise, and I_{IMI}^2 is the noise current due the intermodulation distortion. For lower modulation indices, the RIN is dominant in comparison to the thermal and the quantum noises [23].

The performance of the system is assessed on the basis of SNR or the error vector magnitude (EVM) figures of merit. The EVM expresses the quality of a digital modulated signal and is defined as the difference vector between the measured and the reference signals. Then the SNR can be calculated from the EVM. These figures of merit are calculated as follows:

$$\mathrm{EVM} = \frac{\frac{1}{N_S}\sum_{p=0}^{N_S}\left|S_{p,i} - S_{p,m}\right|^2}{\frac{1}{N_S}\sum_{p=0}^{N_S}\left|S_{p,i}\right|^2} \tag{30}$$

$$\mathrm{SNR} = \left(\frac{1}{\mathrm{EVM}}\right)^2 \tag{31}$$

where S_i and S_m are the ideal and measured constellations, respectively, p is the constellation symbol index, and N_S is the number of constellation symbols.

In Figures 16 to 18, the SNR results for OFDM and SC-FDMA transmissions are represented for three different laser bias currents (I_0= 4, 5, and 6 mA) as a function of the RF signal power, which is defined as:

$$P_{RF} = 10\log_{10}\left(\tau_{i,signal} Z_0\right) + 30\left[\mathrm{dBm}\right] \tag{32}$$

where $\tau_{i,signal}$ is the signal variance and Z_0 is equal to 50 Ω.

From the results presented, it is possible to conclude that there is a good matching between the simulation and the experimental results. It is clear from Figures 16 to 18 that for lower RF power, noise is dominant whereas for higher RF power, the intermodulation distortion, introduced by the VCSEL, becomes the limiting performance factor. The SC-FDMA signal is more sensitive to noise then the OFDM signal for lower RF power, while for higher RF power, the SC-FDMA signal is more robust to the intermodulation distortion. Despite this fact, the maximum SNR values attained are identical in both cases, albeit at a higher RF power for the SC-FDMA case, with no clear performance improvement of the SC-FDMA with respect to the OFDM.

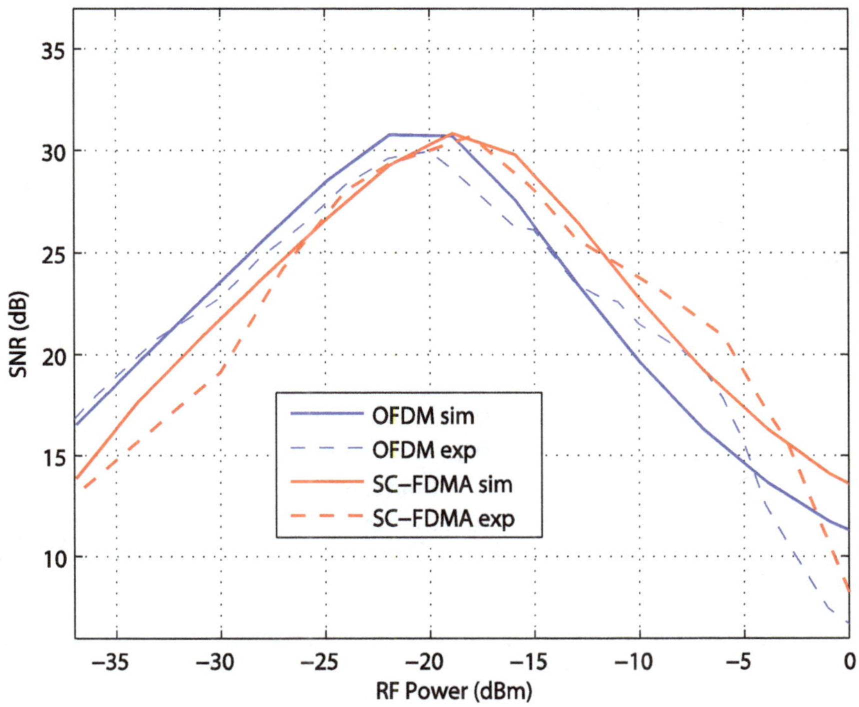

Figure 16. Simulated and experimental SNR for both OFDM and SC-FDMA as a function of the RF power for *Io* = 4mA.

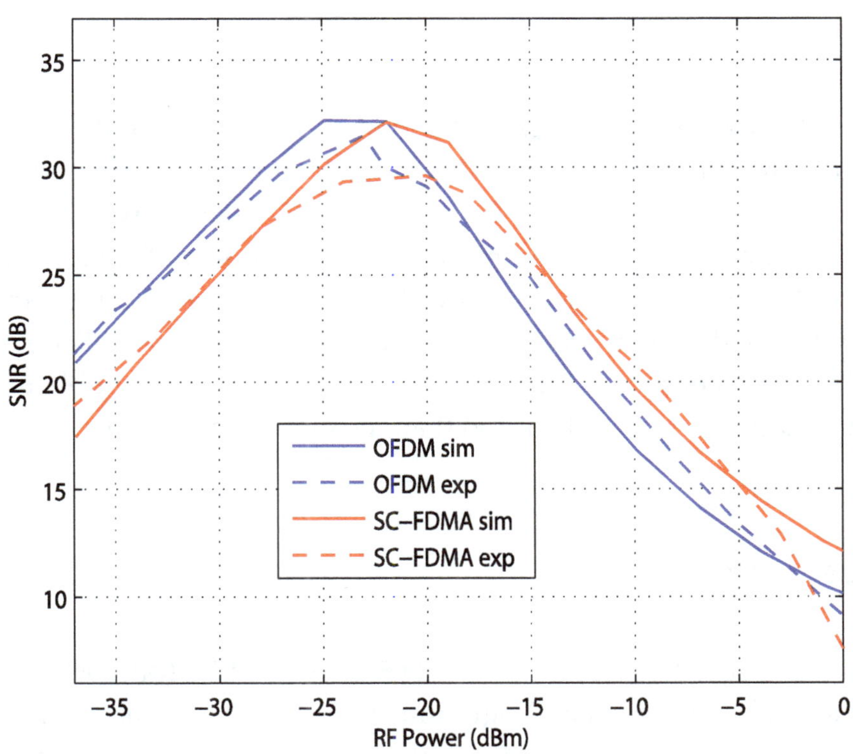

Figure 17. Simulated and experimental SNR for both OFDM and SC-FDMA as a function of the RF power for *I*o = 5mA.

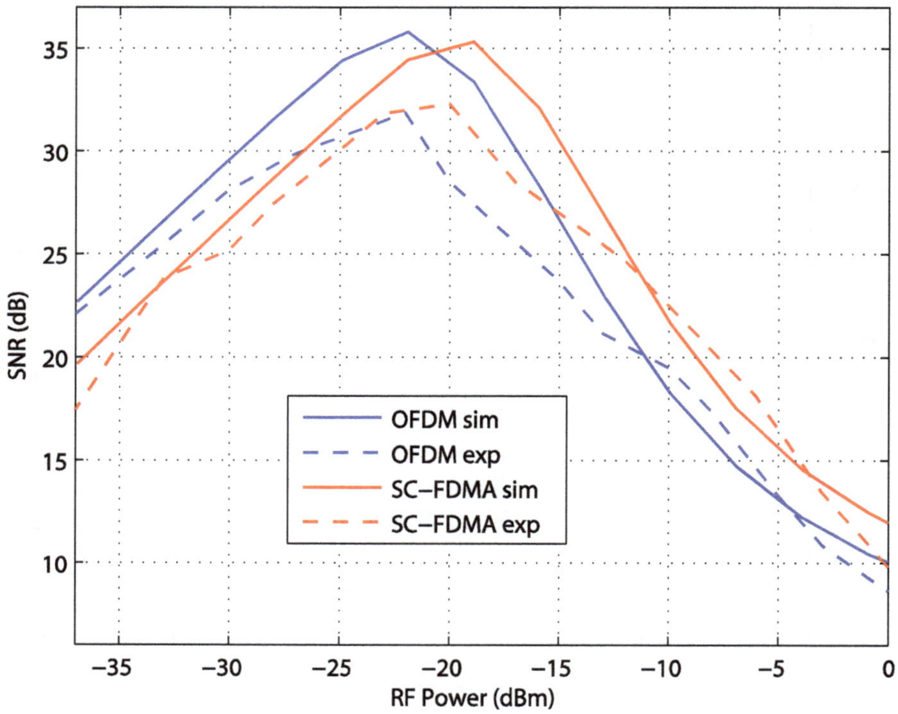

Figure 18. Simulated and experimental SNR for both OFDM and SC-FDMA as a function of the RF power for *Io* = 6mA.

6. Conclusion

We have assessed the performance of OFDM and SC-FDMA in the context of a RoF scenario, based on a directly modulated VCSEL operating at 1550 nm, and direct detection. To that purpose an accurate theoretical model of the laser was presented, for which device parameters were extracted based on fitting the model to experimental data of frequency response and input impedance. The simulation model fits well the experimental results, and we conclude that the SC-FDMA modulation presents a lower PAPR than the OFDM modulation as expected. Moreover, it is observed that the SC-FDMA is more susceptible to noise, yet it is more immune to intermodulation distortion than the OFDM modulation.

Acknowledgements

We acknowledge support from Project "NORTE-07-0124-FEDER-000058" financed by the North Portugal Regional Operational Programme (ON.2 - *O Novo Norte*), under the National Strategic Reference Framework (NSRF), through the European Regional Development Fund (ERDF), and by national funds, through the Portuguese funding agency, *Fundação para a Ciência e a Tecnologia* (FCT). This work was carried out with the support of the TEC4SEA research infrastructure (www.tec4sea.com).

Author details

Henrique M. Salgado[1,2*], Rúben E. Neto[2], Luís M. Pessoa[2] and Pedro J. Batista[1,2]

*Address all correspondence to: hsalgado@inesctec.pt

1 Faculdade de Engenharia da Universidade do Porto, Portugal

2 Instituto de Engenharia de Sistemas e Computadores – Tecnologia e Ciência - INESC TEC, Porto, Portugal

References

[1] Chun-Ting Lin, Chen J, Peng-Chun Peng, et al. Hybrid optical access network integrating fiber-to-the-home and radio-over-fiber systems. IEEE Photonics Technology Letters. 2007;19(8):610-612.

[2] 3GPP. LTE [Internet]. Available from: http://www.3gpp.org/technologies/keywords-acronyms/98-lte [Accessed: January 6, 2014].

[3] Tauber D, Wang G, Geels RS, Bowers JE, Coldren LA. Large and small signal dynamics of vertical cavity surface emitting lasers. Applied Physics Letters. 1993;62(4):325

[4] Marcuse D. Classical derivation of the rate equation. IEEE Journal of Quantum Electron. 1983;19(8):1228-1231.

[5] Westcott J. Investigation of multiple f.m./f.d.m. carriers through a satellite t.w.t. operating near to saturation. Proceedings of the Institution of Electrical Engineers. 1967;114(6):726-740.

[6] Petermann K. Laser Diode Modulation and Noise. Advances in Opto-Electronics (ed.). Netherlands: Springer; 1988, p. 315. DOI: 10.1007/9789400929074

[7] Coldren LA, Corzine SW, Mashonavitch ML. Diode Lasers and Photonic Integrated Circuits (2nd ed.). John Wiley and Sons, Inc.; 2012, p. 744. DOI: 10.1002/9781118148167

[8] Bruensteiner M, Papen GC. Extraction of VCSEL rate-equation parameters for low-bias system simulation. IEEE Journal of Selected Topics in Quantum Electronics. 1999;5(3):487-494.

[9] Salgado H. Performance Assessment of Subcarrier Multiplexed Optical Systems: Implications of Laser Nonlinearities. Ph. D. dissertation. School of Electronic Engineering and Computer Systems, 1993.

[10] Tucker RS, Pope DJ. Microwave circuits models of semiconductor injection lasers. IEEE Transactions on Microwave Theory and Techniques. 1983;31(3):289-294

[11] Tucker RS, Kaminow IP. High-frequency characteristics of directly modulated InGaAsP ridge waveguide and buried heterostructures. Journal of Lightwave Technology. 1984;2(4):385-393.

[12] Agilent Technologies. De-embedding and embedding S-parameters networks using a vector network analyzer: Application Note 1364-1.2004.

[13] Agilent Technologies. In-fixture measurements using vector network analyzers: Application Note AN1287-9. 2006.

[14] Morton PA, Tanbun-Ek T, Logan RA, et al. Frequency response subtraction for simple measurement of intrinsic laser dynamic properties. IEEE Photonics Technology Letters. 1992;4(2):133-136.

[15] Silva S, Salgado HM. VCSEL Laser characterization and modelling for future optical transceiver at the super Large Hadron Collider. In: Transparent Optical Networks, 11th International Conference held on June 28, 2009--July 2, 2009. Azores: IEEE; 2009, p. 1-5.

[16] Ciochina C, Sari H. A review of OFDMA and single-carrier FDMA. In: Wireless Conference (EW), European held from 12-15 April 2010. Lucca: IEEE; 2010, p. 706-710.

[17] Myung HG. Introduction to single carrier FDMA. In: 15th European Signal Processing Conference held from 3-7 September 2007. Poznań, Poland: EURASIP, 2007.

[18] Myung HG, Junsung Lim, Goodman D. Peak-to-average power ratio of single carrier FDMA signals with pulse shaping. In: Personal, Indoor and Mobile Radio Communications, 2006 IEEE 17th International Symposium held from 11-14 September 2006. Helsinki: IEEE; 2006, p. 1-5.

[19] Myung HG, Junsung L, Goodman D. Single carrier FDMA for uplink wireless transmission. IEEE Vehicular Technology Magazine. 2007;1(3):30-38.

[20] Byungjoon Park, Cheon H, Kang C, Daesik Hong. Novel timing estimation method for OFDM systems. IEEE Communications Letters. 2003;7(5):239-241.

[21] Chu D. Polyphase codes with good periodic correlation properties. IEEE Transactions on Information Theory. 1972;18(4):531-532.

[22] Keiser G. Optical Fiber Communication (4th ed.). New York: McGraw-Hill Science/Engineering/Math; 2010, p. 688.

[23] Coelho D, Oliveira JMB, Pessoa LM, Salgado M, Castro JCS. Performance analysis of WDM-PON architecture for UWB distribution in aircraft networks. i-ETC:ISEL Academic Journal of Electronics, Telecommunications and Computers. 2013;2(1):ID16.

8

Novel Floating and Auto-stereoscopic Display with IRLED Sensors Interactive Virtual Touch System

Jian-Chiun Liou

Abstract

A wide range of the types of interactive virtual touch system have been in research and development. As displayed by the trends, users do not need any special equipment, can interact with the images, and under normal circumstances of the interactive nature, saves trouble. We have been studying relevant video interactive systems in which a virtual image, like in the real world, exists to display objects. We developed a floating display and the principle, which is based on an interactive video system, to enable more realistic auto-stereoscopic images.

We developed and applied a multi-view auto-stereoscopic concept, and this newly developed system contains static and moving images. To prove that this concept can be applied, we have developed a series of infrared light-emitting diode sensors character interaction system. In the study, the system is presented in any virtual objects displayed on the screen. These virtual objects move around and interact with the user. The user need not key in any input mode of operation and can use body language to achieve interactive function, as if the fictitious image exists in the ambient space. Here, we use infrared LED array sensors to sense human body motion and finger movements, and we use LED modules to transmit and receive arrays. The systems uses a three-dimensional arrangement of the LED array architecture in order to effectively and accurately detect the user's finger movements and body motions.

The purpose of the entire framework is to have the principle of an actual image go through the optical system of the lens from the external surface of the system in 5 cm floating images. The optical lens system frontage of 50 cm can be seen as a pop-up image, and the image shows a 15° angle. People use their forefinger in a non-contact space virtual touch system to control the floating image menu. Touching virtual objects in virtual space will become the next generation human-machine operations. Touching virtual objects architecture will have an effect on other devices. This fictitious space inter-osculate virtual keyboard idea will have a benefit on people's health (not infected) and can be applied to other public equipment like the ATM, hospital elevator (buttons), and so on.

Keywords: Virtual, Interactive, Floating image, Auto-stereoscopic

1. Introduction

This study shows that the interface does not require any keyboard input and/or a high precision interactive device operation. The purpose of the study is for the actual object image through the optical lens system design to go beyond 5 cm from the system of the drift image. Outside the optical system, in about 50 cm in front of the viewer, a pop-up image can be observed at a viewing angle of 15°. Viewers can use a non-contact touch through IR LED arrays shown in Figure 1 A.

The general contact system with touch-sensitive keys may cause contamination, and germs can potentially spread and reproduce through contact. The development of a touch sensing technology and multi-touch operation technology has gradually been used in all kinds of electrical products for everyday use. Because of privacy requirements, research and development staffs have developed personal information security needs and different ways to deal with the different system architectures [1-6].

As a subsequent introduction, we mainly focus on the common use of electronic equipment in public places: ATM buttons, health and other equipment, and hospital elevator buttons; the system does not touch the electronic products since the operation principle of the function keys can be positioned. Different aspects that are addressed further include a non-touch system to explore the given potential. There are now a popular variety of touch devices, which are necessary to control systems through non-touch input texts, which may be useful in their daily lives in the future.

In the current non-contact control system, the system is set up as a matrix and sent by the infrared - receiver modules. The infrared detector arrays are generally formed by interrupting, and the infrared ray is used to deposit the position of the object. However, in the infrared transmitter, detecting the receiver module current means the infrared transmitting and receiving modules continue to be touched because the object interrupts the corresponding position in the infrared[7-9].

The LCD flat panel display has become the world's leading product. In view of this phenomenon, the trend that dominated the stereo system is the LCD type, giving the viewer a disparity in Figure 1B and the final 3D effect. The 3D effect is assigned a different image, a different view of the area showing the direction of the optical device.

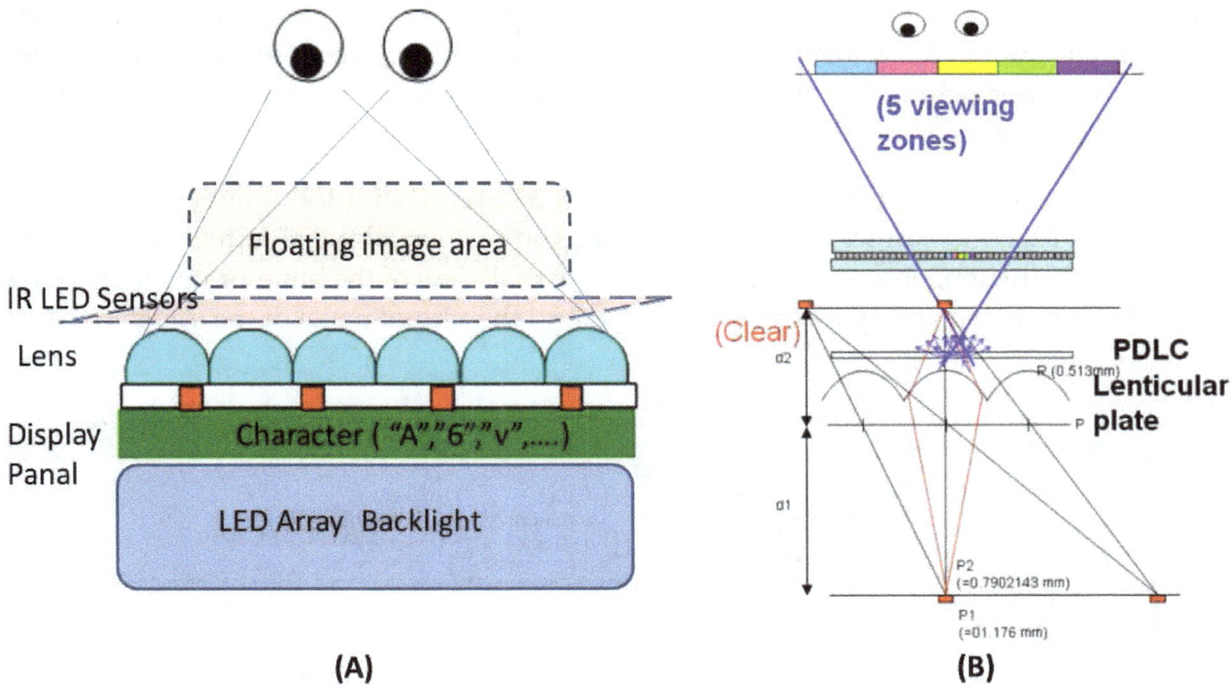

Figure 1. (A). Floating image with interactive touch system. Figure 1(B). Multi-view auto-stereoscopic system.

Some studies on multi-view 3D displays have been developed a while back. Multi-view stereo displays are usually followed up either by the integral imaging method [10-12] or the parallax barrier method [13]. All of these existing methods display a 3D image on a flat screen, and the image cannot be superimposed in real space.

Our interactive imaging studies may be used to pop up in real space through a floating 3D image, and the user can see the floating images without using glasses. Superimposing the projection methods using a real space image projection display has been proposed [14-16]. Based on a non-planar image, users can actually touch the projected image device using these methods, thereby allowing the user objects to experience the real projection surface interaction. However, such a 2D image projection of nature cannot produce multi-view parallax images. In this study, our goal is to create a real image of the touch system and air-infrared, interactive, virtual image with LED sensor.

2. Architecture

Since the fictitious appearance is formed in air, the user can locate the three-dimensional position through the direct contact with the air touch virtual panel. Infrared LED illumination is made in a cylindrical lens mounted in front of a fictitious image as shown in Figure 2.

The fictitious appearance is located in a plurality of sub-regions, which belongs to the operator interface. The object is touched in a sub-region that corresponds by send - receiving. For the corresponding operation function block to perform, the sub-region must be touched. A synchronous processing system is electrically connected to the corresponding transmission - reception. The system is arranged in the same direction of the object sensing delay enabled devices, and the synchronous processing system is configured to synchronize the control settings for each unit.

To solve mishandling, which continuously trigger a current with the same key issues, the operator was asked to correct the perceived image on the virtual panel in the right viewing area. Out of the region, observers perceive only a partial view of the blind or image map. We use the delay function and synchronization to control the virtual panel of the system.

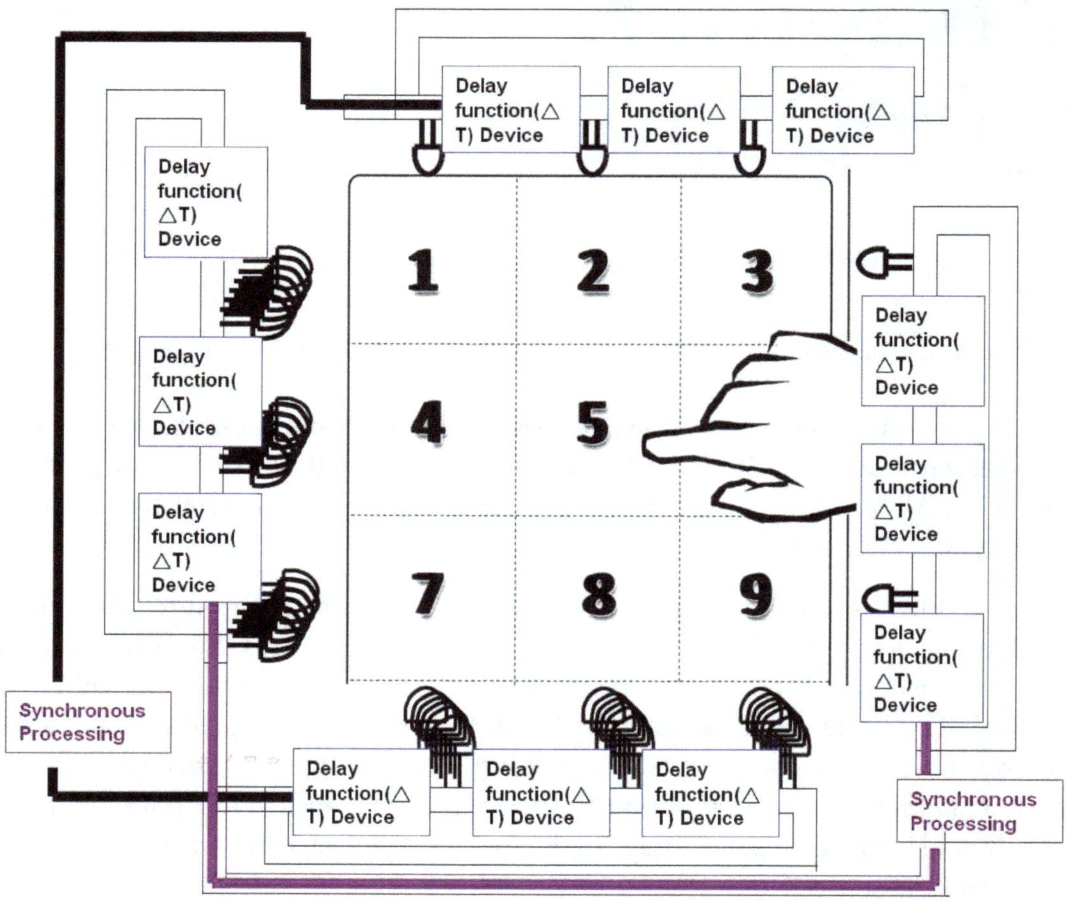

Figure 2. Virtual Touch system with IR LED device array

Novel floating and auto-stereoscopic display with IR LED sensors interactive virtual touch system include two major parts. The first part is the floating image and auto-stereoscopic display. The second part is the design and fabrication of the IR LED sensors multiplexer(mux) system.

2.1. Floating image and auto-stereoscopic display

In recent years, the three-dimensional television market grew quickly. To add, it is included the other three-dimensional floating image systems. The stereo image is shown in Figure 3.

For this reason, the system constructed a floating color image projecting device. According to principles of geometric optics, through a set of lenses or concave mirror and 3D objects that project true 3D images, we can achieve the principles of purpose. Therefore, it uses a display method to replace so that the image is changing the wishes of necessary real 3D objects. However, the two-dimensional image floating image, rather than the three-dimensional real image, has been achieved. Therefore, we propose the free space of the three strategic projects with a wide-angle auto-stereoscopic floating picture below.

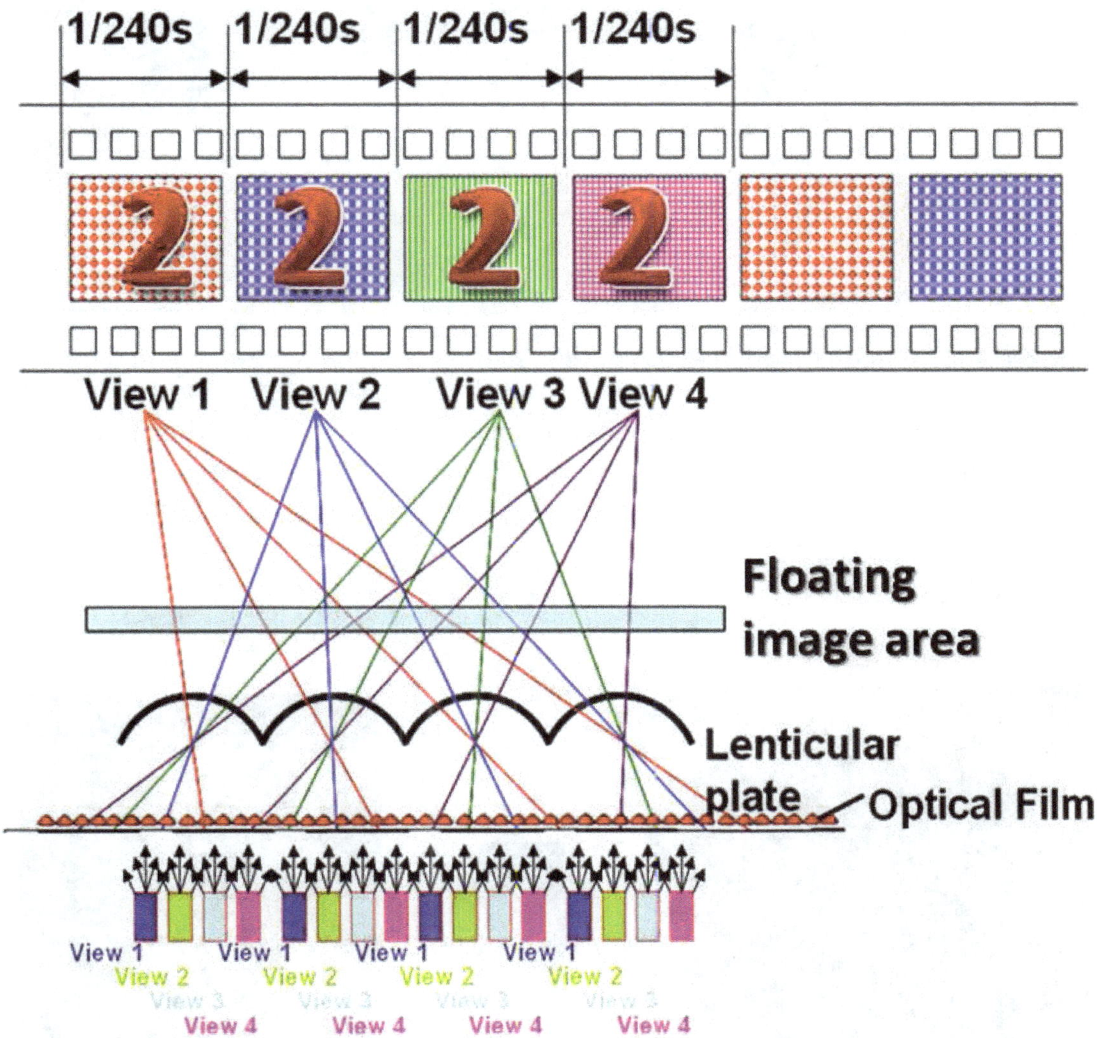

Figure 3. The floating image products with a virtual image.

We propose a floating image display apparatus in Figure 4 that has two main parts: (1) an image appearance system, and (2) an auto-stereoscopic display.

An optical projection system of this type comprises a projection lens, illumination optics in front of the projection lens, and one light modulator, such as a digital mirror device (DMD). The light modulator allows the reflection of the light in a controllable manner to different directions, such as when the ON state light cones, in which the light is projected onto the

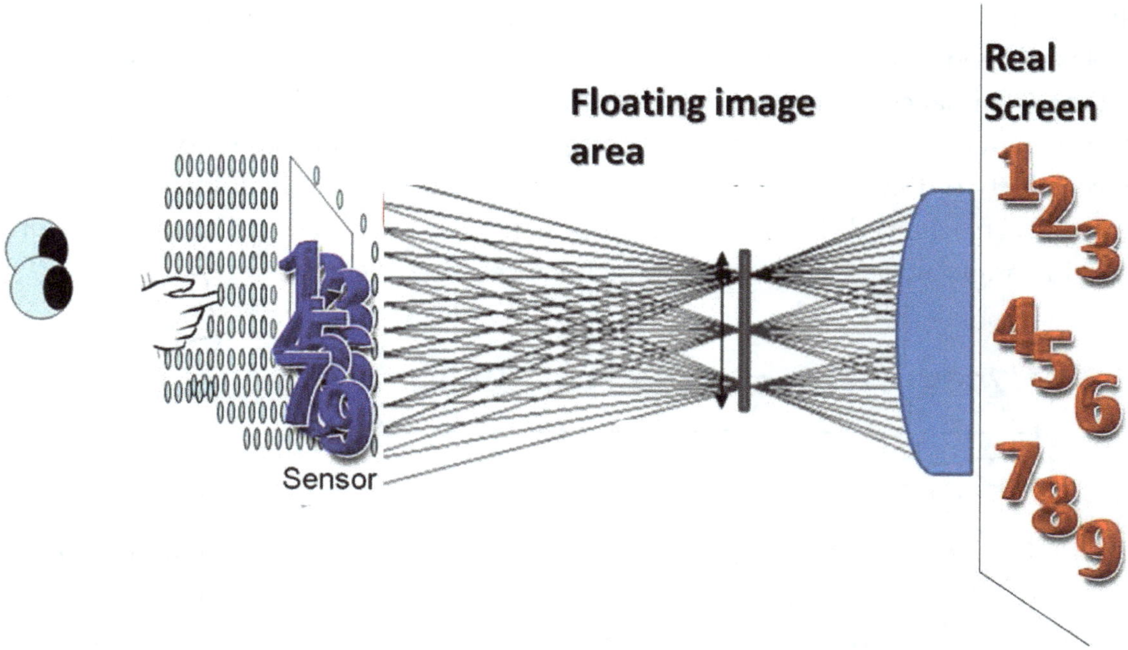

Figure 4. Sensors array in the appearance image system.

projection lens, and the OFF state light cones are formed. In this research, the optical projection system is shown in Figure 5.

Figure 5. The projection lens and one light modulator.

The designed microlens is shown in Figure 6 such that the width of the region through the microlens at a certain viewing zone may be observed as equal to or smaller than the LCD pixel array in a sub-pixel. You can predict high resolution image when viewed from a viewing area to the center of the area when the rays pass through the lenticular center.

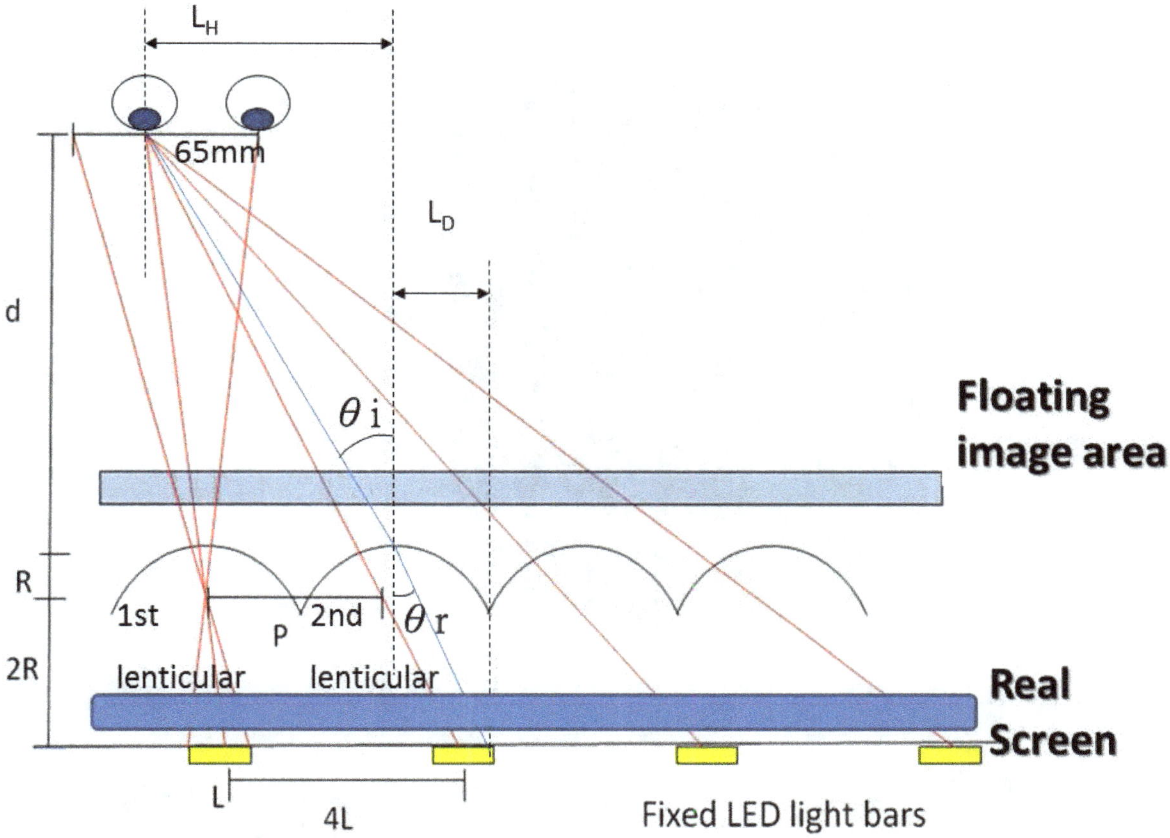

Figure 6. Designed microlens.

The image generating auto-stereoscopic display and its parameters include the viewing distance and the viewing area. In general, the pitch of the viewport to match the eyes and the viewing distances of the monitor determines the purpose. In this apparatus, in particular, the stereoscopic image is viewed by the projection lens shown in Figures 7, 8, 9, and 10. Thus, it illustrates the details of the theory and methods.

A floating real image between the projection lens and the user was set. Moreover, by using the depth detecting module to identify the user position and the control unit electrically connected to the image generator, the projection lens is set. And the depth detecting module, the image generator, and the projection lens set can adjust the position of the floating real image according to the user position.

Regarding the keypad module, it should be noted that by utilizing the control unit and the depth detecting module, the display apparatus in the disclosure can provide a user-friendly operation and a realistic interaction for the user as shown in Figure 10. Specifically, the control unit steers the movement of the image generator according to the position of the user detected by the depth detecting module, so as to adjust the relative positions of the image generator and the projection lens set, the position of the floating real image, and the size of the floating real image. The depth detecting module detects the position of the user, and may also detect the position of the body of the user, or the position of the fingers of the user touching the floating real image.

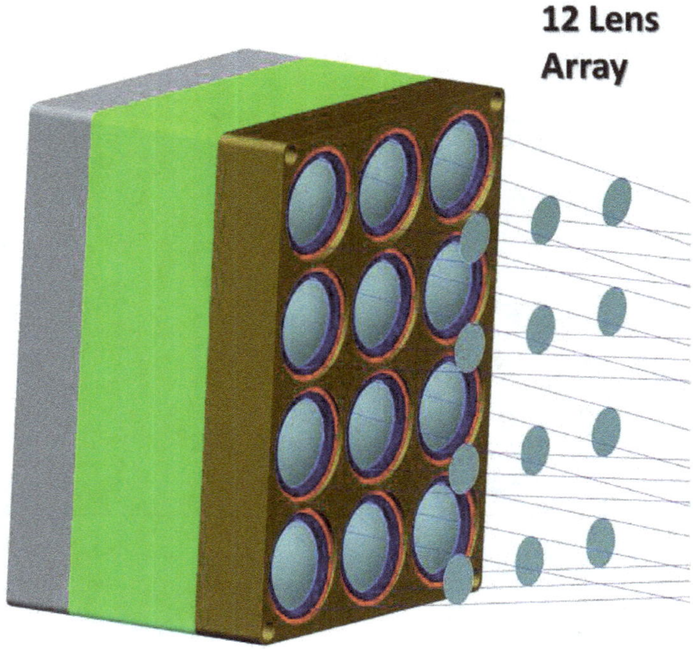

Figure 7. Twelve lens array.

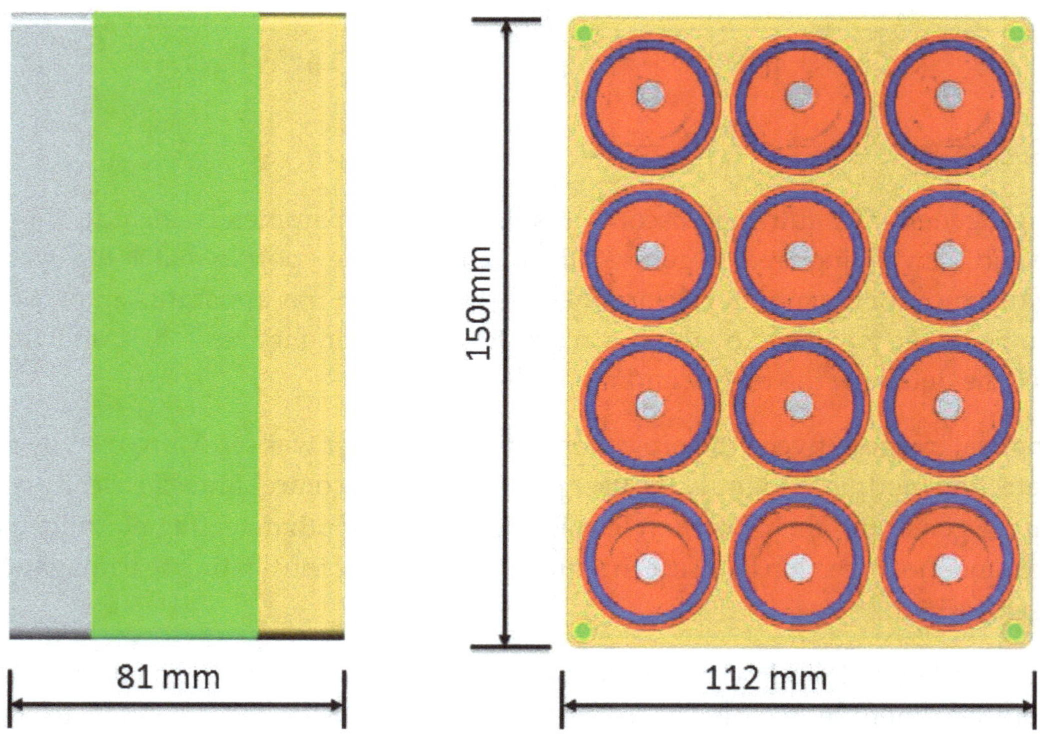

Figure 8. Top view of twelve lens array.

The depth detecting module feeds the detected position information of the user. The control unit can calculate the position of the user, the position of the floating real image, and the required size of the floating real image. Accordingly, the image generator and/or the projection lens set are shifted by corresponding distances to achieve the needed image variation effects.

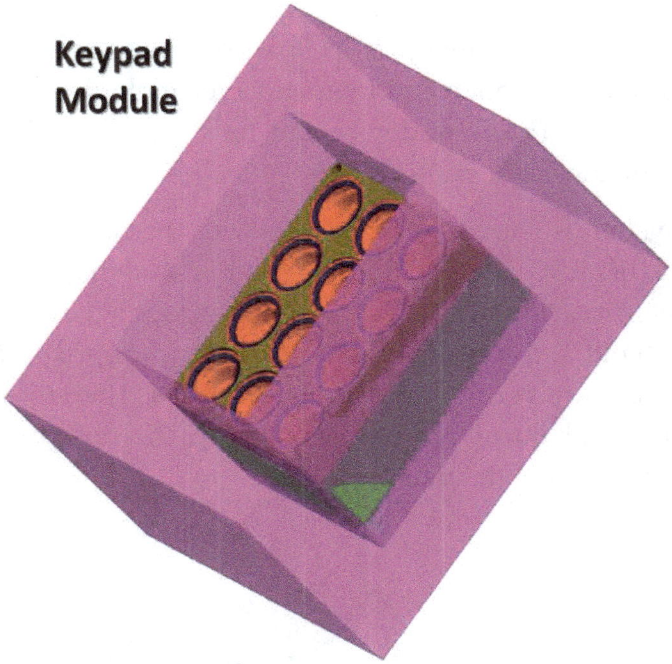

Figure 9. Keypad module.

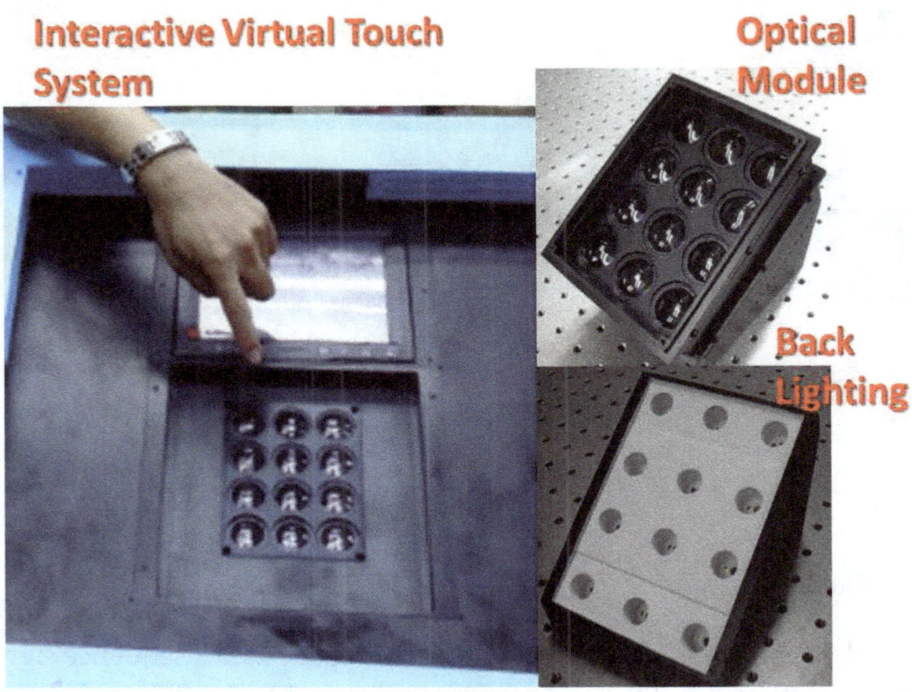

Figure 10. The stereoscopic images through the projected lenses.

2.2. Design and fabrication of IR LED sensors multiplexer (mux) system

We propose an auto-stereoscopic display and an interactive technology locker door floating image device as shown in Figure 8. The pop-up appearance system has two main parts: (1) an image pop-up system, and (2) two sights of auto-stereoscopic display. In these two parts, the audiences can observe the appearance of 3D images in the air with a disparity from a wide viewing angle. The purpose of the image is to float in a free space that is accessible to the viewer's hand. Accordingly, the touch and interactive technologies can include free space to enhance system capabilities. The new device can have many applications, especially for products and information kiosks in public places.

The IR LED sensors system is shown in Figure 11, with the non-touch positioning sensing devices using infrared (IR) LED and proximity sensor. The signals, from a set of proximity sensors, are used to determine the position of the hand through the trilateration method. The distances between the hand and sensors can be obtained from proximity sensors. The positioning results of the hand are verified over mobile devices. Therefore, the system allows a user to intuitively interact with the mobile device real time. This design can waive the necessity of touching the panel or screen of the mobile device.

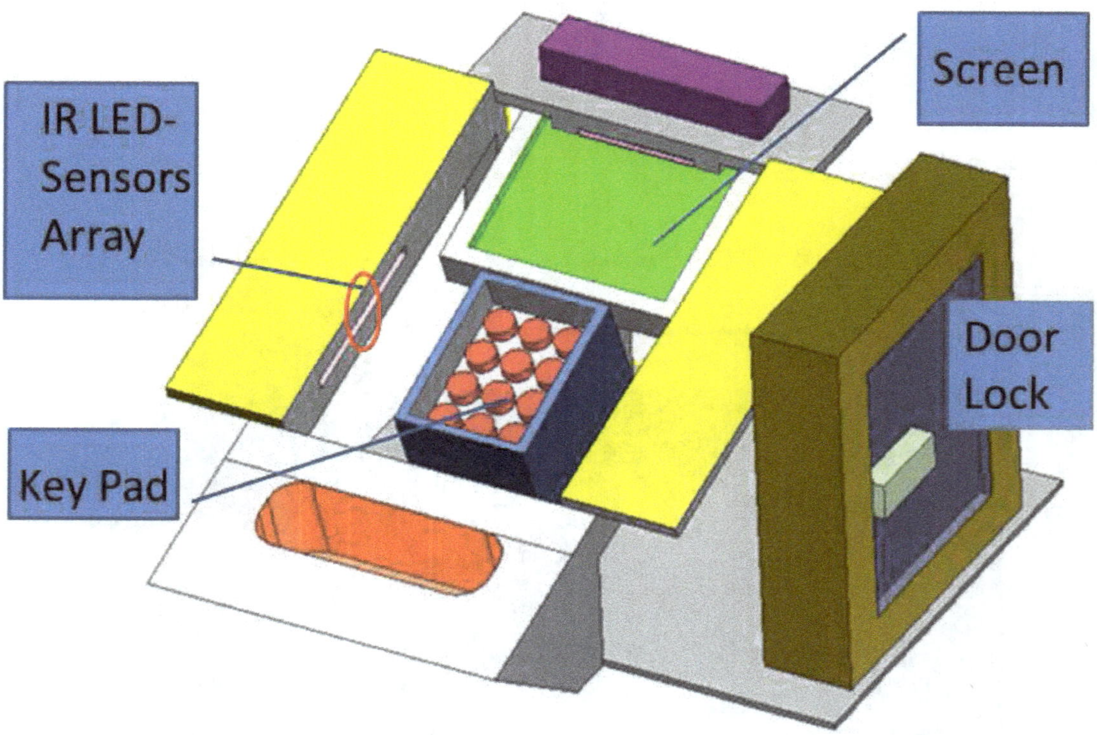

Figure 11. IR LED sensors system.

The multiplexer data control system of the shift register provided a sequence signal. All of the IR LED sensors architecture methods are shown in Figure 12. The floating image is a pop-up space located between the IR LEDs array and sensors array.

Given several mode designs, we have to solve the noise cross-talk between channels and the energy consumption. On the first mode, we designed an "even group and odd group" circuit architecture that is shown in Figure 13. Additionally, the waveform is shown in Figure 14.

The second addressable type of the design is an interval of two elements on different time sequence driving architecture. The architecture shows a sequence that involves the 1st, 4th, 7th;

2nd, 5th, 8th; 3rd, 6th, 9th cyclical signal output controlled elements as shown in Figure 15. The waveform is shown in Figure 16.

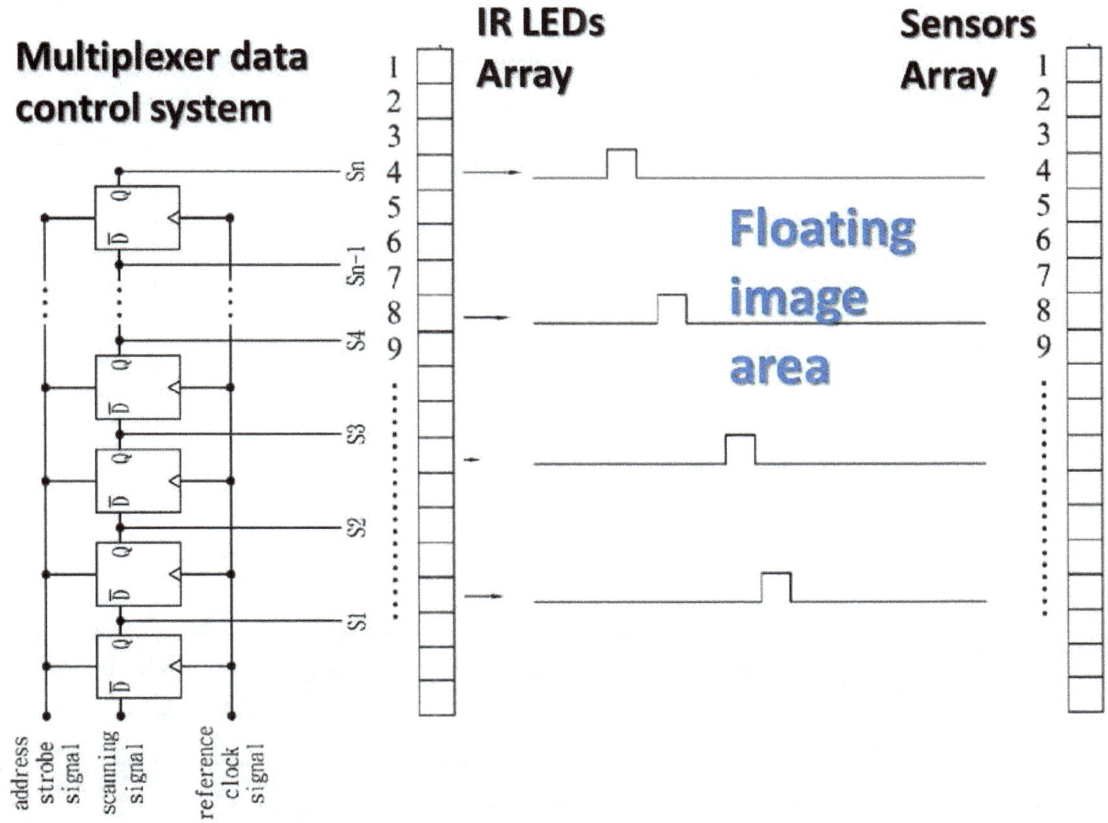

Figure 12. IR LED Sensors multiplexer (mux) system.

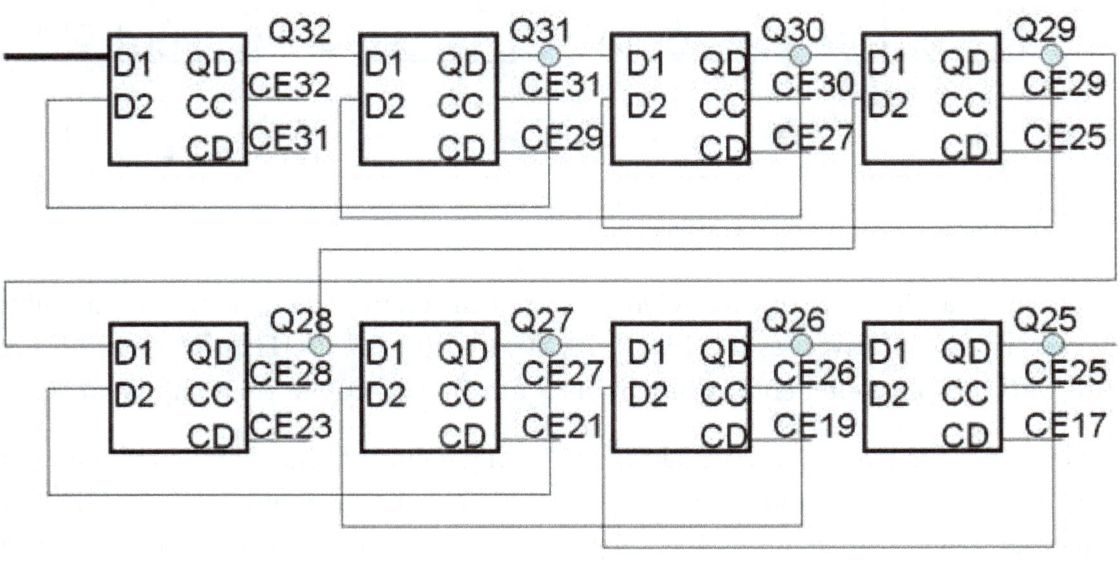

Figure 13. Even and odd group output in different time cycle.

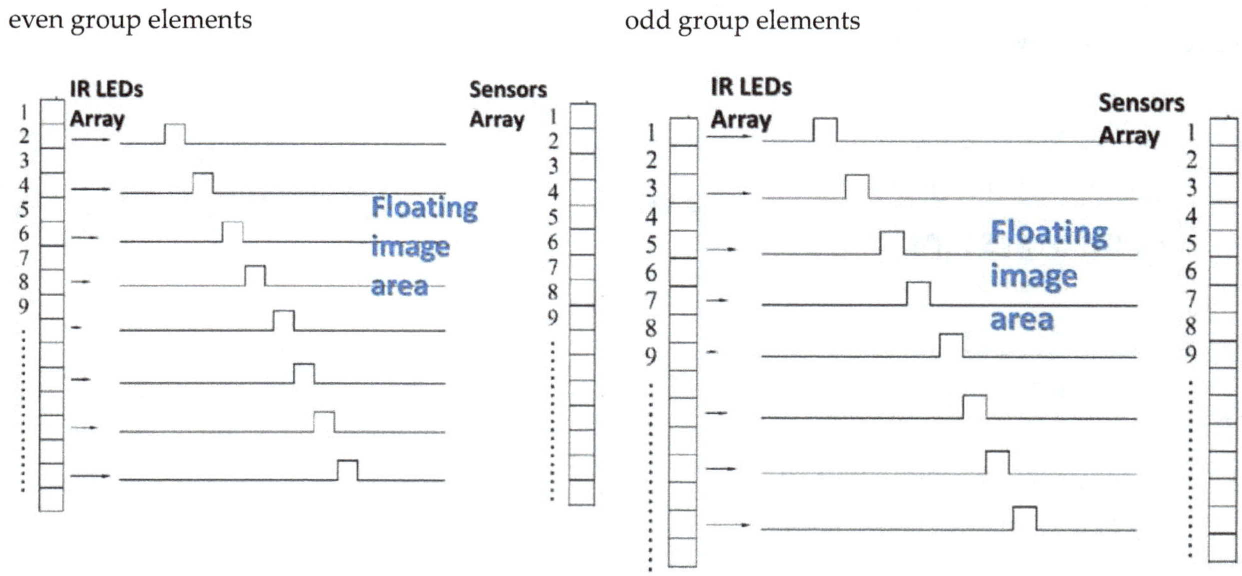

Figure 14. The waveform of even and odd elements.

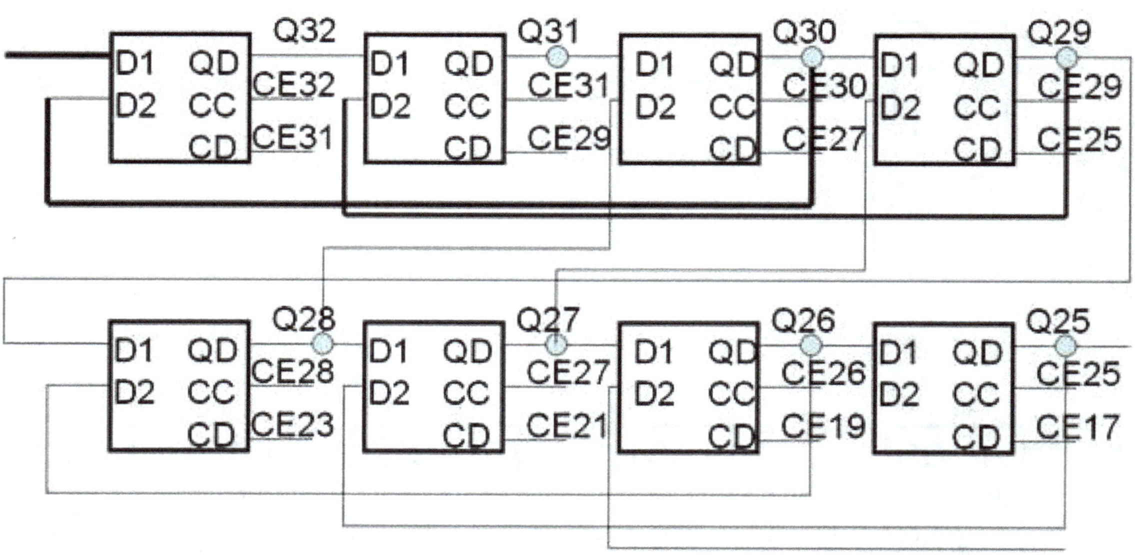

Figure 15. An interval of two elements on a different time driving.

Thirdly, there is an interval of three elements on a different time driving. The architecture shows a sequence that involves the 1st, 5th, 9th; 2nd, 6th, 10th; 3rd, 7th, 11th; 4th, 8th, 12th cyclical signal output controlled elements as shown in Figure 17. The waveform is shown in Figure 18.

Every shift register is used to receive part of the timing counting signal and combined with an odd-even number selection mechanism or a specific number sequence (e.g. 1, 4, 7) element mechanism to generate a set of enabling signals as shown in Figure 19. In this manner, the driving control of the present element circuits can be reached via any random combination of the address signal and each set of enabling signals, i.e., the driving of each element circuit is

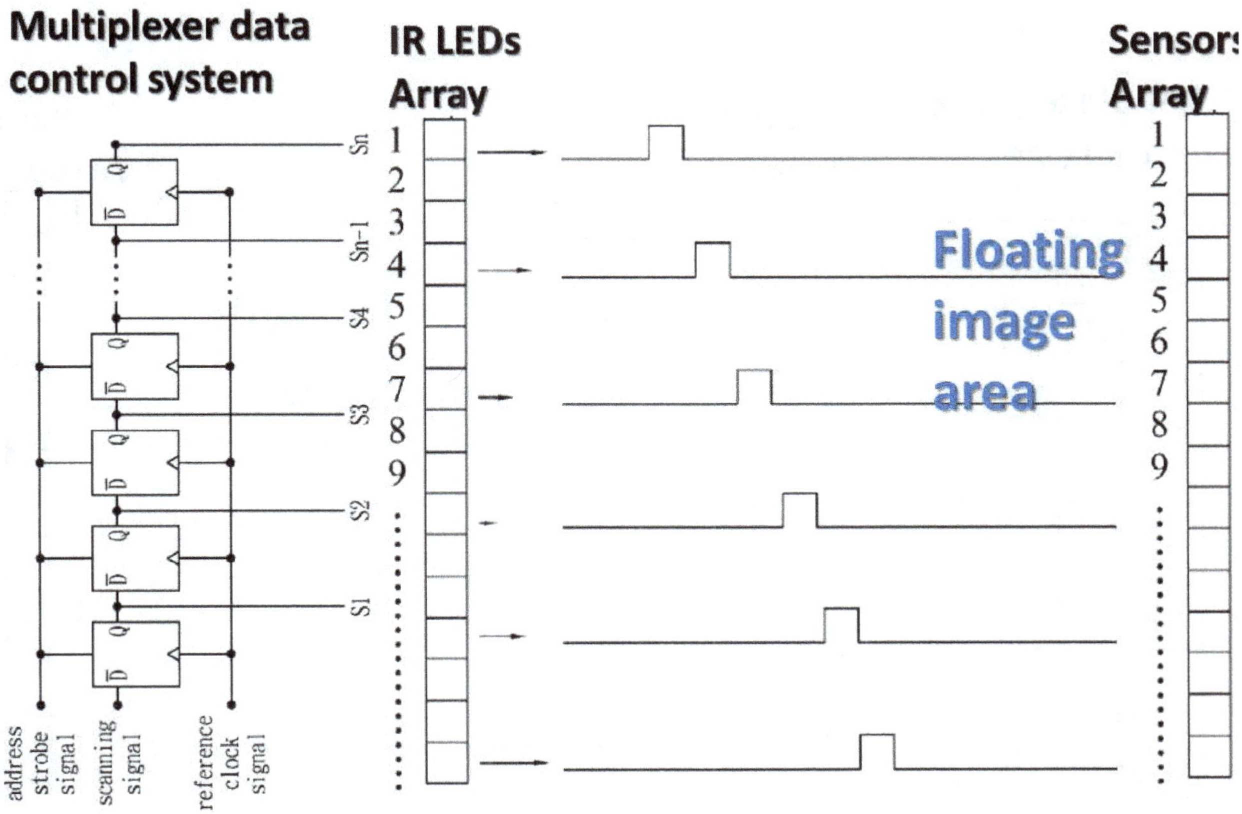

Figure 16. The waveform of interval of two elements.

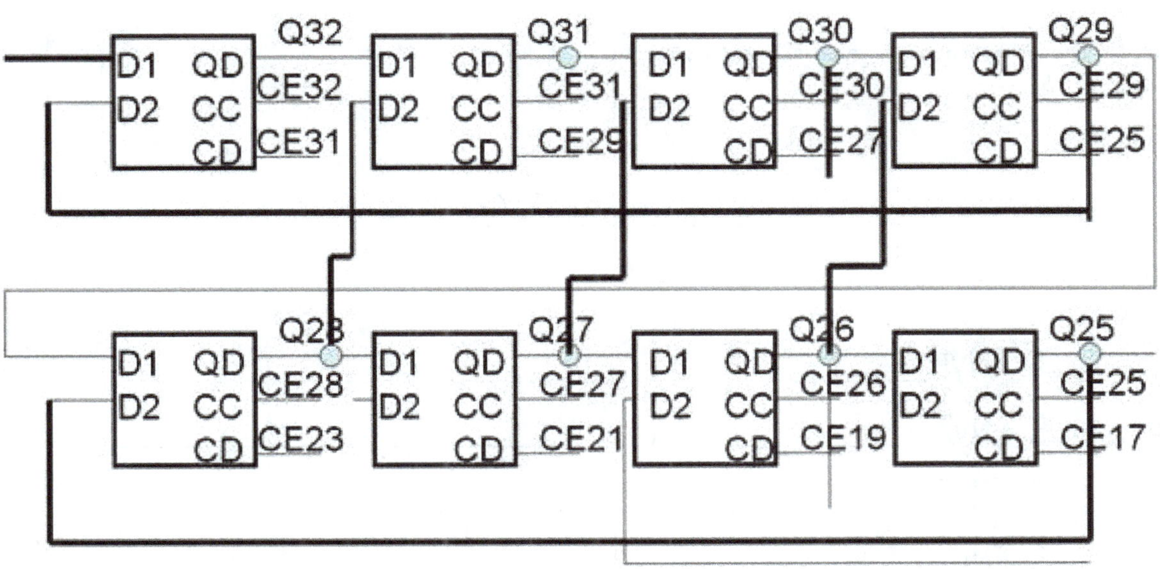

Figure 17. An interval of three elements on a different time driving.

fulfilled via the control of an address signal and one set of the enabling signals. This arrangement is able to prevent generating erroneously triggering signals in the original due to the mutual interference of the IR LED sensors circuits.

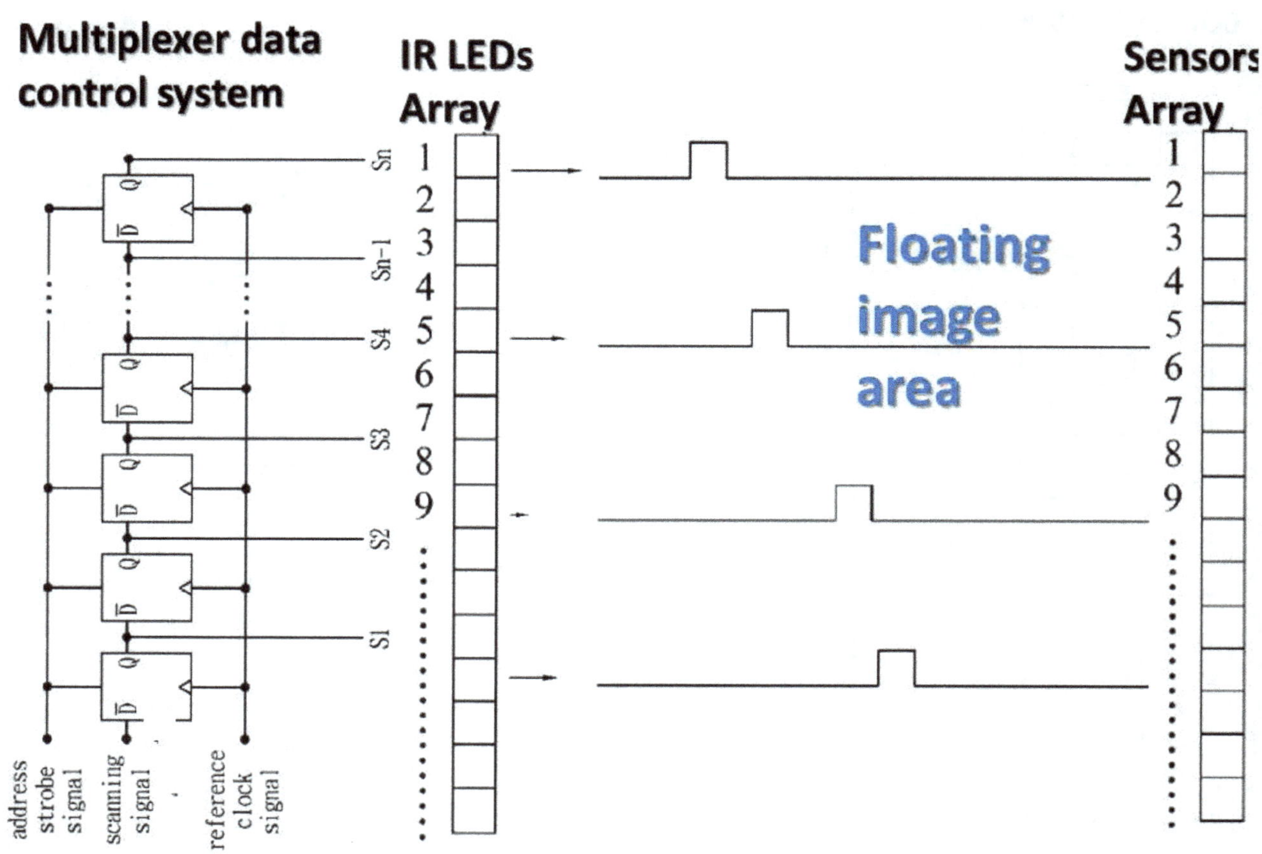

Figure 18. The waveform of the interval of three elements.

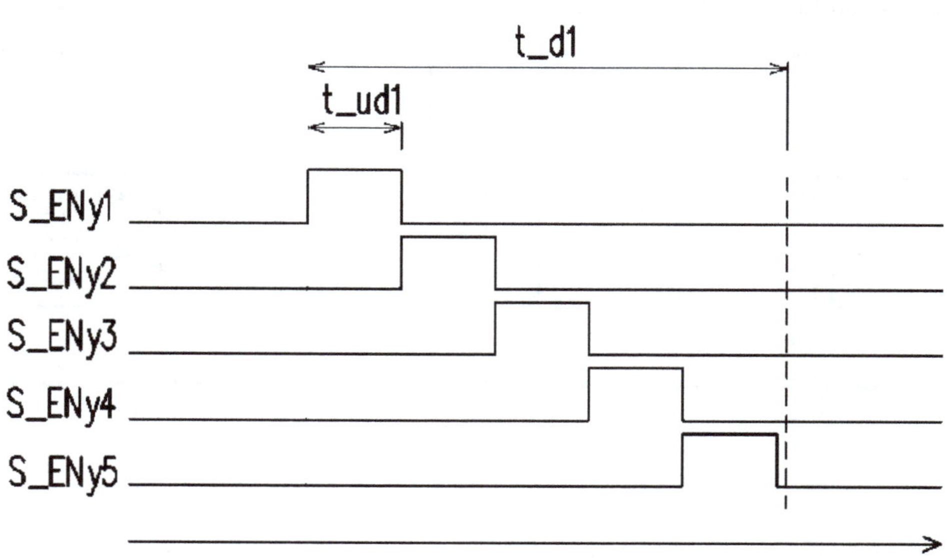

Figure 19. A set of enabling signals.

3. Simulation results

The scanning sequence of the multiplexer is shown in Figure 20. An interleaving data encoding sequence is labeled in red and blue arrows respectively. The inference between neighboring elements is also simulated in Figure 20. The output voltage noise of the third element is lower than half of the level of the second element. The error signal excited by the first elements is significantly reduced.

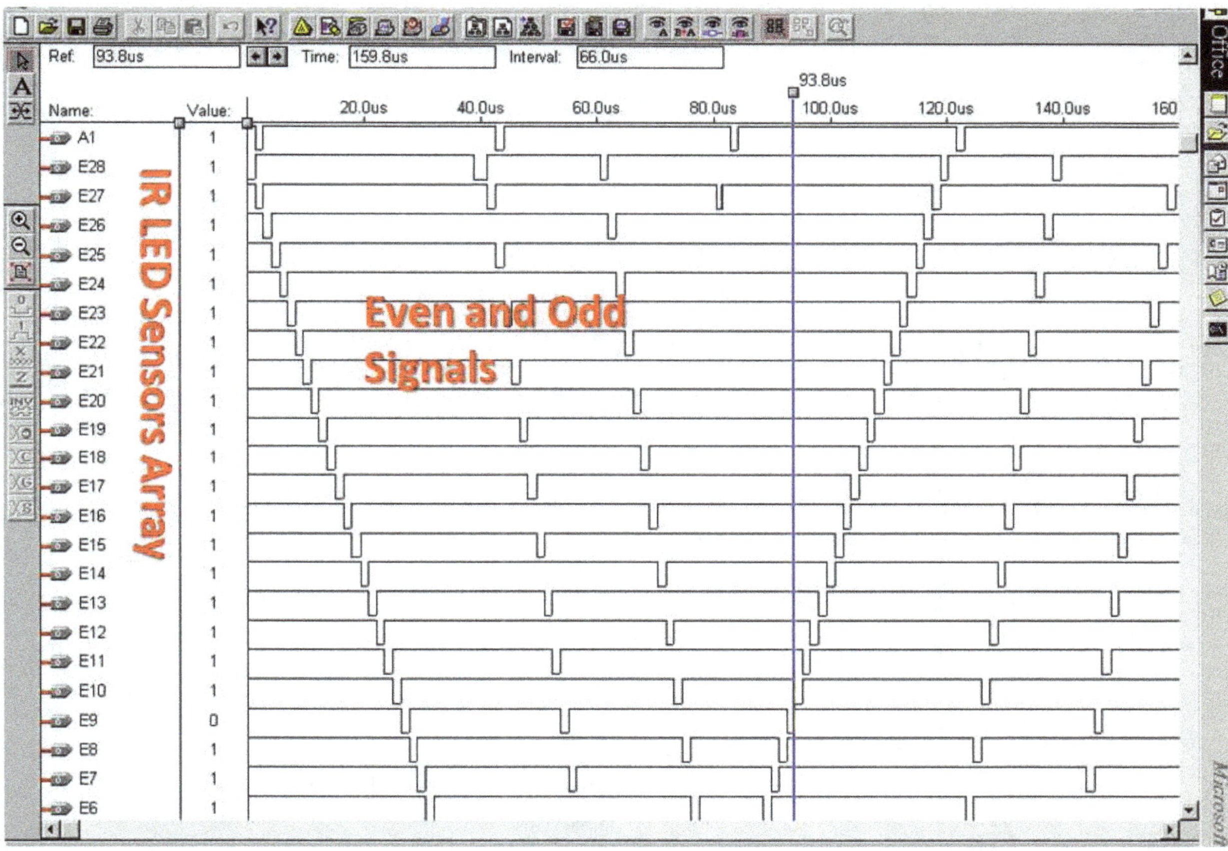

Figure 20. Simulated data encoding sequence for medical ultrasound multiplexer.

The control circuit on IR LED sensor elements verifies addressing elements by a multiplexer logic simulation. There are four kinds of printing tape: type 1 is a sequence of data input for IR LED sensor elements 1, 2, 3,...up to IR LED sensor elements n; type 2 is a sequence of data encoding for IR LED sensor elements n, (n-2), (n-4),...down to final IR LED sensor elements 2; type 3 is a sequence for IR LED sensor elements 1, 3, 5,...up to final IR LED sensor elements; and type 4 is a sequence for IR LED sensor elements (n-1), (n-3), (n-5),....down to final IR LED sensor elements. The driver system design has to simulate the function work, and we used FPGA to verify this.

A novel IR LED sensor elements multiplexer data registration system has been designed and simulated. The controller can be easily adapted to different sizes of the IR LED sensor elements without much hardware arrangement. The simulated results show that the proposed multiplexer controller can reduce the noise interference caused by the excitation of neighboring channels as shown in Figure 21.

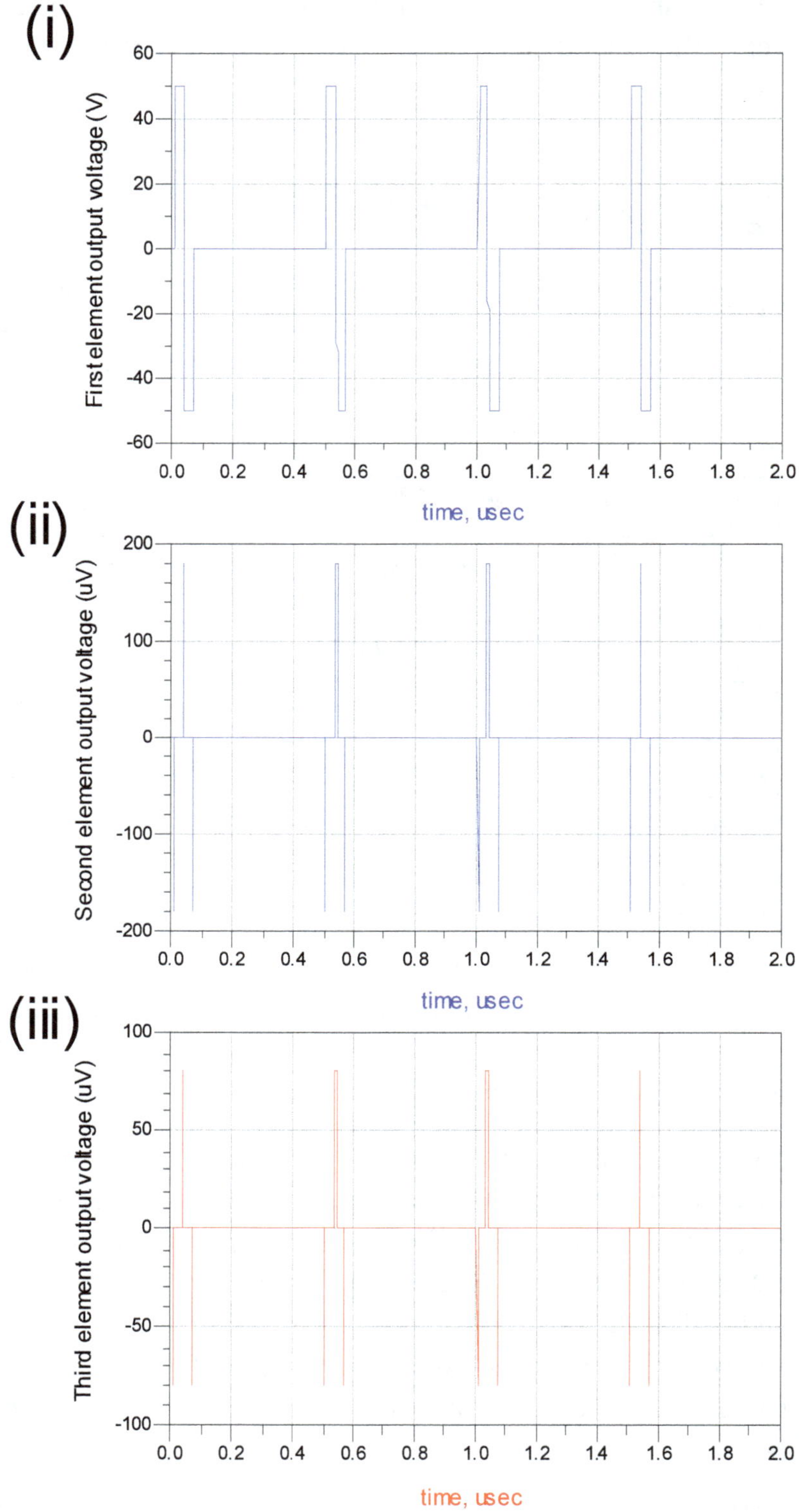

Figure 21. Simulated voltage inferences among the IR LED sensor elements.

A two-dimensional data control system is shown in Figure 22. Under the two-dimensional circuit configurations, a shifting sub-circuit uses a special control method to generate the enabling signals in a serial-in and parallel-out manner. The special control method connects an output of a n-th D flip-flop from the output end to a trigger end of a (n+2)-th D flip-flop.

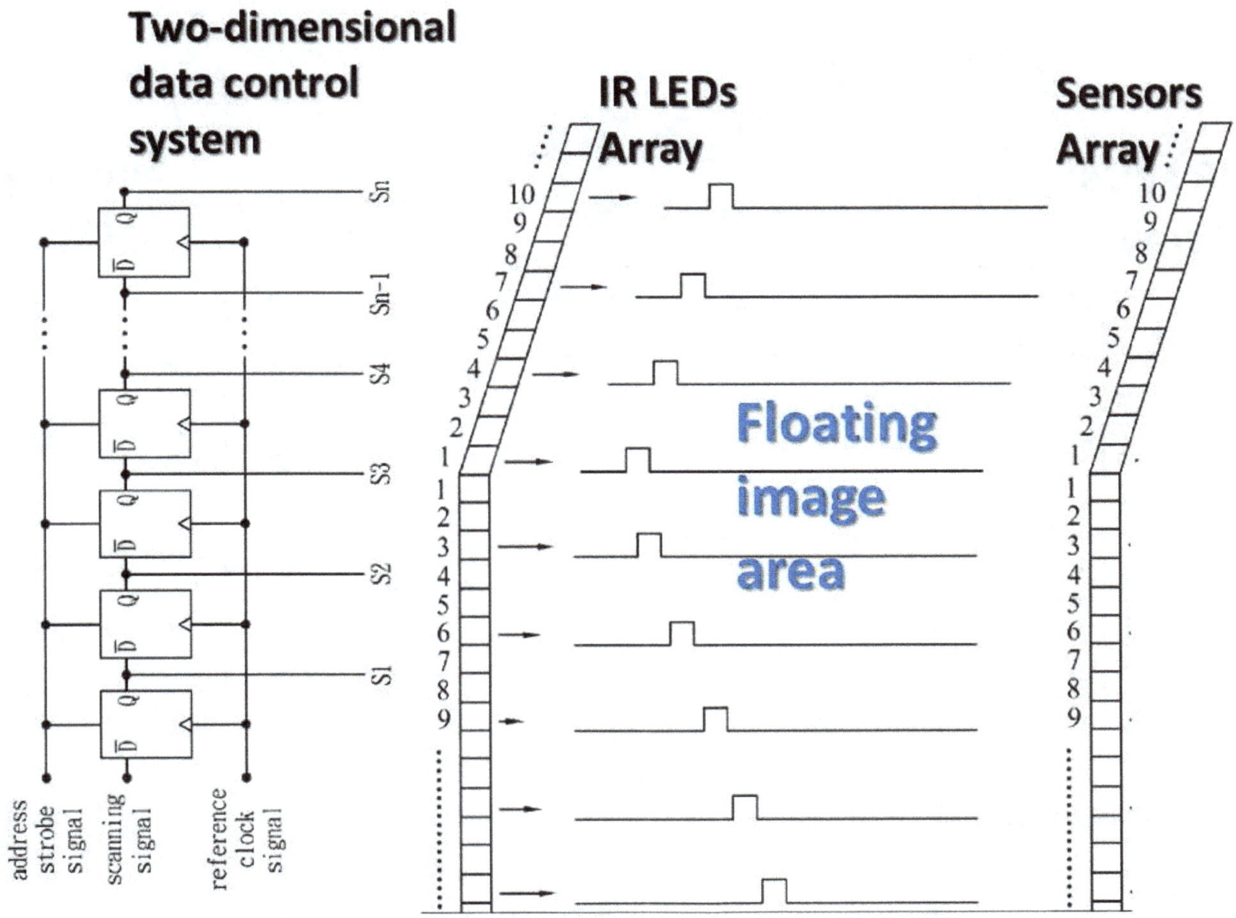

Figure 22. Two-dimensional data control system.

The multi-dimensional data control system is shown in Figure 23. For the same architecture, the three-dimensional data control system method of the shifting circuit connects an output of the n-th D flip-flop from the output end to a trigger end of a (k+3)-th D flip-flop.

4. Integrated system

Figure 24 shows a floating image of the virtual touch system photo. It is the top view of the floating image system. Figure 25 is a cross view of the floating image system. The fictitious image is formed by a focus lens.

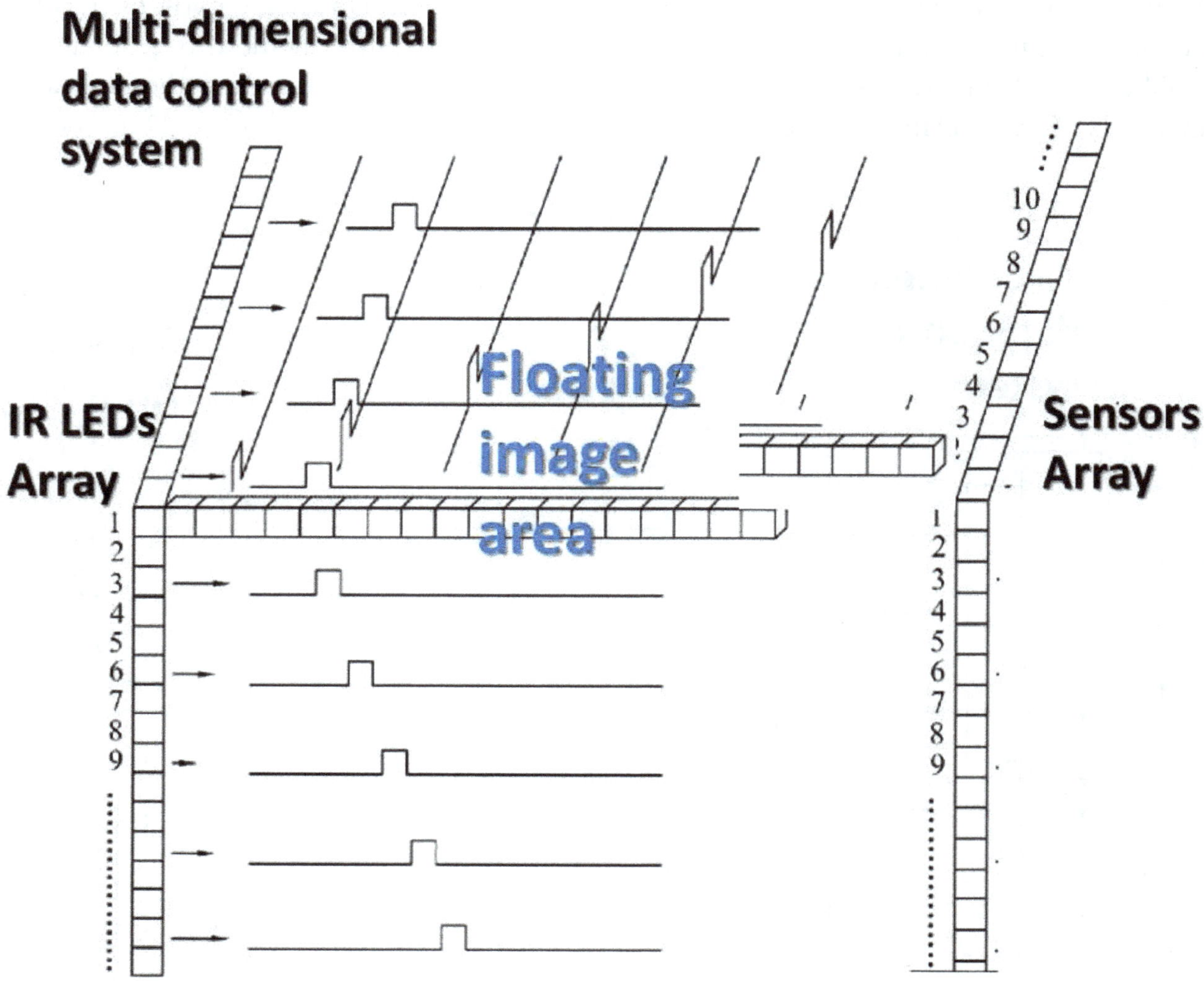

Figure 23. Multi-dimensional data control system.

Figure 24. Top view of the floating image system.

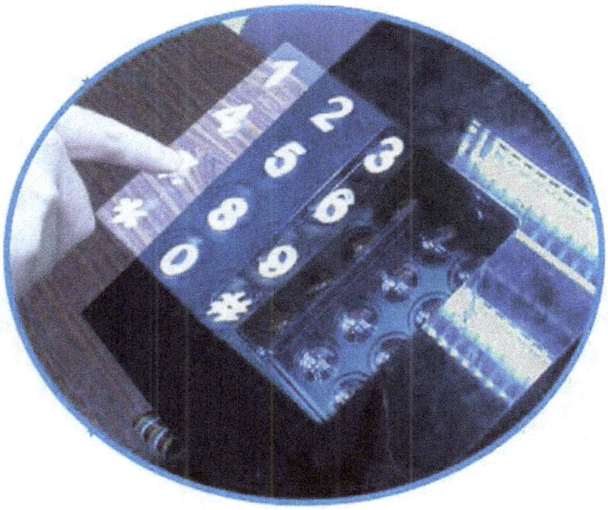

Figure 25. Cross view of the floating image system.

5. Discussion and result

The integrated infrared light sensor module, a floating 3D image architecture and image display is shown in Figures 26 and 27. The image is 20 cm away from the system, and viewers could observe the image to be about 50 cm within the best viewing distance. In addition to displaying three-dimensional images, the application is also a virtual keyboard for the telephone, as shown in Figure 28. Thus, the device can be used to interact with the user interface function. The virtual system displays its application to the locker room door as shown in Figure 29.

Figure 26. IR LED sensors module.

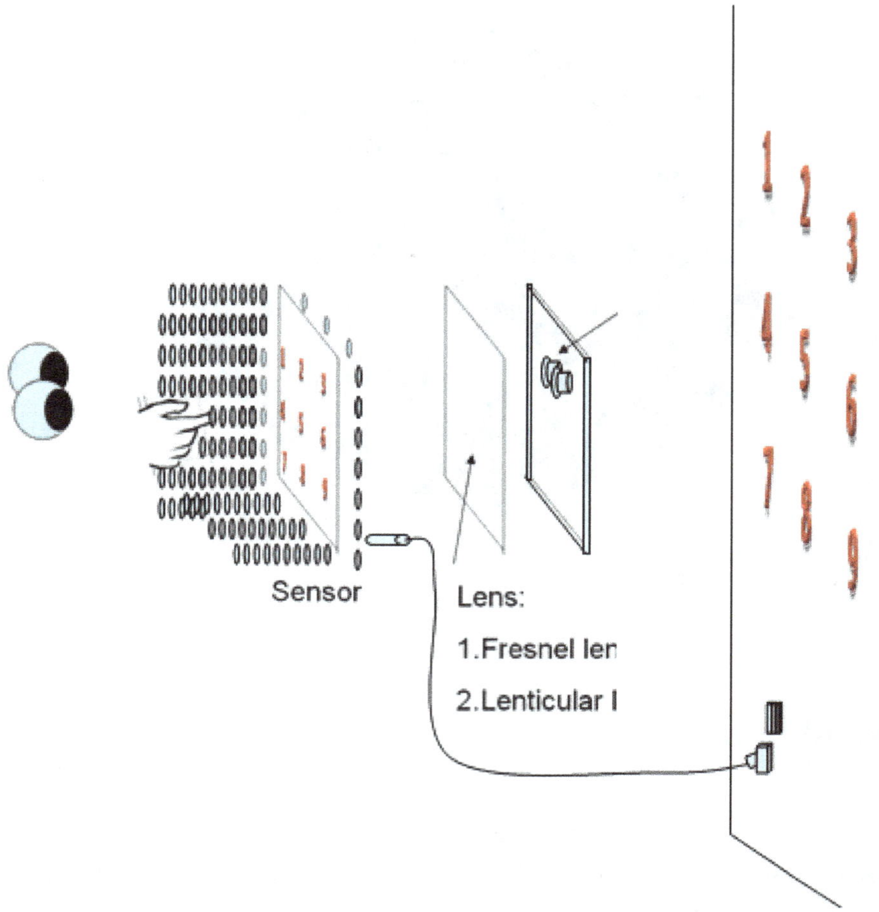

Figure 27. Floating 3D image architecture.

Figure 28. Virtual keypad for telephone.

Figure 29. Virtual systems for the door locker.

Author details

Jian-Chiun Liou

Address all correspondence to: jcliou@kuas.edu.tw

National Kaohsiung University of Applied Sciences (KUAS), The Department of Electronic Engineering, Taiwan

References

[1] Han, J.Y., Low-cost multi-touch sensing through frustrated total internal reflection. ACM UIST, pp. 115-118,(2005).

[2] Kitamura, Y., Nakayama, T., Nakashima, T., and Yamamoto, S., The illusionhole with polarization filters. ACM VRST, p. 244-251, (2006).

[3] Chan, L.W., Chuang, Y.F., Chia, Y.W., Hung, Y.P., and Jane Hsu., A new method for multi-finger detection using a regular diffuser. International Conference on Human-Computer Interaction, (2007).

[4] Chan, L.W., Chuang, Y.F., Yu, M.C., Chao, Y.L., Lee, M.S., Hung, Y.P. and Jane Hsu, Gesture-based interaction for a magic crystal ball. ACM VRST, pp. 157-164(2007).

[5] Wilson, A. TouchLight: An Imaging Touch Screen and Display for Gesture-Based Interaction. International Conference on Multimodal Interfaces, pp. 69-76, (2004).

[6] Morris, M.R., Morris, D., and Winograd, T., Individual audio channels with single display groupware: effects on ommunication and task strategy. ACM CSCW, p. 242-251,(2004).

[7] Sugimoto, M., Hosoi, K., and Hashizume, H., Caretta: system for supporting face-to-face collaboration by integrating personal and shared spaces ACM CHI, p. 41-48, (2004).

[8] Kitamura, Y., Osawa, W., Yamaguchi, T., Takemura, H., and Kishino, F., A display table for strategic collaboration preserving private and public information. International Conference on Entertainment Computing, pp. 167-179,(2005).

[9] Kakehi, T., Iida, M., Naemura, T., Shirai, Y., and Matsushita, M., Lumisight table: an interactive view-dependent tabletop display. IEEE Computer Graphics and Applications, January/February, 25(1), p. 48-53,(2005).

[10] G. Lippmann: La photographie integrale, Comptes-Rendus 146, pp. 446.451, (1908).

[11] Roberto Lopez-Gulliver, Shunsuke Yoshida, Sumio Yano, and Naomi Inoue: gCubik : Real-time Integral Image Rendering for a Cubic 3D Display, ACM SIGGRAPH 2009 Emerging Technologies, New Orleans, USA, August 3-7 (2009).

[12] Koike, T., Sakai, H., Shibahara, T., Oikawa, M., Yamasaki, M., and Utsugi, K., Light field copy machine. In ACM SIGGRAPH ASIA 2009 Emerging Technologies: Adaptation (Yokohama, Japan, December 16 – 19, SIGGRAPH ASIA '09,(2009).

[13] Sakamoto, K., Kimura, R., and Takaki, M.: Parallax polarizer barrier stereoscopic 3D display systems, Active Media Technology, 2005. (AMT 2005). Proceedings of the 2005 International Conference on, pp.469- 474, 19-21 May (2005).

[14] Daisuke Kondo, Toshiyuki Goto, Makoto Kouno, Ryugo Kijima, and Yuzo Takahashi: A Virtual Anatomical Torso for Medical Education using Free From Image Projection, Proc. of the 10th International Conference on Virtual Systems and Multimedia, pp678-685, (2004).

[15] A. Cassinelli and M. Ishikawa: Khronos projector. In SIGGRAPH 2005, Emerging Technologies, (2005).

[16] Y. Kawaguchi: The Art of Gemotion in Space, Information Visualization, 2006. IV 2006. Tenth International Conference on, pp.658-663, 5-7 July (2006).

9

Dye-Sensitized Solar Cells with Graphene Electron Extraction Layer

Lung-Chien Chen

Abstract

This work aims to improve the conversion efficiency of dye-sensitized solar cells (DSSCs) by introducing a new material, graphene, into the DSSC structure. Graphene is a potential material for many applications due to their high electron mobility, outstanding optical properties, and thermal, chemical, and mechanical stability. Therefore, this study changes several parameters, structures, and methods to optimize and compare with the traditional DSSCs. There are three major respects about with or without graphene, the method of plating or sputtering, and the structure of graphene/TiO_2 or TiO_2/graphene/TiO_2 in DSSCs solar cells. Finally, this research knows that the method of sputtering is much better than plating; the conversion efficiency of solar energy with graphene/TiO_2 was increased from 1.45 % to 3.98 %, and the conversion efficiency with TiO_2/graphene/TiO_2 sandwich structure was increased from 1.38 % to 3.93 %. It means that the new material, graphene, works in enhancing the conversion efficiency of DSSCs.

Keywords: DSSCs, Solar cell, Graphene, Sandwich structure, TiO_2

1. Introduction

This chapter aims to review the dye-sensitized solar cells (DSSCs) with graphene structure. DSSCs have been under extensive research. Since the color of the device can be easily varied by choosing different dyes and cells on flexible substrates have been already demonstrated, DSSCs are especially attractive for building integrated photovoltaics. The cell concept can reduce the production costs and energy payback time significantly compared to standard silicon cells or other thin film cells.

However, one of the major issues hindering the rapid commercialization of DSSCs is their lower conversion efficiency compared to conventional p-n junction solar cells [1]. That may be attributed to poor charge separation in DSSC structure. Therefore, charge transfer structure, such as Au nanoparticles and quantum dots, has been employed in a DSSC to improve the device performance through charge separation in the photoelectrodes [2-5].

Graphene is a potential material for many applications due to their high electron mobility, outstanding optical properties, and thermal, chemical, and mechanical stability [6-10]. Therefore, the second section in this article illustrates the principle of electron extraction layer. TiO_2 plays an important role on the electron-extraction layer. We will discuss the electron transmission on dye-sensitized solar cells. The effect of the electron-transporting layer of the solar cell is very important. Therefore, we show the I-V characteristics of the DSSCs. The cell performance was measured, which had different electron-extraction layer structures.

The third section discusses the preparation method of the graphene. Graphene is a potential material for many applications due to their high electron mobility, outstanding optical properties, and thermal, chemical, and mechanical stability.

In the fourth section, the graphene was introduced into the DSSC structure to improve electron conversion efficiency. This study investigates the effect on the graphene layer as electron transport layer in the DSSC structure deposited by the magnetron sputtering method; in particular, it examines the performance of the DSSCs with the graphene electron transport layer.

The fifth section reveals a new DSSC structure. The structure has provided excellent performance and higher photoelectric conversion efficiency by DSSC with the TiO_2/graphene/TiO_2 sandwich structure. This section focuses on the improvement that is associated with the increase in electron transport efficiency and the absorption of light in the visible range.

The concluding paragraphs will summarize some parameters of DSSC with or without a graphene layer which was prepared by sputtering and then discuss the DSSC's parameters and reasons which have different preparation methods of graphene layer. Finally, there are some concluding remarks.

2. Principle of electron transport layer

Figure 1 sketches the structure of the basic DSSC which can be divided into several parts. They have a basic structure that comprises two conductive substrates (one is photoelectrode and the other is counter electrode), an absorbing layer of semiconductor materials, dye molecules, and a redox electrolyte. The basic principle of operation of DSSCs includes the following: (1) the light irradiates on the DSSC and the photons will pass through the photoelectrode to the dye layer which is absorbed by the photosensitizer dye molecule. (2) The photosensitizers are excited from the ground state (S) to the excited state (S*). The excited electrons are injected into the conduction band of the TiO_2 electrode. This results in the oxidation of the photosensitizer (S^+). (3) Electrons are injected from the photoexcited dye into the conductive band of the semiconductor. The electrons will pass from the electric transport layer to the external circuit. (4) The oxidized photosensitizer (S^+) accepts electrons from the I^- ion redox mediator, leading to regeneration of the ground state (S), and the I^- is oxidized to the oxidized state, I_3^-, and transports the positive charges to the counter electrode. (5) The oxidized redox mediator, I_3^-, diffuses toward the counter electrode and then it is reduced to I^- ions.

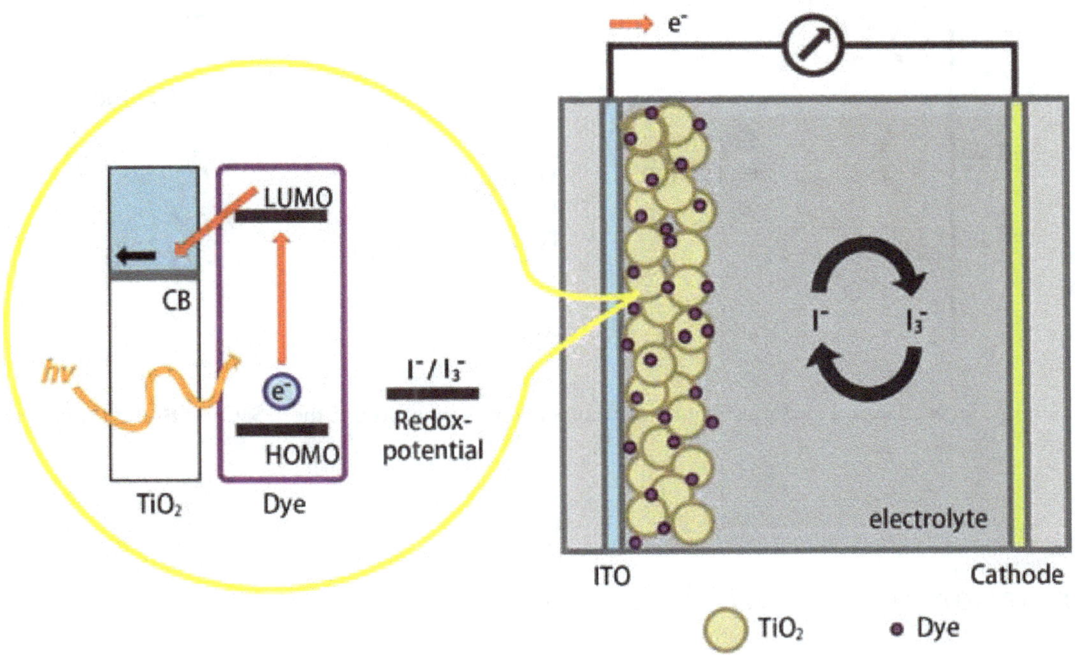

Figure 1. Schematic cross section of the completed structure

The principle of electron transport (or extraction) layer inserted in the traditional DSSC structure had been reported [11-17]. Figure 2 shows the energy level diagram and mechanism of photocurrent generation in TiO_2 DSSCs with the graphene layer. The work function of the graphene layer is around 4.5 eV [18,19]. Graphene has a work function similar to that of the indium tin oxide (ITO) (4.8 eV) electrode. The graphene layer does not prevent the flow of injected electrons down to the ITO electrode because its work function exceeds that of the ITO electrode [20-22]. Therefore, the brief operating process is as follows. Dye N719 was excited by incident light, and electrons transit from HOMO to LUMO. The LUMO and HOMO are the lowest unoccupied molecular orbit and highest occupied molecular orbit, respectively. Electrons are injected into the graphene electron transport layer via the TiO_2 photoelectrode. The electrons transferred to the graphene electron transport layer were collected at the back contact to generate a photocurrent. Therefore, the inserted graphene layer collects electrons and acts as a transporter in the effective separation of charge and rapid transport of the photogenerated electrons.

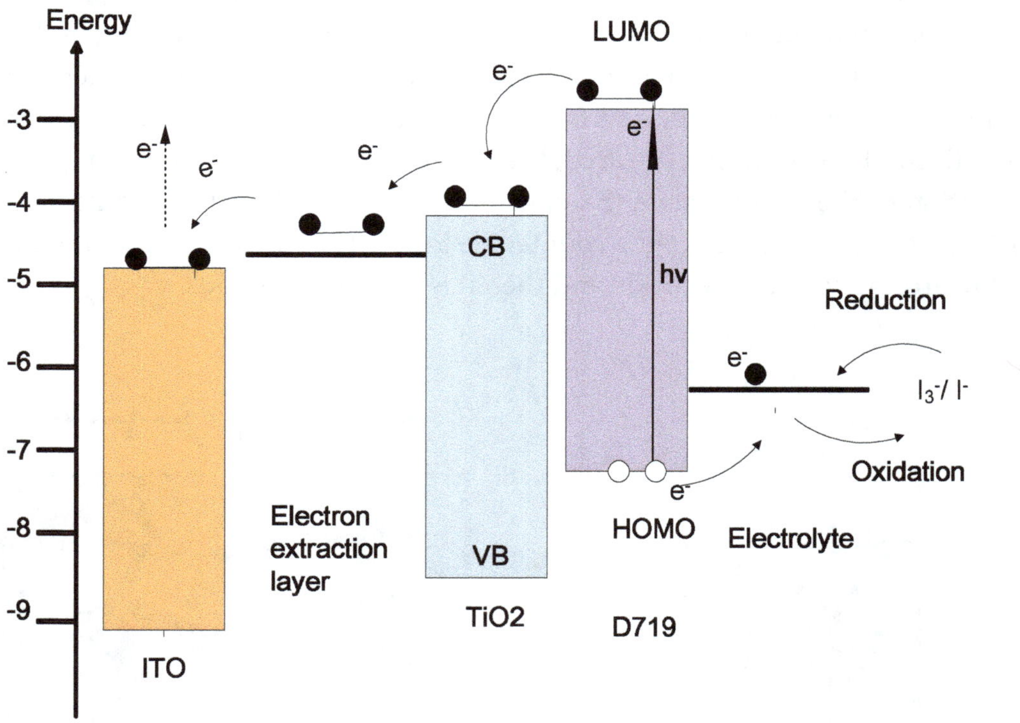

Figure 2. Energy level diagram and mechanism of photocurrent generation in the DSSCs with the graphene electron transfer layer

3. Preparation and characteristics of graphene

Graphene is a potential material for many applications such as extensively utilized in organic photovoltaic (PV) cells. It has excellent optical and electrical characteristics which are exploited in transparent conductive films or electrodes by their high electron mobility [6-8,10,17], outstanding optical properties, and thermal, chemical, and mechanical stability [6–10,12]. However, it is hard to produce high-quality graphene to use in the sputter deposition method. Therefore, this study uses plating method to plate graphene and compare with the traditional method.

First, acetylacetone and Triton X-100 were added into 10 ml water by using syringe. The TiO_2 compound solution was stirred for 24 h using a magnetic stirrer. After mixing the TiO_2 compound solution, the TiO_2 colloid is obtained. The graphene was stacked on the ITO substrate by electroplating process. The plating solution is graphene dispersion. The plating solution was injected into the beaker and stirred with an air pump. Figure 3 shows the electroplating process. The anode connected to the graphite, and the cathode connected to the ITO substrate. The speeds of the coating process were 500 rpm for 20 s and 2,000 rpm for 60 s. The thickness of TiO_2 is about 13 μm. After the annealing process by using RTA (rapid thermal annealing) at 450 °C for 30 min, the strength of the anatase structure would be enhanced. When the samples cool down to room temperature, they were soaked into the N719 solution; the

N719 solution is mixed ethanol and N719 powder. The samples will produce electrons when they are illuminated with light after they are soaked into the N719 solution.

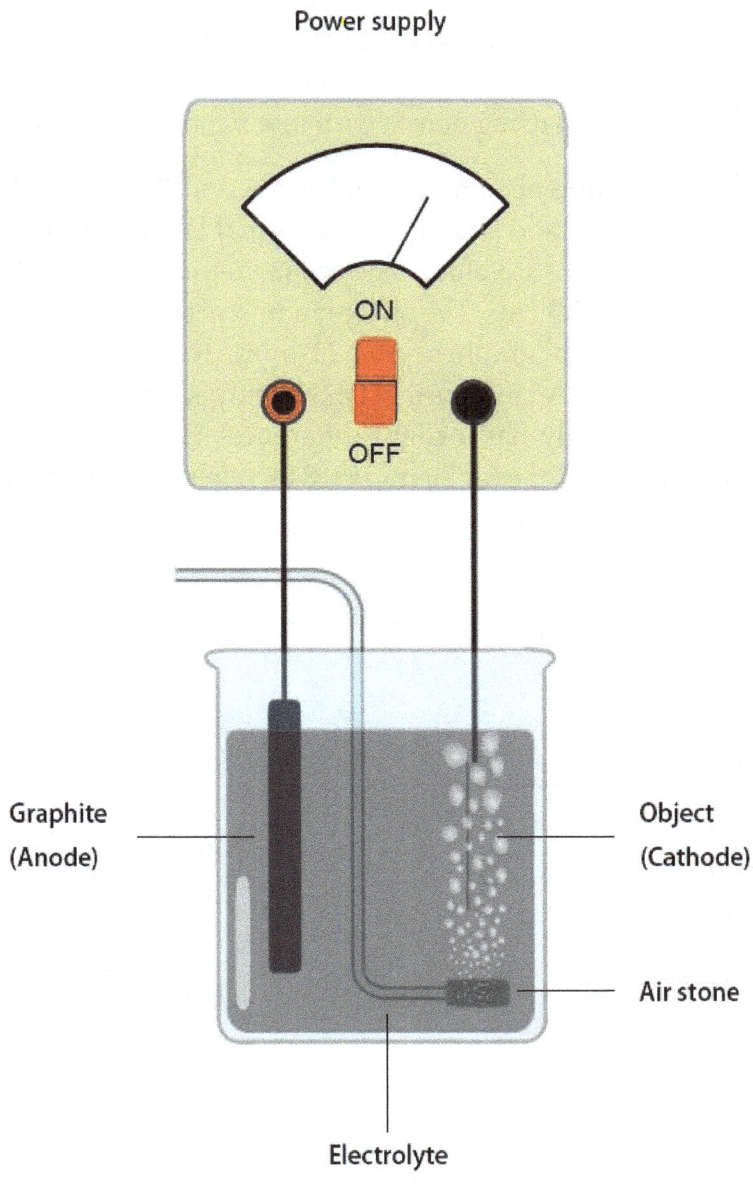

Figure 3. The graphene electroplating methods

A graphene layer was sputtered on indium tin oxide (ITO) conductive glass substrate by radio-frequency magnetron sputtering with a graphite target. It is the electron transport layer that improves the electron transfer in the DSSC structure.

First, the solution consisting of TiO_2 nanocrystalline powder, Triton X-100, acetic acid, and deionized water was mixed as a colloidal solution, and the colloidal solutions were daubed uniformly onto the graphene electron transfer layer to form a thick film. After annealing, the photoelectrode with the graphene layer was immersed in N719 dye absorption $((Bu4N)_2$-$[Ru(dcbpyH)2(NCS)_2]$ complex) in ethanol for 24 h. To increase its anatase content, the samples were sintered at 450 for 30 min. The electrolyte was composed of iodide and lithium iodide with and without 4-tertbutylpyridine (TBP) in propylene carbonate. Then a thick layer of platinum was sputtered onto ITO substrate as a counter electrode. Cells were fabricated by placing sealing films (SX1170-60, Solaronix) between the two electrodes and leaving just two via-holes for injection of electrolyte. The sealing process was carried out on a hot plate. Then the electrolyte was injected into the space between the two electrodes through the via-holes. Finally, the via-holes were sealed using epoxy with low vapor transmission rate.

This study fabricated three different samples: samples A and B are plated with graphene for 20 min and 30 min, respectively, and sample C is a normal DSSC. Figure 4 and Table 1 show the I-V curves and the measurement values. The cell is measured under AM 1.5 illumination at 25 °C. The active area is 0.3×0.3 cm^2. The short-circuit current densities of the samples are 4.97 mA/cm^2 (electroplated with graphene for 20 min), 5.42 mA/cm^2 (electroplated with graphene for 30 min), and 11.2 mA/cm^2 (normal DSSC), respectively. The value of open-circuit voltage between the samples only has a slight difference. The efficiency of the samples are 0.796 % (electroplated with graphene for 20 min), 0.844 % (electroplated with graphene for 30 min) and 3.93 % (normal DSSC), respectively.

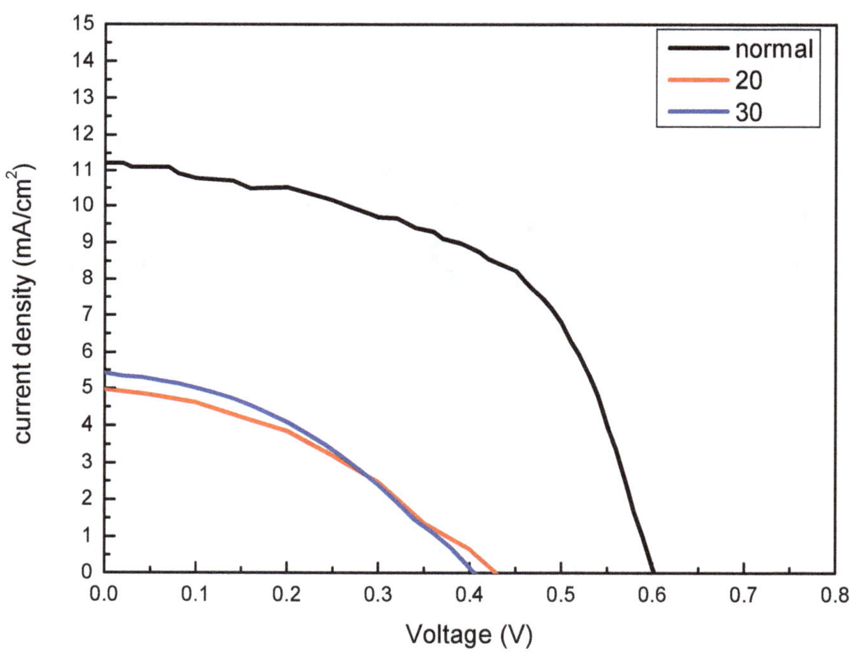

Figure 4. I-V curves of three samples

Sample	Jsc (mA/cm^2)	Voc (mV)	FF%	Efficiency (%)
A(20 min)	4.97	0.4	0.401 %	0.796 %
B(30 min)	5.42	0.4	0.389 %	0.844 %
C (normal DSSC)	11.2	0.6	0.585 %	3.93 %

Table 1. The measurement values of this study

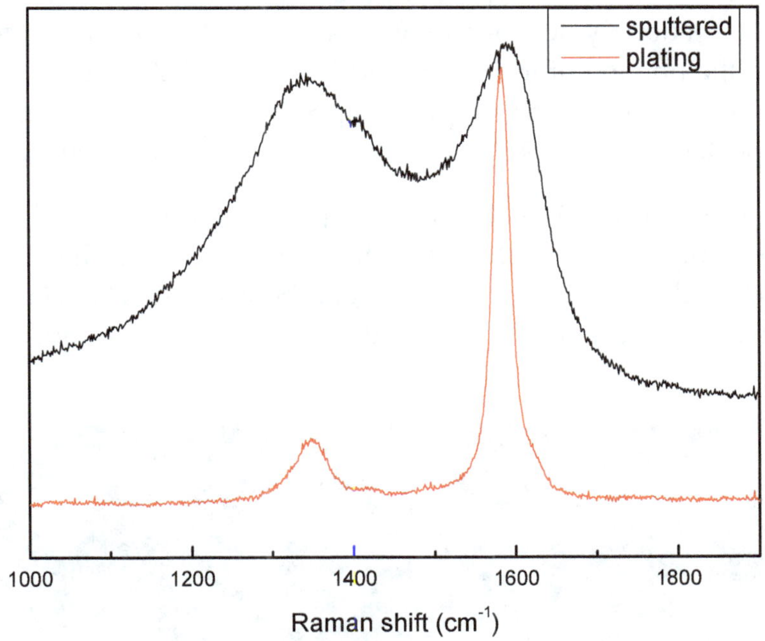

Figure 5. The Raman spectra of plated graphene and sputtered graphene in G-band and D-band

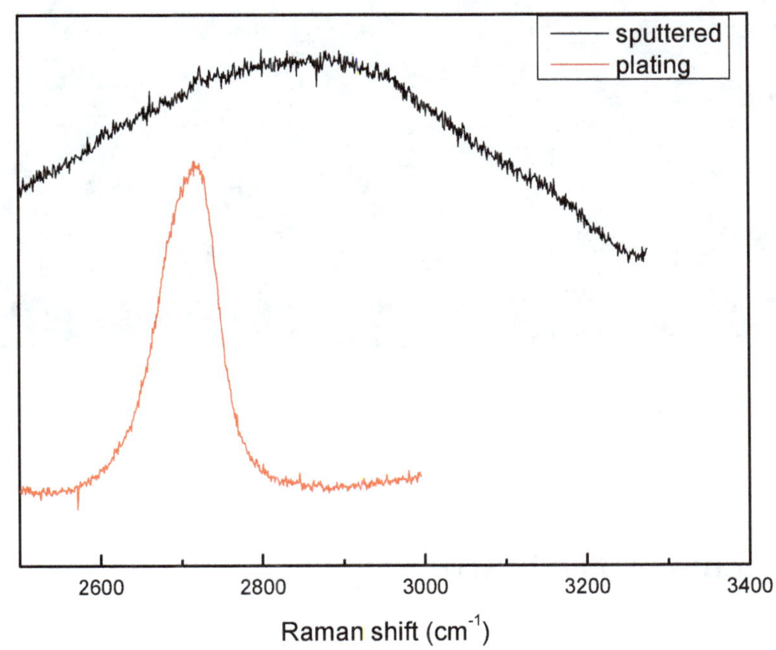

Figure 6. The Raman spectra of plated graphene and sputtered graphene in 2D-band

Figures 5 and 6 show the Raman spectra of electroplated graphene and sputtered graphene (normal DSSC). As shown in Figure 5, electroplated graphene and sputtered graphene have Raman peaks at 1,350 cm^{-1} and 1,580 cm^{-1} [23–25]. The ID/IG of plating graphene is 0.52, and the sputtered graphene is 0.96. The higher value of ID/IG shows good preservation of the highly crystalline structure of graphene. Figure 5 compares the D-band of two different procedures [26]; it shows that using the plating process is much better than using the sputtering process. The G-band is a doubly degenerate phonon mode at the Brillouin zone center; the D-band is a defect and the phonon branches around K point [27]. As shown in Figure 6, the 2D-band is a two-phonon double-resonance process [28] and is similar to the G-band but has a more complicated peak structure [29-31]. It depends on the photon energy and polarization.

Figure 7 shows the top-view SEM image of electroplating graphene on ITO glass. As shown in Figure 7, the graphene flakes with 10 μm width were stacked on the ITO glasses. Figure 8 shows the cross-sectional SEM image of plating graphene on the ITO glass. The graphene is successfully plated on the ITO glasses, and the thickness is around 8 μm.

Figure 7. Top view of electroplating graphene on ITO glass surface

4. DSSCs with graphene/TiO2 active layer

Because graphene has high electron mobility, we use graphene as an electron transport layer to improve the electron transfer in the DSSC. That is DSSCs with graphene/TiO_2 active layer. The graphene flakes prepared by using the electroplating method have demonstrated a superior graphene property by Raman scattering. However, the DSSCs with graphene flakes exhibited poor power conversion efficiency, owing to the high series resistance caused by the discontinue graphene flakes. Therefore, sputtered graphene was employed to replace the graphene flakes prepared by electroplating to improve the electrical properties of the DSSCs even the sputtered graphene including graphene oxide.

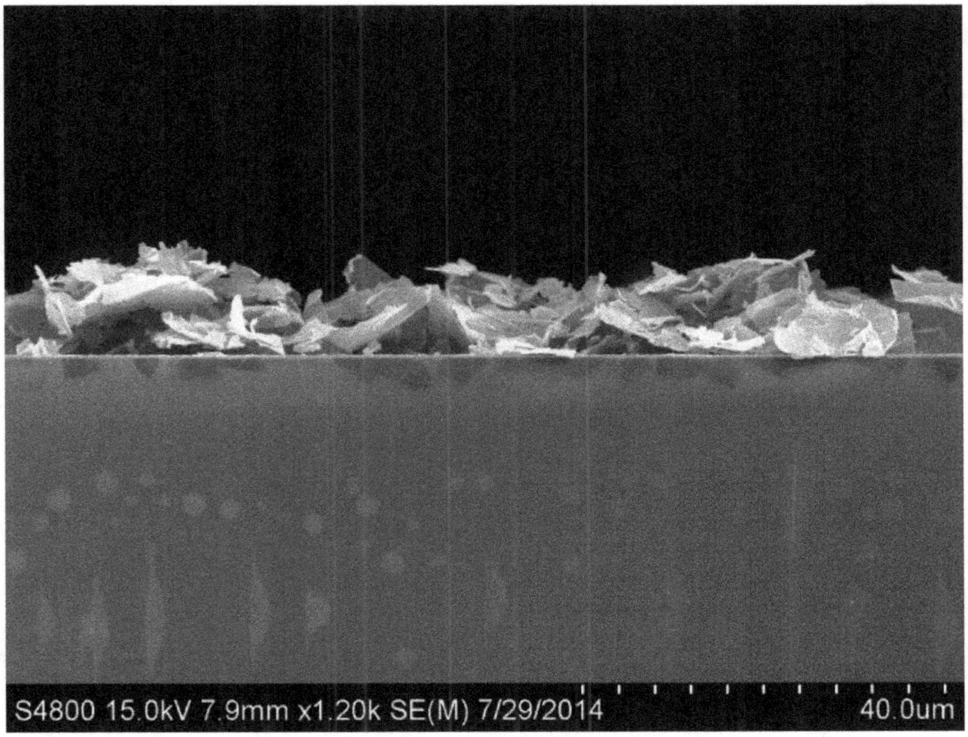

Figure 8. Cross-section SEM image of the plated graphene

First of all, a 60-nm-thick graphene layer was sputtered on indium tin oxide (ITO) conductive glass substrate by radio-frequency magnetron sputtering as an electron transport layer. Next, the solution consisting of TiO_2 was mixed as a colloidal solution which was daubed uniformly onto the graphene electron transfer layer to form a thick film. Then a 100-nm-thick layer of platinum was sputtered onto ITO substrate as a counter electrode. Cells were fabricated by placing sealing films between the two electrodes and leaving just two via-holes for injection of electrolyte. Then, the electrolyte was injected into the space between the two electrodes through the via-holes. Finally, the via-holes were sealed using epoxy with low vapor transmission rate. Figure 9 shows the cross section of the completed structure.

Afterward, we began examining its results by comparing the 60-nm-thick graphene electron transport layer with the 100-nm-thick graphene layer. Figure 10 shows the absorption of TiO_2 DSSCs with and without the graphene electron transfer layer in visible range. As shown in Figure 10, the graphene electron transport layer has an increased absorption coefficient in the range of 310–400 nm. Therefore, the graphene electron transport layer is also an absorption layer to improve the absorption of the solar cells.

Figure 11 shows the *I-V* characteristics of the DSSCs. This figure shows cell performance between TiO_2 DSSCs and TiO_2/graphene under AM 1.5 illumination with a solar intensity of 100 mW/cm^2 at 25°C. The cell has an active area of 3×3 mm^2 and no antireflective coating.

Finally, we examine its result by measuring the cell parameters, open-circuit voltage (V_{oc}), short-circuit current (J_{sc}), fill factor (FF), and energy conversion efficiency (E_{ff}) which are summarized in Table 2 [12].

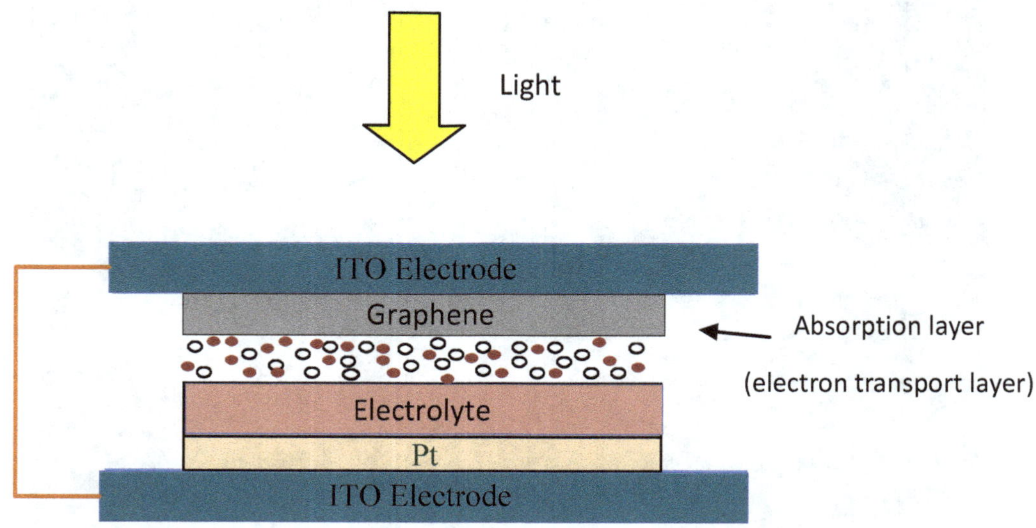

Figure 9. Schematic cross section of the completed structure

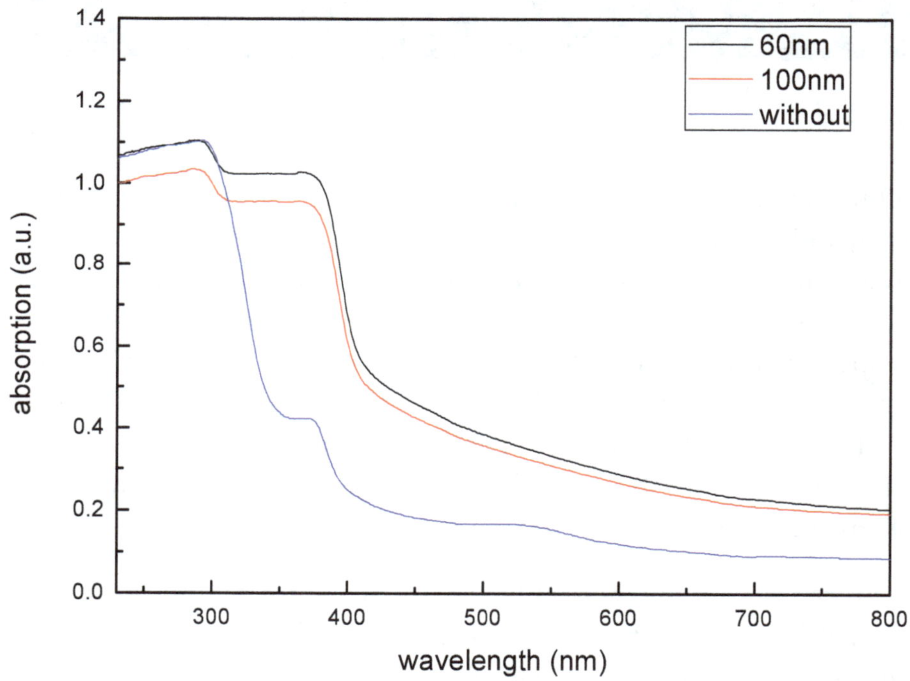

Figure 10. Absorption spectra of the DSSCs with and without the graphene electron transfer layer [12]

	TiO_2	Graphene+TiO_2
Jsc (mA/cm^2)	6.9	17.5
V_{oc} (V)	0.5	0.5
FF	0.419	0.456
η (%)	1.45	3.98

Table 2. The parameters of TiO2 DSSCs with and without graphene electron transport layer [12]

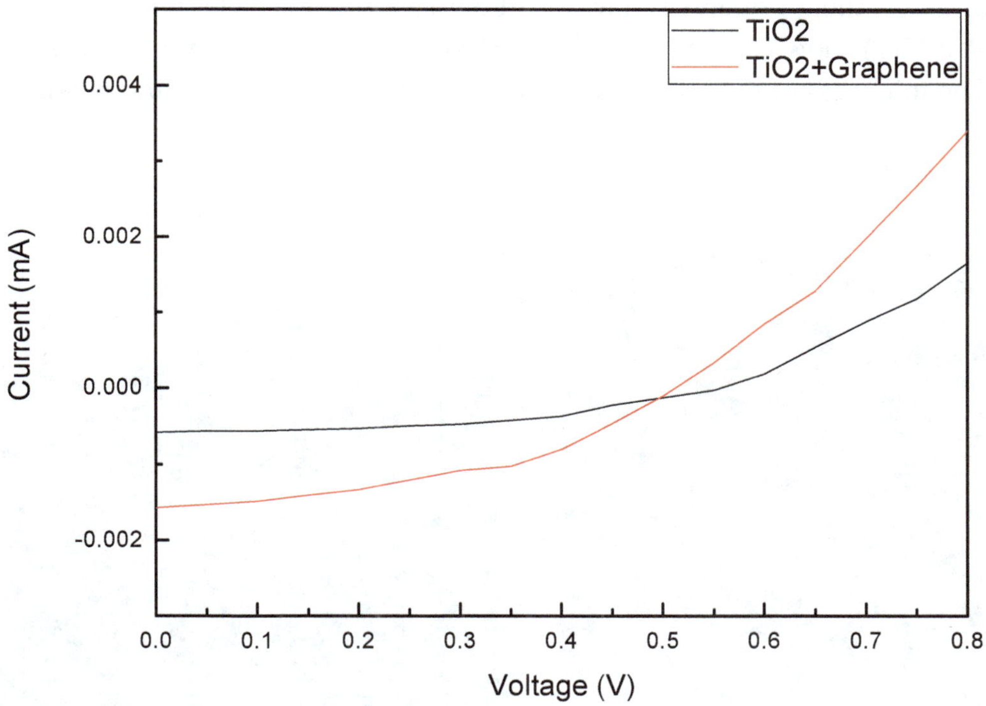

Figure 11. *I-V* curves of the DSSCs with and without the graphene electron transfer layer under illumination [12]

According to Figures 10 and 11 and Table 2, the short-circuit current rises up to 17.5, fill factor to 0.456, and energy conversion efficiency to 3.98 %. The enhanced performance of DSSCs with a graphene was attributed to the increase in electron transport efficiency and light absorption in visible range.

5. DSSCs with TiO_2/Graphene/TiO_2 sandwich structure

Because of the TiO_2/graphene sandwich structure, the efficiency on traditional DSSCs improved. As a result, we use three sandwich structures to achieve the desired outcomes of the following experiment. The enhanced performance of DSSCs with the sandwich structure can be attributed to an increase in electron transport efficiency and in the absorption of light in the visible range. The preparation of TiO_2 photoelectrodes is done by the following: the TiO_2 slurry was prepared by mixing 6 g of nanocrystalline powder, 0.1 mL Triton X-100, and 0.2 mL acetylacetone. The graphene film is deposited on the surface of the first photoelectrode layer, a single TiO_2 photoelectrode layer. This is spin-coated with the rate of rotation of 2,000 rpm, a sandwich structure with three rotational speed to 4,000 rpm, in the present experiment for comparison.

In summary, the DSSC with the sandwich structure in this study exhibited a Voc of 0.6 V, a high Jsc of 11.22 mA cm^{-2}, a fill factor (FF) of 0.58, and a calculated η of 3.93 %, which is 60 % higher than that of a DSSC with the traditional structure.

Figure 12(a) shows the top-view SEM image of the TiO_2 nanoparticles with mean diameter of 50 nm. Figure 12(b) shows the cross-sectional SEM image of a TiO_2/graphene/TiO_2 sandwich structure. The thickness of the graphene electron extraction layer is around 60 nm.

Figure 12. SEM images of (a) TiO2 nanoparticles and (b) TiO_2/graphene/TiO_2 sandwich structure [17]

Figures 13(a) and 13(b) present the Raman scattering spectra of the graphene film that was deposited on the glass substrate using the process that was described in the section on the preparation of graphene. The spectra include important peaks that correspond to the D-band (approximately 1,350 cm^{-1}), the G-band (approximately 1,580 cm^{-1}), and the 2D-band (approximately 2,700 cm^{-1}).

Figure 14 displays the UV-vis spectra of photoelectrodes with different structures before and after they were loaded with dye. Clearly, the photoelectrode with the TiO_2/graphene/TiO_2 sandwich structure has a higher absorption than those with the traditional structure both before and after loading with dye.

Figure 15 presents the energy level diagram of the DSSC with the TiO_2/graphene/TiO_2 sandwich structure. Under illumination, electrons from the photoexcited dye are transported to the conduction band (CB) of TiO_2 via the CB of the graphene and TiO_2. The transportation path via the CB of graphene is in addition to the traditional path. Owing to the excellent electrical conduction of the graphene, the graphene layer bridges behave as a channel for transferring electrons and rapidly transport the photoexcited electrons. The graphene is homogeneous throughout the system, and the excited electrons are captured by the graphene without any obstruction. The collected electrons can be rapidly and effectively transported to the CB of TiO_2 through graphene bridges. In the interface of graphene and TiO_2, the resistance through which charges are transported is reduced relative to the DSSC without graphene electron transport layer, and the recombination and back-reaction processes are suppressed.

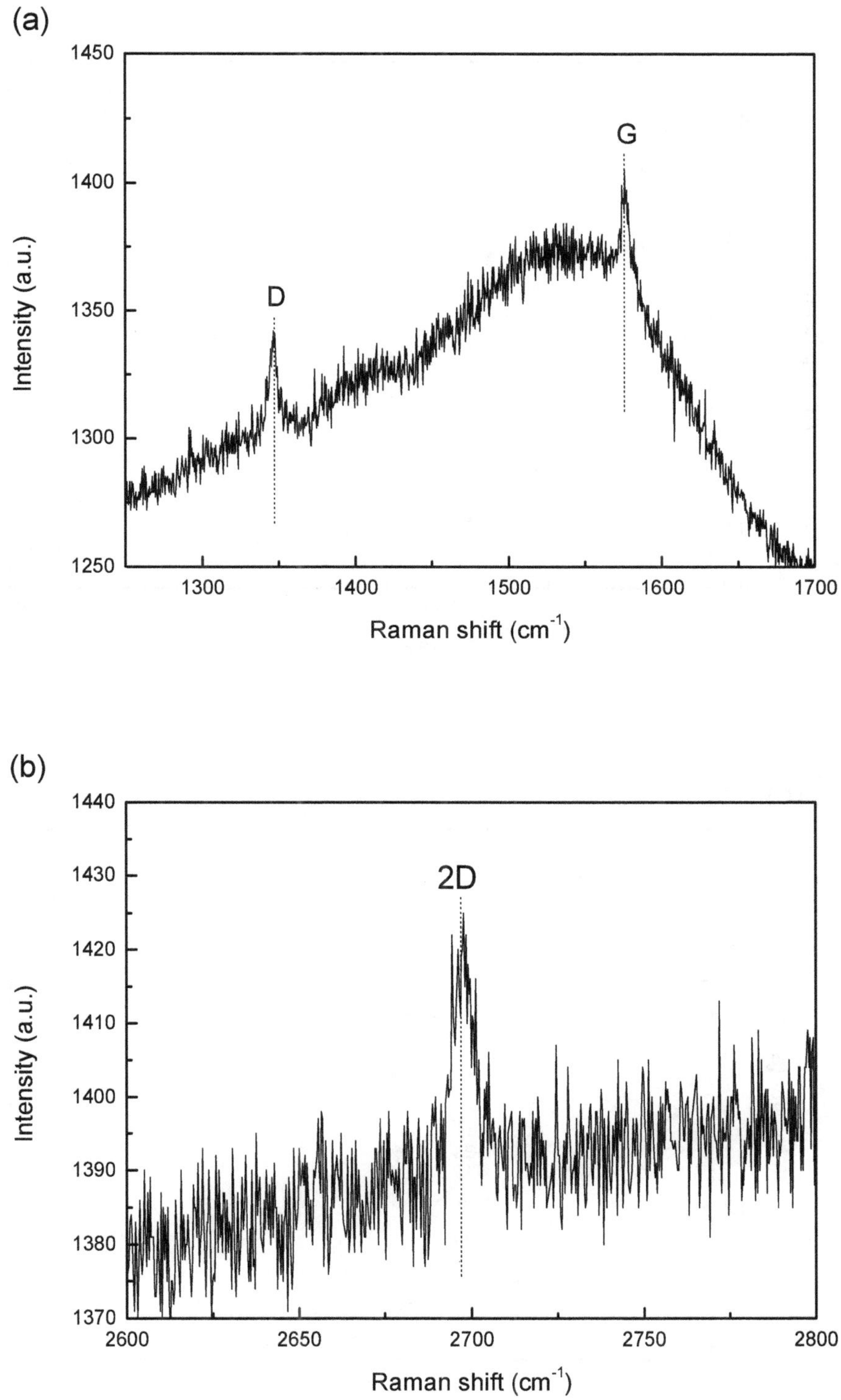

Figure 13. Raman scattering spectra of graphene film deposited on glass substrate. The spectra include important peaks that correspond to the D-band (1,350 cm^{-1}), the G-band (1,580 cm^{-1}), and the 2D-band (2,700 cm^{-1}) [17]

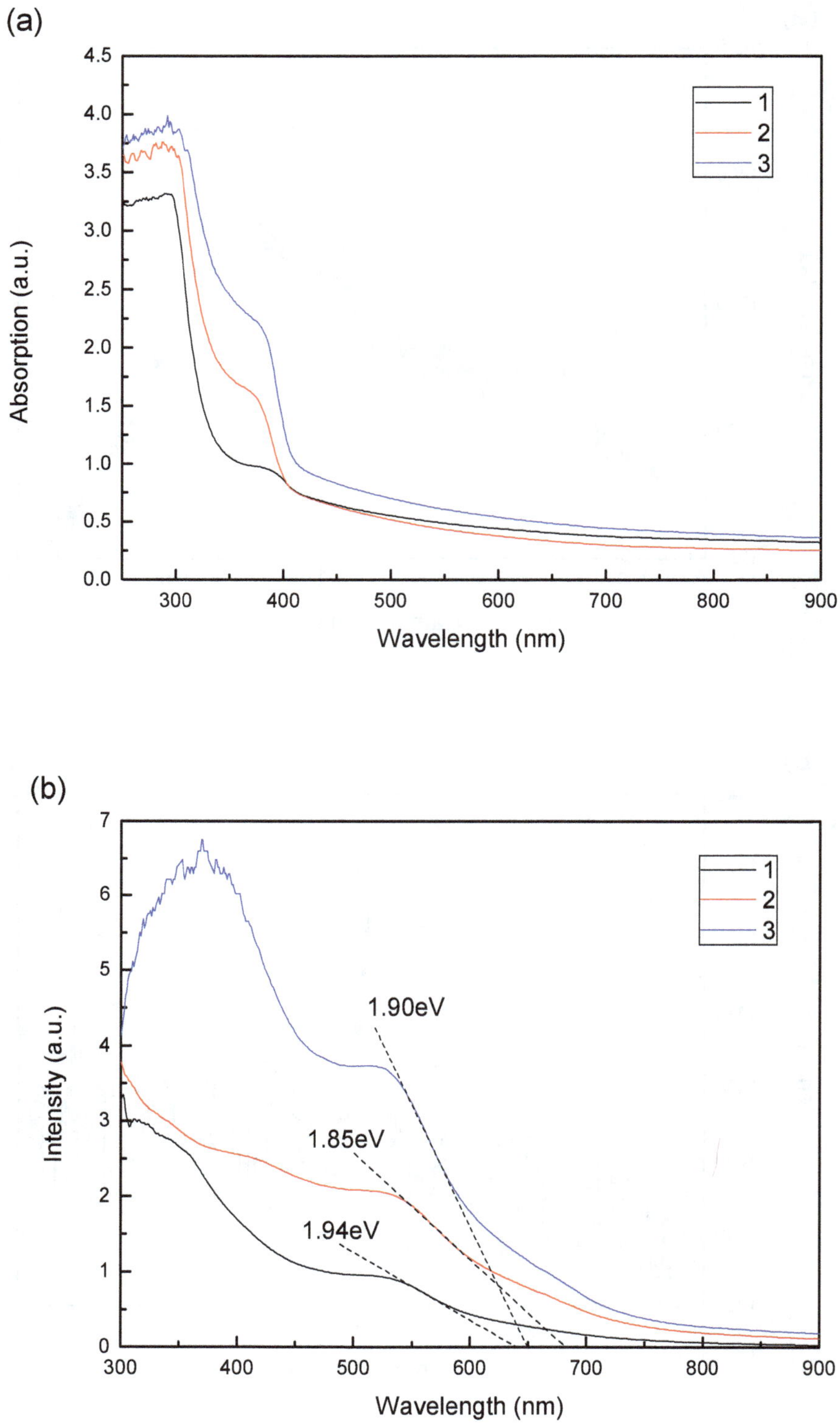

Figure 14. UV-vis absorption spectra of DSSCs with different structures (a) before and (b) after dye loading [17]

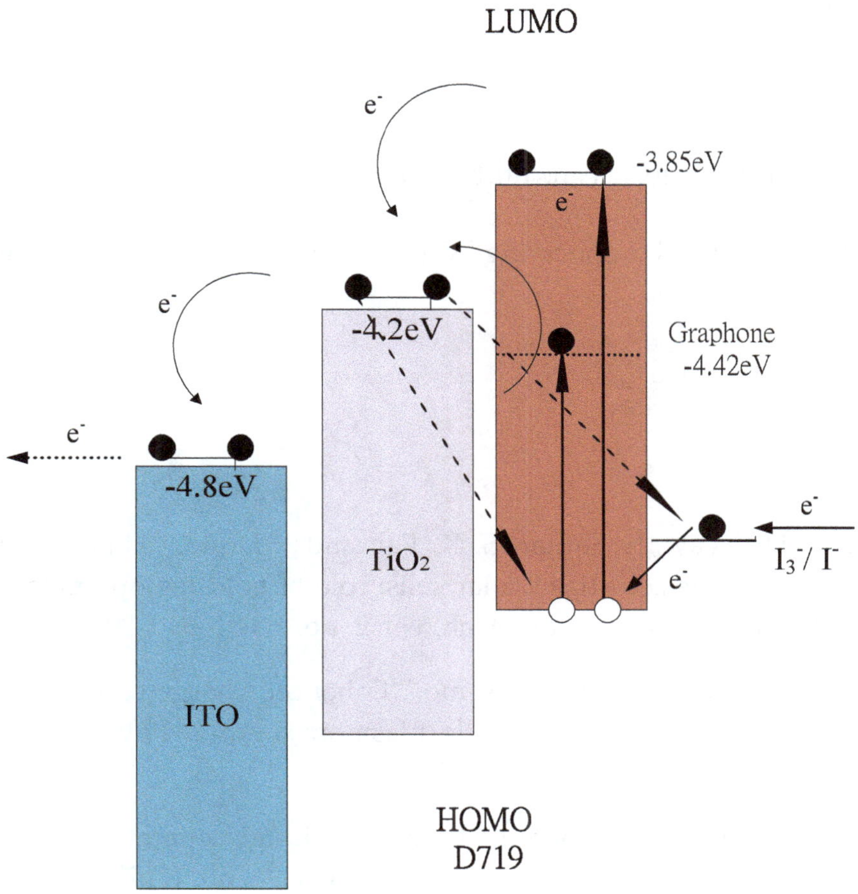

Figure 15. Energy level diagram and mechanism of photocurrent generation in DSSCs with TiO_2 /graphene/TiO_2 sandwich structure [17]

6. Conclusion

To enhance the performance of DSSCs, this work used nanostructure graphene electron transfer layer by plating or sputtering and compared the difference between the DSSC structure with graphene/TiO_2 and with TiO_2/graphene/TiO_2. From the I-V curves, sputtered graphene is much better than plated graphene because the plated graphene has a scattered distribution of ITO. The enhanced performance of DSSCs with a graphene may be attributed to the increase in electron transport efficiency and light absorption in visible range, especially in the range of 310–400 nm. Therefore, the efficiency of conversion of solar energy with graphene+TiO_2 to electricity were increased from 1.45 % to 3.98 %, and the efficiency of conversion of solar energy with TiO_2/graphene/TiO_2 sandwich structure to electricity was increased from 1.38 % to 3.93 %, respectively, under simulated full-sun illumination. This improvement in performance is associated with an increase in the absorption of light, a wide range of absorption wavelengths, shorter charge transportation distances, and the suppression of charge recombination when the graphene is applied.

Author details

Lung-Chien Chen

Address all correspondence to: ocean@ntut.edu.tw

Department of Electro-optical Engineering, National Taipei University of Technology, Taipei, Taiwan

References

[1] T. Bora, H. H. Kyaw, S. Sarkar, S. K. Pal, and J. Dutta, "Highly efficient ZnO/Au Schottky barrier dye-sensitized solar cells: role of gold nanoparticles on the charge-transfer process," Beilstein J Nanotech, vol. 2, no. 1, 681–690, 2011.

[2] C. Hagglund, M. Zach, and B. Kasemo, "Enhanced charge carrier generation in dye sensitized solar cells by nanoparticle plasmons," Appl. Phys. Lett., vol. 92, no. 1, 2008.

[3] S. Barazzouk and S. Hotchandani, "Enhanced charge separation in chlorophyll a solar cell by gold nanoparticles," J. Appl. Phys., vol. 96, no. 12, 7744–7746, 2004.

[4] S. W. Tong, C. F. Zhang, C. Y. Jiang et al., "Improvement in the hole collection of polymer solar cells by utilizing gold nanoparticle buffer layer," Chem. Phys. Lett., vol. 453, no. 1–3, 73–76, 2008.

[5] L. C. Chen, C. C. Wang, and B. S. Tseng, "Enhancement in nano-crystalline TiO2 solar cells sensitized with ZnPc by nanoparticles, " J Optoelectron Biome.,vol.1, no.3, 249–254, 2009.

[6] X. Du, I. Skachko, A. Barker, and E. Y. Andrei, "Approaching ballistic transport in suspended grapheme," Nat Nanotechnol., vol. 3, no. 8, 491–495, 2008.

[7] R. R. Nair, P. Blake, A. N. Grigorenko et al., "Fine structure constant defines visual transparency of graphene," Science, vol. 320, no. 5881, 1308, 2008.

[8] X. Wang, L. Zhi, N. Tsao, Z. Tomovic, J. Li, and K. M. Mullen, "Transparent carbon films as electrodes in organic solar cells," Angew Chem. Int. Edit., vol. 47, no. 16, 2990–2992, 2008.

[9] X. Wang, L. Zhi, and K. Mullen, "Transparent, conductive graphene electrodes for dye-sensitized solar cells," Nano Lett., vol. 8, no. 1, 323–327, 2008.

[10] S. Bae, H. Kim, Y. Lee et al., "Roll-to-roll production of 30-inch graphene films for transparent electrodes," Nat. Nanotechnol, vol. 5, no. 8, 574–578, 2010.

[11] D. W. Chang, H. J. Choi, A. Filer, and J. B. Baek, "Graphene in photovoltaic applications: Organic photovoltaic cells (OPVs) and dye-sensitized solar cells (DSSCs)", J. Mater. Chem. A, vol. 2,12136–12149, 2014.

[12] C. H. Hsu, J. R. Wu, L. C. Chen, P. S. Chan, and C. C. Chen, "Enhanced performance of dye-sensitized solar cells with nanostructure graphene electron transfer layer," Adv. Mater. Sci. Eng, 107352, 2014.

[13] H. Zhang, W. Wang, H. Liu, R. Wang, Y. Chen, and Z. Wang, "Effects of TiO2 film thickness on photovoltaic properties of dye-sensitized solar cell and its enhanced performance by graphene combination," Mater. Res. Bull., vol. 49, 126–131, 2014.

[14] H. N. Kim, H. Yoo, and J. H. Moon, "Graphene-embedded 3D TiO2 inverse opal electrodes for highly efficient dye-sensitized solar cells: Morphological characteristics and photocurrent enhancement," Nanoscale, vol. 5, 4200–4204, 2013.

[15] C. Yang, H. Bi, D. Wan, F. Huang, X. Xie, and M. Jiang, "Direct PECVD growth of vertically erected graphene walls on dielectric substrates as excellent multifunctional electrodes," J. Mater. Chem. A, vol. 1, 770–775, 2013.

[16] B. Tang and G. Hu, "Two kinds of graphene-based composites for photoanode applying in dye-sensitized solar cell," J. Power Source, vol. 220,95–102, 2012.

[17] L. C. Chen, C. H, Hsu, P. S. Chan, X. Zhang, and C. J. Huang, "Improving the performance of dye-sensitized solar cells with TiO2/graphene/TiO2 sandwich structure," Nanoscale Res. Lett., vol. 9, 380, 2014.

[18] M. Koshino and T. Ando, "Electronic structures and optical absorption of multilayer graphenes," Solid State Commun., vol. 149, no. 27-28, 1123–1127, 2009.

[19] G. Giovannetti, P. A. Khomyakov, G. Brocks, V. M. Karpan, J. van den Brink, and P. J. Kelly, "Doping graphene with metal contacts," Phys. Rev. Lett., vol. 101, no. 2, 2008.

[20] R. Czerw, B. Foley, D. Tekleab, A. Rubio, P. M. Ajayan, and D. L. Carroll, "Substrate-interface interactions between carbon nanotubes and the supporting substrate," Phys. Rev. Lett. B, vol. 66, 2002.

[21] A. Kongkanand, R. Martinez-Dominguez, and P. V. Kamat, "Single wall carbon nanotube scaffolds for photoelectron chemical solar cells. Capture and transport of photogenerated electrons," Nano. Lett., vol. 7, no. 3, 676–680, 2007.

[22] F. Xu, J. Chen, and X. Wuetal, "Graphene scaffolds enhanced photogenerated electron transport in ZnO photoanodes for high- efficiency dye-sensitized solar cells," J Phys. Chem. C, vol. 117, 8619–8627, 2013.

[23] A. C. Ferrari, J. C. Meyer, V. Scardaci, C. Casiraghi, M Lazzeri, F. Mauri, S. Piscanec, D. Jiang, K. S. Novoselov, S. Roth, and A. K. Geim, "Raman spectrum of graphene and graphene layers," Phys. Rev. Lett., vol. 97, 187401, 2006.

[24] X. L. Fang, M. Y. Li, K. M. Guo, Y. D. Zhu, Z. Q. Hu, X. L. Liu, B. L. Chen, and X. Z. Zhao, "Photoelectrodes modification by N doping for dye-sensitized solar cells," Electrochim. Acta 65, 174–178, 2012.

[25] X. L. Fang, M. Y. Li, K. M. Guo, X. L. Liu, Y. Zhu, B. Sebo, and X. H. Zhao, "Graphene-compositing optimization of the properties of dye-sensitized solar cells Solar," Energy, vol. 101, 176–181, March 2014.

[26] Y. Y. Wang, Z. H. Ni, T. Yu, Z. X. Shen, H. M. Wang, Y. H. Wu, W. Chen, and A. T. S. Wee, "Raman studies of monolayer graphene," The Substrate Effect J. Phys. Chem. C, vol. 112, no. 29, 10637–10640, 2008.

[27] J. N. Zheng, S. S. Li, F. Y. Chen, N. Bao, A. J. Wang, J. R. Hen, and J. J. Feng, "Facile synthesis of platinum–ruthenium nanodendrites supported on reduced graphene oxide with enhanced electrocatalytic properties," J Power Sources, vol. 266, 259–267, 2014.

[28] AZoM, "Characterization of graphene using Raman spectroscopy," http://www.azom.com/article.aspx?ArticleID=6271.

[29] M. Huang, H. Yan, T. F. Heinz, and J. Hone, "Probing strain-induced electronic structure change in graphene by Raman spectroscopy," Nano Lett., vol. 10, 4074–4079, 2010.

[30] O. Frank, M. Mohr, J. Maultzsch, C. Thomsen, I. Riaz, R. Jalil et al. "Raman 2D-band splitting in graphene," Theory Expt ACS Nano., vol. 5, no. 3, 2231–2239, 2011.

[31] D. Yoon, Y. W. Son, and H. Cheong, "Strain-dependent splitting of the double-resonance Raman scattering band in graphene," Phys. Rev. Lett., vol. 106, no. 15, 155502-1–4, 2011.

10

All-Optical Waveguide-Type Switch using Saturable Absorption in Graphene

Misaki Takahashi, Hiroki Kishikawa,
Nobuo Goto and Shin-ichiro Yanagiya

Abstract

An all-optical switch using saturable absorption that accompanies refractive-index change such as in graphene films is proposed. The switching conditions are theoretically derived for arbitrary values of absorption and refractive-index change. It is found that switching can be performed by weaker control light when the refractive-index change is accompanied in saturable absorption. Switching conditions for various combinations of absorption and refractive-index change are confirmed by FD-BPM simulation. As an experimental demonstration of saturable absorption in graphene, we discuss measured nonlinear absorption in a graphene-loaded waveguide. The saturable absorption and refractive-index change through vertically placed multilayered graphene sheets are also discussed based on reported experimental results.

Keywords: Graphene, integrated optics, optical switch, saturable absorption, nonlinear refractive-index change

1. Introduction

Fast-response optical switches are one of the key devices in photonic routers and in optical signal processors. The switching response depends on the control mechanism of the switch. All-optical switching using optical nonlinear effects has a potential for picosecond-order response. Optically controlled switches include devices using phase shift induced by optical nonlinear Kerr effect [1] and phase shift in semiconductor optical amplifiers (SOAs) [2] and in quantum dots [3].

We have proposed an all-optical switch controlled by Raman amplification [4,5] or by saturable absorption [6-8], where optical signal amplitude is controlled instead of the signal phase. In the previously proposed switches, refractive-index change induced by all optical nonlinear interactions was assumed to be small enough to be negligible [4,6] or to be adjusted at an integer multiple of 2π [7]. When we assume graphene sheets as the saturable absorbing material, the induced refractive-index change cannot be ignored [9]. In this book chapter, we theoretically analyze the proposed switch to derive switching conditions in general when both the absorption and the index change are induced by nonlinear interaction [8]. It is found that, with arbitrary combination of absorption and induced phase shift, optical switching can be performed by employing appropriate fixed attenuator and fixed phase-shifter in the switching device. We confirm the switching performance by finite-difference beam-propagation-method (FD-BPM) simulation. As an experimental demonstration of saturable absorption in graphene, we also discuss measured nonlinear absorption in a graphene-loaded waveguide [7]. Nonlinear refractive-index change in graphene is also discussed to evaluate optical nonlinear phase change through vertically placed multilayer graphene based on the reported experimental results [9].

2. Graphene for optical processing

Graphene is a two-dimensional (2D) monolayer of sp^2 -bonded carbon atoms and has a dense honeycomb crystal structure. This carbon material has been attracting increasing interest, in particular, since the Nobel Prize was awarded to Novoselov and Geim for their significant contribution in graphene researches [10]. The application areas of graphene are shown in Fig. 1 [11]. Unique characteristics of graphene are driven by its linear massless band structure. In electronic devices, electrons in graphene move at ultrafast speed, and various applications have been investigated. In the field of optics, a monolayer graphene shows large broadband optical absorption of 2.3%, and excellent saturable absorption phenomena having ultrafast carrier dynamics. The saturable absorption has been applied to mode-locked fiber lasers [12-14].

In our proposed optical switch, the saturable absorption characteristics are employed to control optical signal light. In such optical waveguide devices, two schemes, loaded type and vertically inserted type, to introduce graphene in the optical waveguide can be considered as shown in Fig. 2, where graphene is introduced in one of the two arms of a Mach-Zehnder interferometer (MZI) as an example. The former is easy to fabricate in integrated-optic device structure. On the other hand, the latter is easy to evaluate the effect of graphene, whereas optical scattering at the inserted plane of the waveguide may induce insertion loss. We consider all-optical control of signal light by intense control light through saturable absorption.

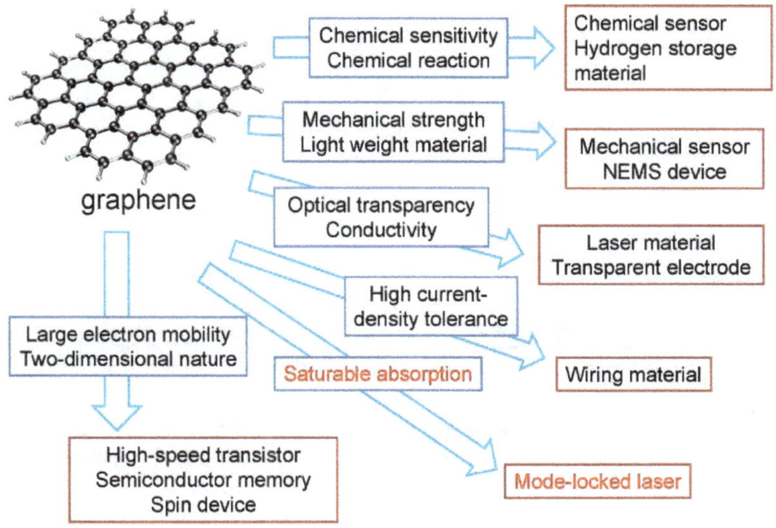

Figure 1. Application area of graphene.

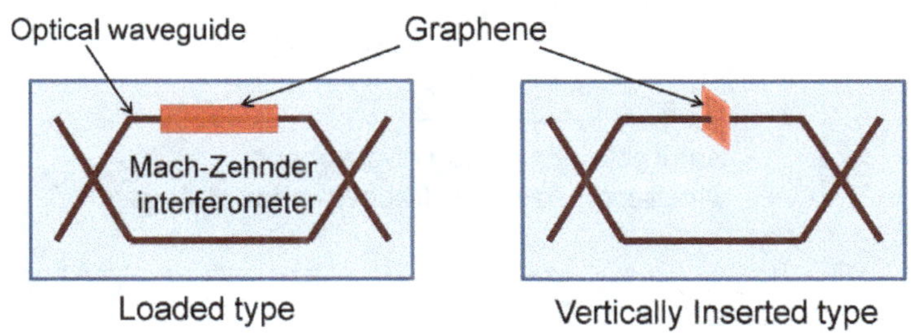

Figure 2. Introduction of graphene in optical waveguide.

We consider a model for controlling the transmittance through a graphene-introduced waveguide. The signal light with electrical-field amplitude $E_{s,in}$ attenuates to be $E_{s,out}$ through the vertically inserted graphene sheets or the graphene-loaded waveguide as shown in Fig. 3(a). The intense control light with electrical-field amplitude $E_{c,in}$ also attenuates to be $E_{c,out}$. The attenuation of the signal light can be decreased by the intense control light due to saturable absorption of graphene. This control of absorption can be equivalently represented by a cascade connection of a fixed attenuator with amplitude attenuation coefficient t_{low} and an amplifier controlled by the control light whose amplitude amplification coefficient is α as shown in Fig. 3(b). The amplification corresponds to decrease of the absorption due to the saturable absorption induced by the intense control light. An induced nonlinear phase shift through saturable absorption is denoted by ϕ. We denote the coefficient including amplification and phase shift as $\alpha' = \alpha \exp(j\ \phi)$. These coefficients are defined by

$$t_{low} = \frac{E_{s,out}(E_{c,in} = 0)}{E_{s,in}}, \tag{1}$$

and

$$\alpha' = \alpha e^{j\phi} = \frac{E_{\mathrm{s,out}}(E_{\mathrm{c,in}} \neq 0)}{E_{\mathrm{s,out}}(E_{\mathrm{c,in}} = 0)}. \tag{2}$$

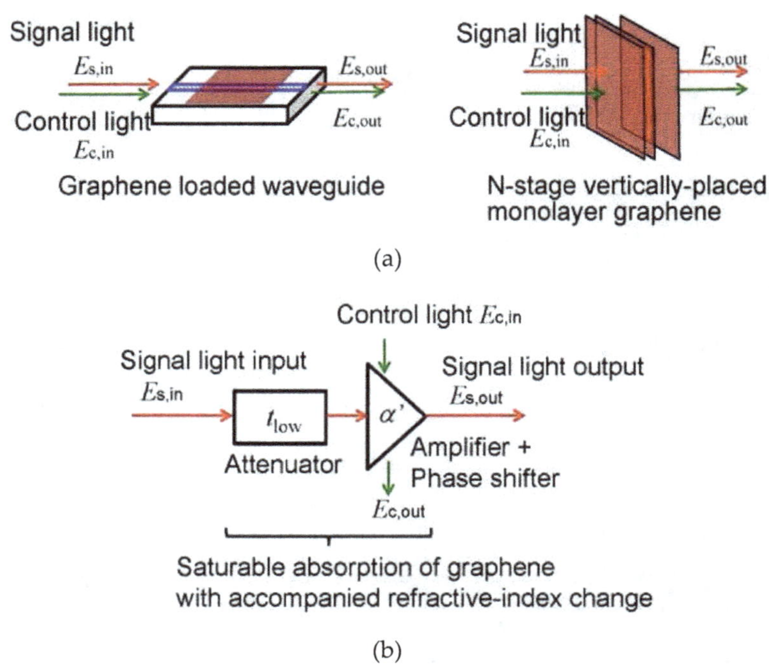

Figure 3. Modeling of light control by graphene; (a) signal and control lights in graphene, and (b) a model with fixed attenuator and variable amplifier and phase-shifter controlled by control light.

3. Optical switch configuration

The proposed switch consists of two cascaded MZIs connected with asymmetric X-junction couplers, as shown in Fig. 4 [8]. We consider two types of configuration, Type A and B, depending on whether a fixed attenuator is employed in the upper arm of the second MZI or in the lower one, respectively. It is noted that a conventional switch using phase control can be formed with a single MZI. Optical signals in the first MZI are controlled by vertically inserted or overlaid graphene. Control light at a different wavelength or in an orthogonally different polarization from the signal light is coupled in both arms of the first MZI. By inserting the control light, the absorption in the graphene film is reduced due to saturable absorption, which is accompanied by the refractive-index change in the graphene film. Thus, the attenuation and the phase of the signal light along the first MZI arm are controlled by the control light. The transmittance for optical field amplitude in each arm is given by $t_{\mathrm{low}}\ \alpha_i \exp(j\phi_i) = t_{\mathrm{low}}\ \alpha_i'$, i=A and B. In either arm of the second MZI, a fixed attenuator with a fixed phase-shifter has the transmittance given by $\beta_{\mathrm{fix}} = |\beta_{\mathrm{fix}}| \exp(j\ \arg(\beta_{\mathrm{fix}}))$.

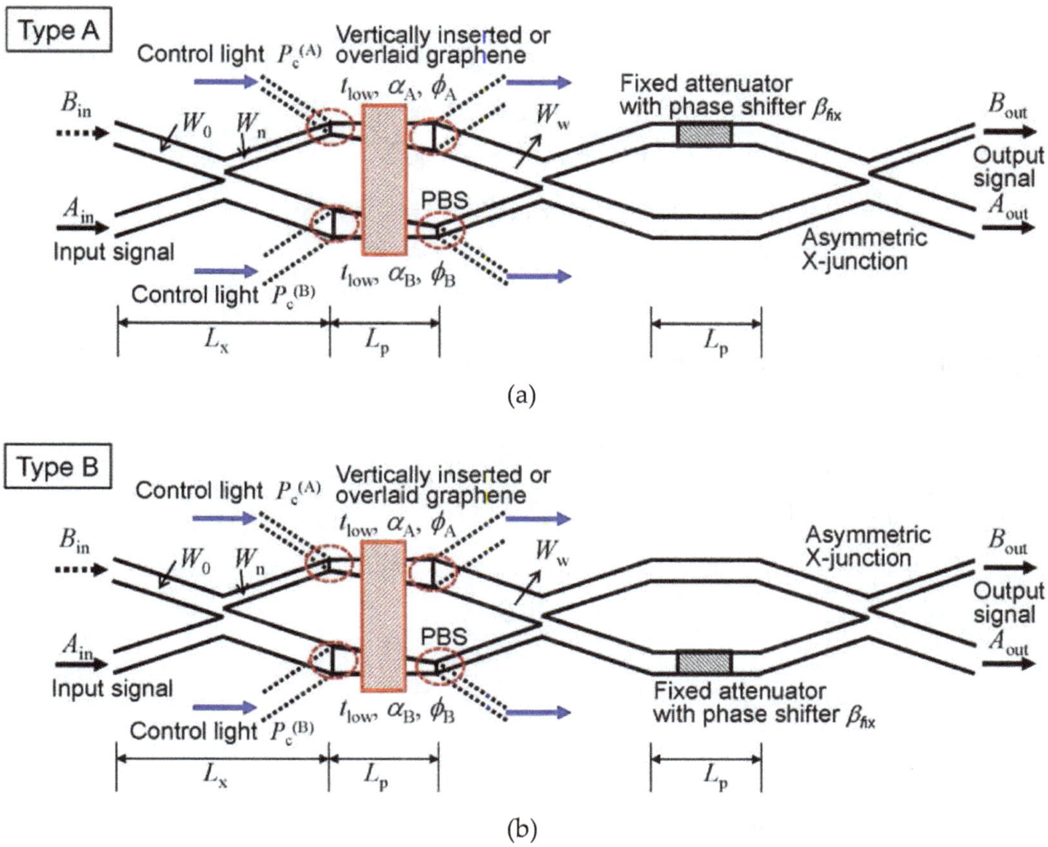

Figure 4. Switch structure; (a) Type A and (b) Type B.

Switching response time is mainly limited by the response of the saturable absorption in graphene. It is reported that, from pump-probe experiments, the carrier relaxation time and carrier–carrier intraband scattering time-constant are less than 1 ps [13]. A response time in a 20-μm-long graphene-coated microfiber was also measured to be 2.2 ps [15]. Since the switching is kept as on-state during the control signal existence, optical control pulse with rise and fall time less than 1 ps can result in picosecond switching.

4. Switching conditions

The input-to-output relation for optical field amplitudes in case of Type A shown in Fig. 4(a) is given by

$$\begin{aligned}\begin{bmatrix} A_{out} \\ B_{out} \end{bmatrix} &= \left(\frac{1}{\sqrt{2}}\right)^3 \begin{bmatrix} 1 & 1 \\ -1 & 1 \end{bmatrix} \begin{bmatrix} 1 & 0 \\ 0 & \beta_{fix} \end{bmatrix} \begin{bmatrix} 1 & 1 \\ -1 & 1 \end{bmatrix} \begin{bmatrix} t_{low}\alpha_B' & 0 \\ 0 & t_{low}\alpha_A' \end{bmatrix} \begin{bmatrix} 1 & 1 \\ -1 & 1 \end{bmatrix} \begin{bmatrix} A_{in} \\ B_{in} \end{bmatrix} \\ &= \frac{t_{low}}{2\sqrt{2}} \begin{bmatrix} a_{11} & a_{12} \\ a_{21} & a_{22} \end{bmatrix} \begin{bmatrix} A_{in} \\ B_{in} \end{bmatrix}\end{aligned} \tag{3}$$

where

$$\begin{cases} a_{11} = \alpha_A^{'} - \alpha_B^{'} - \beta_{\text{fix}}(\alpha_A^{'} + \alpha_B^{'}) \\ a_{12} = \alpha_A^{'} + \alpha_B^{'} + \beta_{\text{fix}}(-\alpha_A^{'} + \alpha_B^{'}) \\ a_{21} = -\alpha_A^{'} + \alpha_B^{'} - \beta_{\text{fix}}(\alpha_A^{'} + \alpha_B^{'}) \\ a_{22} = -\alpha_A^{'} - \alpha_B^{'} + \beta_{\text{fix}}(-\alpha_A^{'} + \alpha_B^{'}) \end{cases}. \tag{4}$$

Similar equations for input-to-output relation in Type B shown in Fig. 4(b) are derived as

$$\begin{aligned} \begin{bmatrix} A_{\text{out}} \\ B_{\text{out}} \end{bmatrix} &= \left(\frac{1}{\sqrt{2}}\right)^3 \begin{bmatrix} 1 & 1 \\ -1 & 1 \end{bmatrix} \begin{bmatrix} \beta_{\text{fix}} & 0 \\ 0 & 1 \end{bmatrix} \begin{bmatrix} 1 & 1 \\ -1 & 1 \end{bmatrix} \begin{bmatrix} t_{\text{low}}\alpha_B' & 0 \\ 0 & t_{\text{low}}\alpha_A' \end{bmatrix} \begin{bmatrix} 1 & 1 \\ -1 & 1 \end{bmatrix} \begin{bmatrix} A_{\text{in}} \\ B_{\text{in}} \end{bmatrix} \\ &= \frac{t_{\text{low}}}{2\sqrt{2}} \begin{bmatrix} a_{11} & a_{12} \\ a_{21} & a_{22} \end{bmatrix} \begin{bmatrix} A_{\text{in}} \\ B_{\text{in}} \end{bmatrix} \end{aligned} \tag{5}$$

where

$$\begin{cases} a_{11} = -\alpha_A^{'} - \alpha_B^{'} + \beta_{\text{fix}}(\alpha_A^{'} - \alpha_B^{'}) \\ a_{12} = -\alpha_A^{'} + \alpha_B^{'} + \beta_{\text{fix}}(\alpha_A^{'} + \alpha_B^{'}) \\ a_{21} = -\alpha_A^{'} - \alpha_B^{'} - \beta_{\text{fix}}(\alpha_A^{'} - \alpha_B^{'}) \\ a_{22} = -\alpha_A^{'} + \alpha_B^{'} - \beta_{\text{fix}}(\alpha_A^{'} + \alpha_B^{'}) \end{cases}. \tag{6}$$

We consider the case that an optical signal is incident only at the lower port, that is, $A_{\text{in}}=E_{\text{in}}$ and $B_{\text{in}}=0$. Then, eqs. (3) and (5) are simplified to be

$$\begin{bmatrix} A_{\text{out}} \\ B_{\text{out}} \end{bmatrix} = \frac{t_{\text{low}}E_{\text{in}}}{2\sqrt{2}} \begin{bmatrix} a_{11} \\ a_{21} \end{bmatrix}. \tag{7}$$

We now derive the conditions required for complete switching. We assume that switching is operated by feeding a control light with power $P_c^{(A)}$ in the upper arm of the first MZI to switch to the lower output port as A_{out}, and by feeding a control light $P_c^{(B)}$ in the lower arm to switch to the upper output port as B_{out}. Therefore, the parameters $\alpha_A^{'}$ and $\alpha_B^{'}$ due to saturable absorption are to be set as $\alpha_A^{'}=1$ to switch to the upper output port as B_{out} and $\alpha_B^{'}=1$ to switch to the lower output port as A_{out} because no control light is fed in these arms.

Switching conditions for β_{fix} and the output field B_{out} in the case of Type A are obtained from eqs. (4) and (7) as follows:

1. Switching to B_{out}:

$$\beta_{\text{fix}} = \frac{\alpha_B' - 1}{\alpha_B' + 1} \text{ and } B_{\text{out}} = \frac{t_{\text{low}} E_{\text{in}} (1 - \alpha_B')}{\sqrt{2}}. \tag{8}$$

2. Switching to A_{out}:

$$\beta_{\text{fix}} = \frac{\alpha_A' - 1}{\alpha_A' + 1} \text{ and } A_{\text{out}} = \frac{t_{\text{low}} E_{\text{in}} (1 - \alpha_A')}{\sqrt{2}}. \tag{9}$$

It is found from these conditions that, when α_A' for switching to A_{out} is equal to α_B' for switching to B_{out}, switching can be operated with a fixed β_{fix}.

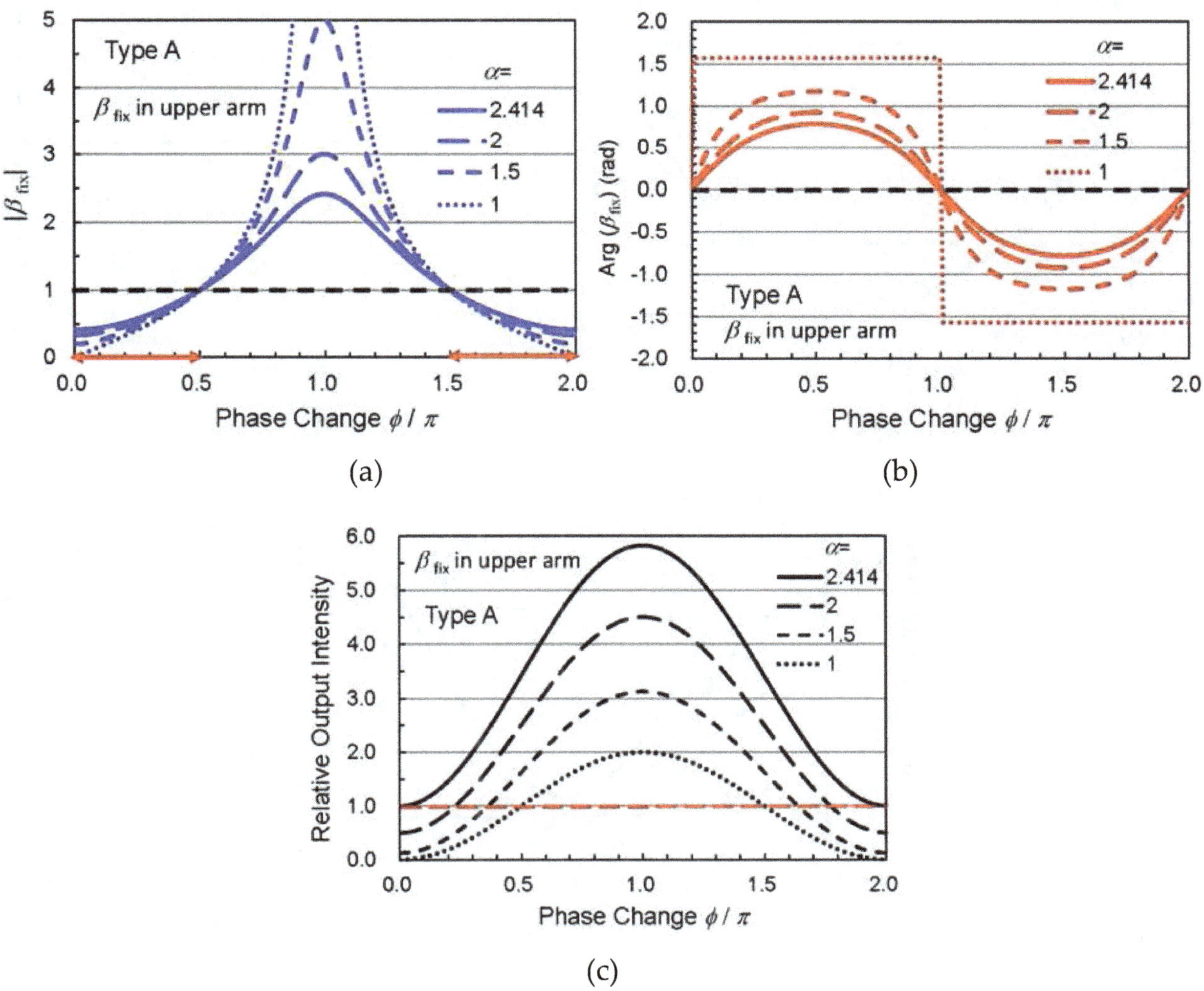

Figure 5. Switching conditions in switch Type A; (a) the fixed attenuator and (b) phase-shifter in the second MZI required for switching and (c) the output intensity.

Similarly, switching conditions in the case of Type B are obtained from eqs. (6) and (7) as follows:

1. Switching to B_{out} :

$$\beta_{\text{fix}} = \frac{\alpha_B^{'} + 1}{\alpha_B^{'} - 1} \text{ and } B_{\text{out}} = -\frac{t_{\text{low}} E_{\text{in}} (1 + \alpha_B^{'})}{\sqrt{2}}. \tag{10}$$

2. Switching to A_{out} :

$$\beta_{\text{fix}} = \frac{\alpha_A^{'} + 1}{\alpha_A^{'} - 1} \text{ and } A_{\text{out}} = \frac{t_{\text{low}} E_{\text{in}} (1 + \alpha_A^{'})}{\sqrt{2}}. \tag{11}$$

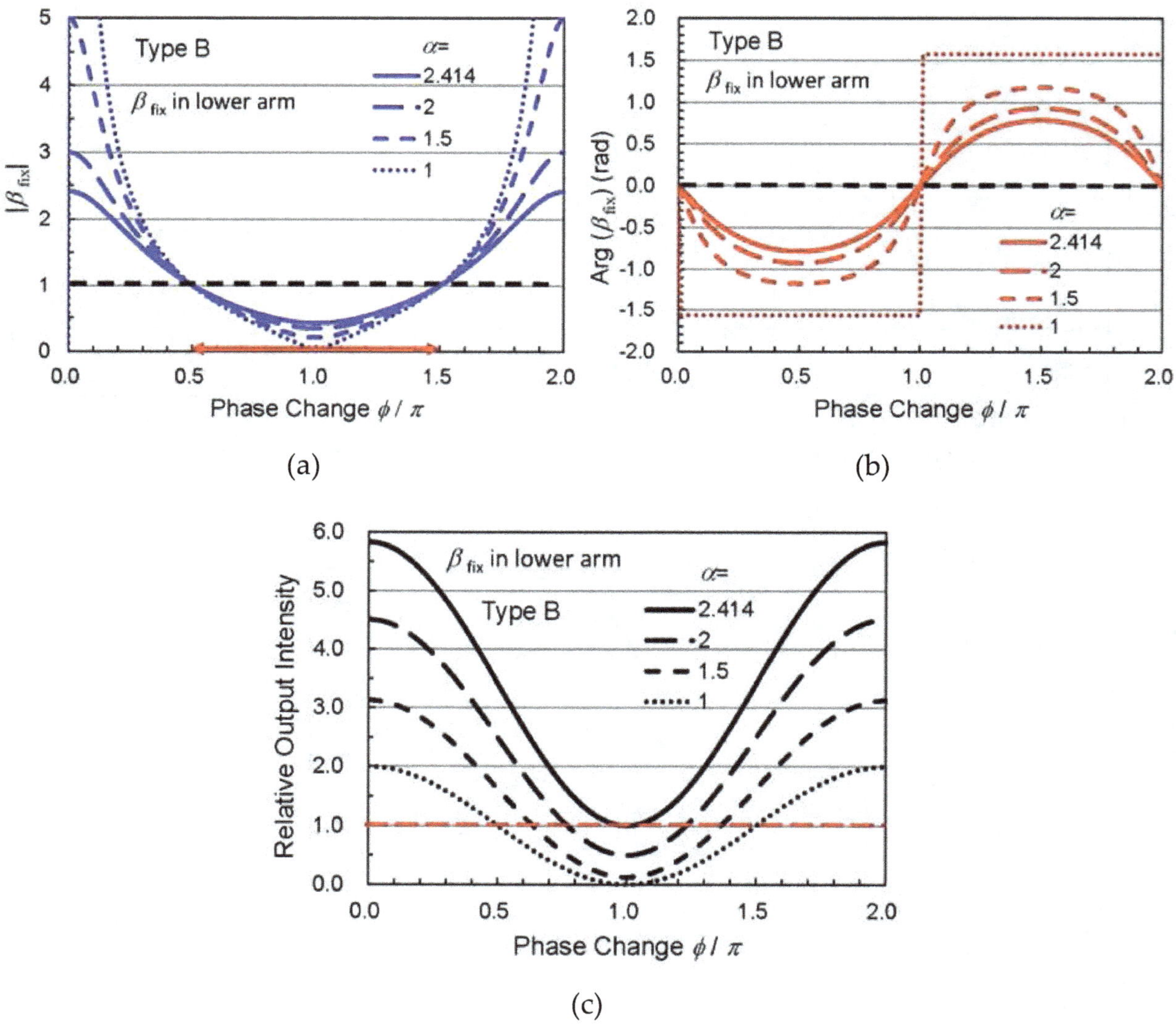

Figure 6. Switching conditions in switch Type B; (a) the fixed attenuator and (b) phase-shifter in the second MZI required for switching and (c) the output intensity.

From these equations, switching conditions required for $\beta_{\rm fix}$ and the relative output intensities $|A_{\rm out}/(t_{\rm low}E_{\rm in})|^2$ and $|B_{\rm out}/(t_{\rm low}E_{\rm in})|^2$ are calculated as a function of phase change $\phi_{\rm A}$ or $\phi_{\rm B}$ as shown in Figs. 5 and 6 for Type A and B, respectively, where $\alpha_i=1+\sqrt{2}$, 2.0, 1.5, and 1.0 are assumed as a parameter. Since $|\beta_{\rm fix}|\le 1$ can be realized by an attenuator, Type A should be selected for $|\phi_i|\le\pi/2$ and $3\pi/2\le|\phi_i|\le 2\pi$ and Type B for $\pi/2\le|\phi_i|\le 3\pi/2$, respectively, in the range of $0\le|\phi_i|\le 2\pi$. An arbitrary combination of α_i and ϕ_i can be used to operate switching by adjusting $|\beta_{\rm fix}|$ and arg($\beta_{\rm fix}$).

We consider three cases as examples of switching. As a first case, case (1), we assume that no phase shift is accompanied in saturable absorption, that is, $\phi_A=\phi_B=0$, and no phase shift is set in $\beta_{\rm fix}$. This corresponds to the case discussed in ref. [7]. The switching conditions in Type A are derived from eqs. (8) and (9) as follows:

1. Switching to $B_{\rm out}$ $(\alpha_A'=1,\ \alpha_B'=1+\sqrt{2})$:

$$\beta_{\rm fix}=\frac{1}{1+\sqrt{2}}\left(\rightarrow \arg\left(\beta_{\rm fix}\right)=0\right)\text{ and }B_{\rm out}=-t_{\rm low}E_{\rm in}. \tag{12}$$

2. Switching to $A_{\rm out}$ $(\alpha_A'=1+\sqrt{2},\ \alpha_B'=1)$:

$$\beta_{\rm fix}=\frac{1}{1+\sqrt{2}}\left(\rightarrow \arg\left(\beta_{\rm fix}\right)=0\right)\text{ and }A_{\rm out}=-t_{\rm low}E_{\rm in}. \tag{13}$$

This case corresponds to $\alpha_i=1+\sqrt{2}\cong 2.414$ and $\phi_i=0$ in Fig. 5(a). The operating point for $\beta_{\rm fix}$ and the output intensity are indicated by red circles on green curves in Fig. 7(a).

As a next case, case (2a), we assume the same $\alpha_i=1+\sqrt{2}$ as case (1), whereas a phase shift $\phi_i=\pi/2$ is accompanied. The switching conditions in Type A are derived from eqs. (8) and (9) as follows:

1. Switching to $B_{\rm out}$ $(\alpha_A'=1,\ \alpha_B=1+\sqrt{2},\ \phi_B=\pi/2)$:

$$\beta_{\rm fix}=\frac{j\left(1+\sqrt{2}\right)-1}{j\left(1+\sqrt{2}\right)+1}\left(\rightarrow\left|\beta_{\rm fix}\right|=1,\arg\left(\beta_{\rm fix}\right)=\frac{\pi}{4}\right)\text{ and }B_{\rm out}=\frac{t_{\rm low}E_{\rm in}[1-j\left(1+\sqrt{2}\right)]}{\sqrt{2}}. \tag{14}$$

2. Switching to $A_{\rm out}$ $(\alpha_A=1+\sqrt{2},\ \phi_A=\pi/2,\ \alpha_B'=1)$:

$$\beta_{\rm fix}=\frac{j\left(1+\sqrt{2}\right)-1}{j\left(1+\sqrt{2}\right)+1}\left(\rightarrow\left|\beta_{\rm fix}\right|=1,\arg\left(\beta_{\rm fix}\right)=\frac{\pi}{4}\right)\text{ and }A_{\rm out}=\frac{t_{\rm low}E_{\rm in}[1-j\left(1+\sqrt{2}\right)]}{\sqrt{2}}. \tag{15}$$

It is noted that, since $\{[1-j(1+\sqrt{2})]/\sqrt{2}\}^2=2+\sqrt{2}$, the output intensity is increased from 1.0 to 3.41 compared with case(1). Thus, when $\phi_i=\pi/2$ is accompanied in saturable absorption, the excess loss through the switch is decreased by $10\log_{10}(2+\sqrt{2})=5.33$ (dB). The operating point for β_{fix} and the output intensity are indicated in Fig. 7(b).

When ϕ_i increases further to $\pi/2 \le \phi_i \le 3\pi/2$, switching in type A cannot be realized with a fixed attenuator because $|\beta_{\text{fix}}| \ge 1$, and the switch structure of Type B has to be used to realize with a fixed attenuator. In this case, $|\beta_{\text{fix}}|$, $\arg(\beta_{\text{fix}})$ and the normalized output are found in Fig. 6. We consider a case of $\phi_i=\pi$ as case (2b). The switching conditions in Type B are derived from eqs. (10) and (11) as follows:

1. Switching to B_{out} $(\alpha_A'=1,\ \alpha_B=1+\sqrt{2},\ \phi_B=\pi)$:

$$\beta_{\text{fix}} = \frac{1}{1+\sqrt{2}} \left(\rightarrow \arg(\beta_{\text{fix}})=0\right) \text{ and } B_{\text{out}} = t_{\text{low}} E_{\text{in}} . \tag{16}$$

2. Switching to A_{out} $(\alpha_A=1+\sqrt{2},\ \phi_A=\pi,\ \alpha_B'=1)$:

$$\beta_{\text{fix}} = \frac{1}{1+\sqrt{2}} \left(\rightarrow \arg(\beta_{\text{fix}})=0\right) \text{ and } A_{\text{out}} = t_{\text{low}} E_{\text{in}} . \tag{17}$$

The same switching conditions are found for $\phi_i=(2n-1)\pi,\ n=1, 2, \ldots$. It is noted that, for $\phi_i=2n\pi$, the same switching conditions as for $\phi_i=0$ in Type A are applied. The operating point for β_{fix} and the output intensity are indicated in Fig. 7(c).

We consider improvement of control efficiency by taking the phase shift ϕ_i and $\arg(\beta_{\text{fix}})$ into account. Although $\alpha_i=1+\sqrt{2}$ is required with a fixed β_{fix} with $\arg(\beta_{\text{fix}})=0$ in the second MZI, switching can be performed even with $\alpha_i<1+\sqrt{2}$ as shown in Figs. 5 and 6. We consider a case of $\phi_i=0$. By decreasing α_i from $1+\sqrt{2}$ to 2.0, the required β_{fix} decreases to $1/3$, resulting in decrease of the normalized output from 1 to 1/2 as shown in Fig. 5. Thus, by decreasing α_i from $1+\sqrt{2}$ to 2.0, insertion loss of the switch increases by 3 dB. Next, we consider a case of $\phi_i=\pi/2$ and $\alpha_i=2.0$ as case(3). The switching conditions in Type A are derived from eqs. (8) and (9) as follows:

1. Switching to B_{out} $(\alpha_A'=1,\ \alpha_B=2,\ \phi_B=\pi/2)$:

$$\beta_{\text{fix}} = \frac{3+j4}{5} \left(\rightarrow |\beta_{\text{fix}}|=1,\ \arg(\beta_{\text{fix}})=0.927\right) \text{ and } B_{\text{out}} = \frac{t_{\text{low}} E_{\text{in}}(1-j2)}{\sqrt{2}} . \tag{18}$$

2. Switching to A_{out} $(\alpha_A=2,\ \phi_A=\pi/2,\ \alpha_B'=1)$:

$$\beta_{\text{fix}} = \frac{3+j4}{5} \left(\rightarrow |\beta_{\text{fix}}|=1, \arg(\beta_{\text{fix}})=0.927\right) \text{ and } A_{\text{out}} = \frac{t_{\text{low}} E_{\text{in}}(1-j2)}{\sqrt{2}} . \tag{19}$$

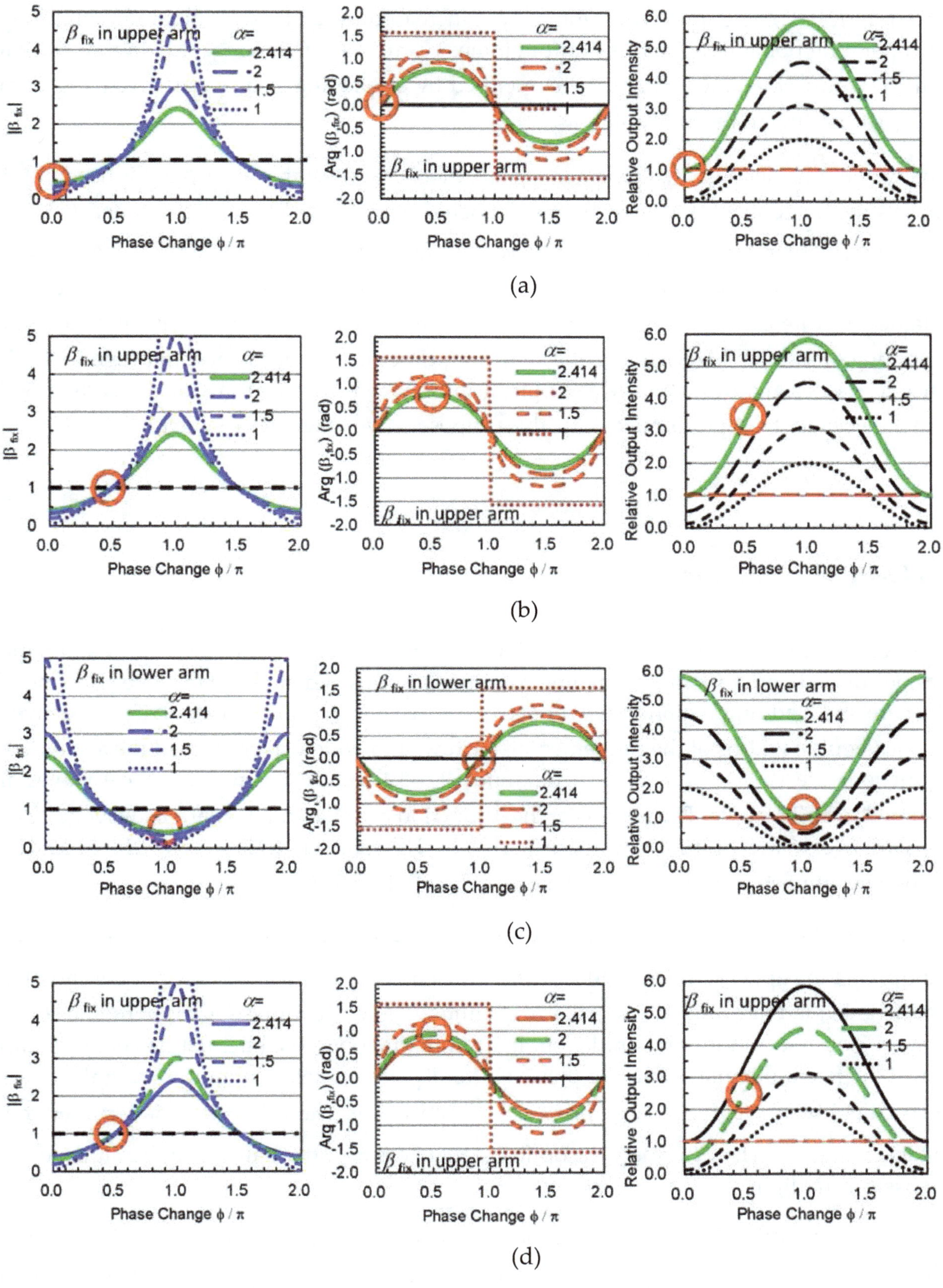

Figure 7. Operation points in the switching conditions for four kinds of switches: (a) case(1), (b) case(2a), (c) case(2b), and (d) case(3).

Since $[(1-j2)/\sqrt{2}]^2$=2.5, the output intensity is increased by 2.5 times compared with case(1). The operating point for β_{fix} and the output intensity are indicated in Fig. 7(d). In the case with a further reduced α_i, switching can be operated by employing appropriate fixed $|\beta_{\text{fix}}|$ and $\arg(\beta_{\text{fix}})$ according to the phase shift ϕ_i.

Finally, we consider a limited case with α_i=1.0, that is, no saturable absorption is used but the phase shift ϕ_i alone is used in switching. When ϕ_i is increased to $\pi/2$, β_{fix} is derived to be $\beta_{\text{fix}}=(-1+j)/(1+j)=e^{j\pi/2}$, that is, a $\pi/2$ phase-shifter is used as β_{fix}. The normalized output intensity is 1.0. When ϕ_i is further increased to π, $|\beta_{\text{fix}}|$ reduced to 0 in Type B. No output is obtained in this case. This switching condition of $\phi_i=\pi$ is used in a conventional switch consisting of a single MZI. That is, the switching is performed at the output of the first MZI. Since β_{fix} is designed to equalize the optical intensity in the two arms of the second MZI, switching cannot be performed with the proposed two-stage MZI switch, for α_i=1.0 and $\phi_i=\pi$.

5. FD-BPM simulation

In order to confirm the switching operation with the proposed switch, FD-BPM simulation was performed under various switching conditions. We consider a 2D slab waveguide model. The core and cladding regions have refractive indices of n_c = 1.461 and n_s = 1.45, respectively. The waveguide widths of the fundamental, narrow, and wide waveguides of the asymmetric X-junctions are W_1=3.0 μm, W_n=2.6 μm, and W_w=3.4 μm, respectively. Optical waves are assumed to be TE mode. The lengths of the X-junction coupler and the parallel waveguides in arms are L_x =16 mm and L_p =1 mm, respectively. The total length is 50 mm. The distance of the two input ports is 23 μm. The optical wavelength is 1550 nm.

In the simulation, α_i and ϕ_i caused by saturable absorption are equivalently simulated just by multiplying the optical electric field by the coefficient $\alpha_i e^{j\phi_i}$ at a plane located at the end of the graphene-loaded or graphene-inserted waveguide. The waveguides for coupling the control light are not modeled in this simulation. In a similar manner, the attenuation corresponding to t_{low}, the fixed attenuator and the phase-shifter β_{fix} are modeled by multiplying the optical electric field by t_{low} and $|\beta_{\text{fix}}|\exp[j\ \arg(\beta_{\text{fix}})]$, respectively. The value of t_{low} is assumed to be an ideal value of $1/(1+\sqrt{2})$, where the optical signal does not attenuate through graphene with intense control light.

Fig. 8(a) shows optical intensities along the switch in case (1) obtained by FD-BPM simulation, where $\alpha_i=1+\sqrt{2}$, ϕ_i=0, and $\beta_{fix}=1/(1+\sqrt{2})$ are assumed. Similarly, optical intensities along the switch in case (2a), case (2b), and case (3) are plotted in Fig. 8(b), (c), and (d), respectively. The switched output intensities are summarized in Table 1. The simulated outputs show good agreement with theoretical results with error less than 3.5%.

6. Measurement of saturable absorption through graphene-loaded waveguide

In this section, we show an experimental result of optical saturable absorption in a monolayer-graphene-loaded waveguide. The optical waveguide was fabricated by K^+ ion exchange on a soda-lime glass substrate. Optical waveguide was formed by K^+ ion exchange through a 30 μm opening in an aluminum mask. Potassium nitrate was used as the ion source and ion exchange was carried out at 370 degrees Celsius for 4 hours. After polishing the end facets of the waveguide, a sheet of monolayer graphene was transferred on the glass substrate surface. The monolayer graphene was formed by CVD on copper foil (iTRIX Corporation). The length of the graphene sheet is about 7 mm. The waveguide structure is shown in Fig. 9(a). A microscopic top-view of the waveguide and a near-field output intensity profile from the waveguide output end facet are also shown in Fig. 9(b) and (c), respectively. Raman spectrum of the transferred graphene on the glass substrate was measured as shown in Fig. 10. In the Raman spectrum, the D peak at 1350 cm^{-1} is correlated with the disorder of the graphene lattice; the G peak at 1580 cm^{-1} and the 2D peak around 2700 cm^{-1} correspond to the phonon excitation at the Brillouin zone center and the second-order of zone-boundary phonons, respectively [16].

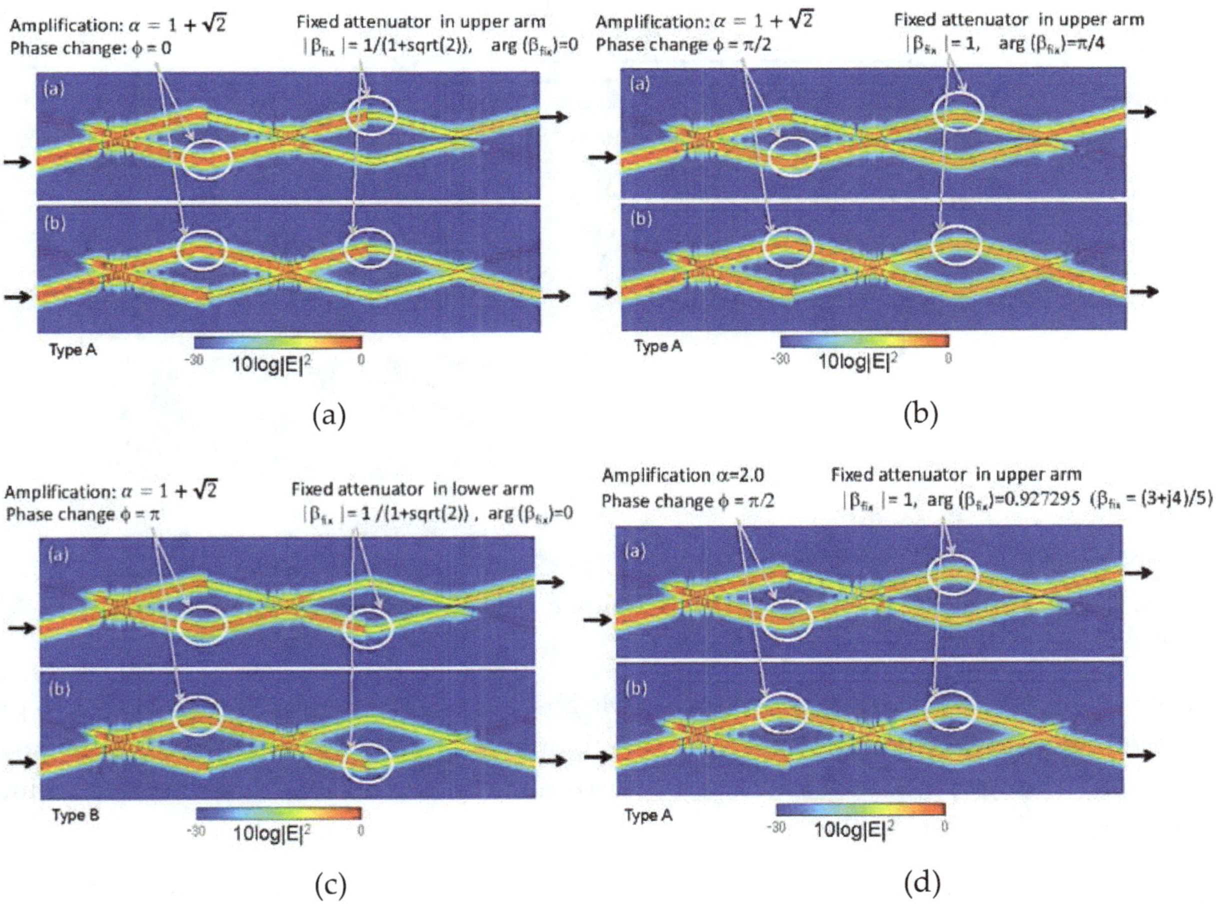

Figure 8. Optical signal intensities along the switches obtained by FD-BPM simulation for (a) case(1), (b) case(2a), (c) case(2b), and (d) case(3).

Case	α_i	ϕ_i	$\lvert\beta_{\text{fix}}\rvert$	$\arg(\beta_{\text{fix}})$	Control arm	Theory		Simulation	
						$\lvert A_{\text{out}}\rvert^2$	$\lvert B_{\text{out}}\rvert^2$	$\lvert A_{\text{out}}\rvert^2$	$\lvert B_{\text{out}}\rvert^2$
(1)	$1+\sqrt{2}$	0	$\frac{1}{1+\sqrt{2}}$	0	B	0	0.1716	6.281×10^{-4}	1.653×10^{-1}
					A	0.1716	0	1.689×10^{-1}	6.195×10^{-4}
(2a)	$1+\sqrt{2}$	$\frac{\pi}{2}$	1	$\frac{\pi}{4}$	B	0	0.5858	1.197×10^{-3}	5.647×10^{-1}
					A	0.5858	0	5.704×10^{-1}	1.168×10^{-3}
(2b)	$1+\sqrt{2}$	π	$\frac{1}{1+\sqrt{2}}$	0	B	0	0.1716	6.158×10^{-4}	1.665×10^{-1}
					A	0.1716	0	1.680×10^{-1}	6.060×10^{-4}
(3)	2	$\frac{\pi}{2}$	1	$\arg\left(\frac{3+j4}{5}\right)$	B	0	0.4289	1.101×10^{-3}	4.135×10^{-1}
					A	0.4289	0	4.173×10^{-1}	1.070×10^{-3}

Table 1. Theoretical and simulated output intensities.

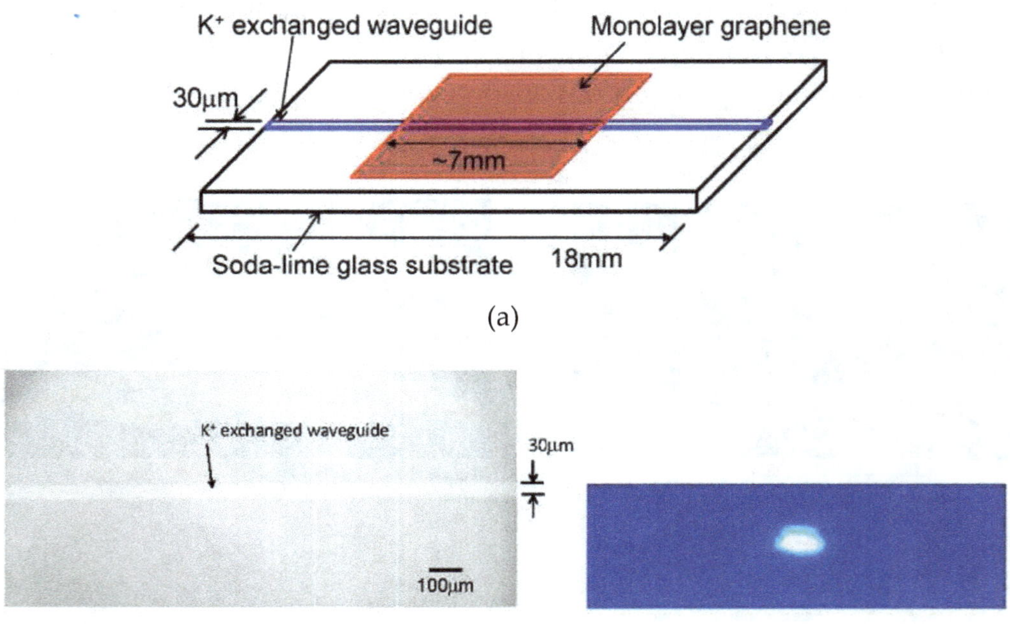

Figure 9. Graphene-loaded optical waveguide; (a) the structure, (b) the microscopic top-view, and (c) the near field pattern from the waveguide.

The experimental setup for measuring saturable absorption is illustrated in Fig. 11. A 1.56 μm femtosecond laser with width of 0.4 ps at repetition rate of 41.96 MHz was used as the laser source. The average power is 4.4 mW. The polarization-controlled laser light was coupled into the waveguide by ×20 objective lens.

Figure 12 shows the measured insertion loss for TE and TM modes as a function of the incident average power at the waveguide input. The measured loss includes coupling loss, waveguide-propagation loss and attenuation loss due to graphene. The insertion loss reduces by more than 10 dB as the incident power increases. This is considered to be caused by saturable

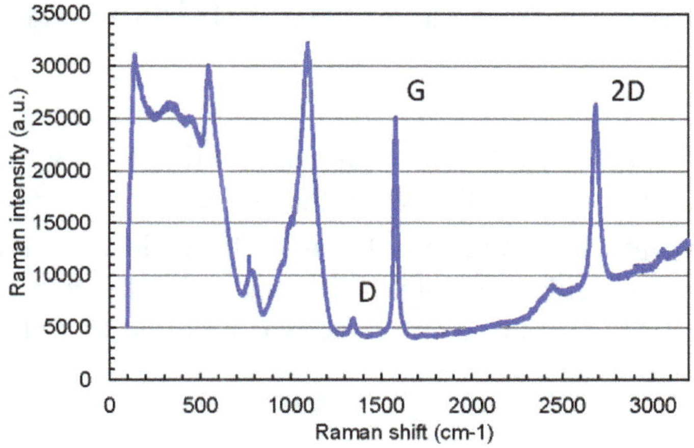

Figure 10. Raman spectrum from the graphene-loaded waveguide.

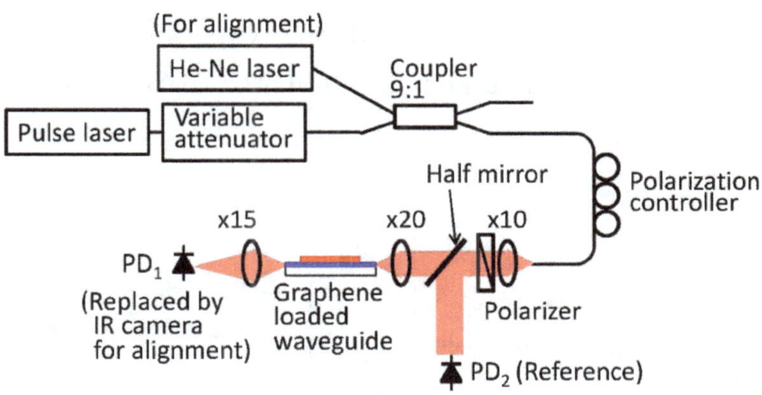

Figure 11. Experimental setup for measurement of nonlinear insertion loss.

absorption of graphene. This modulation depth of the transmittance through the waveguide satisfies the requirement for optical switching. The insertion loss in TM mode is larger than that in TE mode by about 10 dB at low optical power. The difference reduces to about 3dB at high optical power.

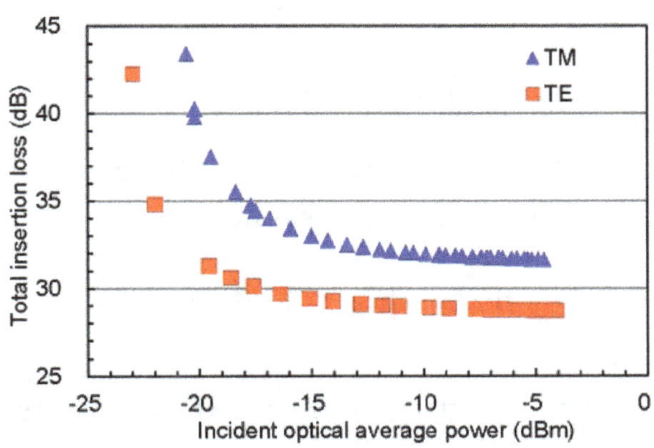

Figure 12. Nonlinear insertion loss through a graphene-loaded optical waveguide.

7. Evaluation of transmittance and phase shift in loosely multilayered graphene

It is found from the discussion in the previous sections that the switch can be operated by adjusting the fixed β_{fix} in the second MZI even if the phase shift ϕ_i is unknown. In this section, we discuss the transmittance and the phase shift in vertically inserted multilayered graphene based on a reported experimental result by H. Zhang et al. [9]. The transmittance in loosely stacked three-layer graphene was measured and was reported to be fitted by the following experimental formula:

$$T(I) = T(0) + \Delta T_0 - \frac{\Delta T_0}{1 + I / I_{sat}} \tag{20}$$

where I is the incident optical peak power density, $T(0)$ is the transmittance at weak optical power density. The parameters fitted with the experimental data are ΔT_0=0.051, I_{sat}=0.074 $\mathrm{GW/cm^2}$. When $T(0)$ is assumed as $T(0)=0.977^3=0.9326$, the transmittance $T(I)$ is plotted as shown in Fig. 13(a).

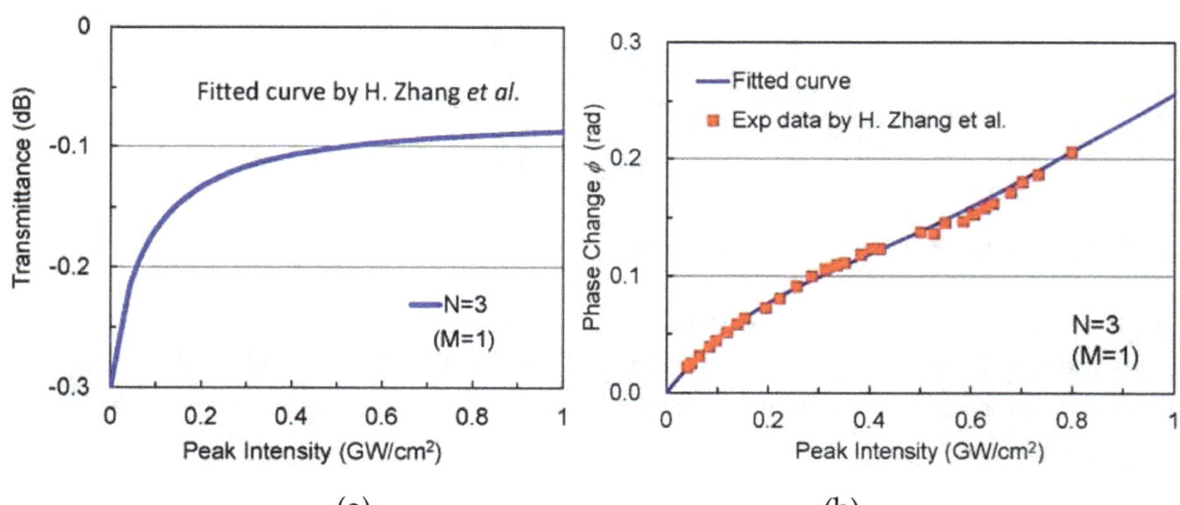

Figure 13. Transmittance and phase change through loosely three-layered graphene.

Reported experimental data of the accompanied phase change are shown in Fig. 13(b) [9]. We find an experimental formula from these data as follows:

$$\phi(I) = k_0 L I \left[a_1 \exp\left(-a_2 I^{a_3}\right) + n_{2,c} \right] \times 10^9 \tag{21}$$

where $k_0=2\pi/\lambda$ with λ=1.55 μm, L =1 nm, and the fitted parameters are $n_{2,c}=6.3\times10^{-8}$ $\mathrm{cm^2/W}$, $a_1=6.75\times10^{-8}$, a_2=6.49, and a_3=1.34. The optical incident peak power density is in unit of $\mathrm{GW/cm^2}$. The fitted curve is also shown in Fig. 13(b). This phase change was derived from nonlinear refractive index which might include parametric process such as four-wave mixing and strong nonparametric process [9].

We now estimate transmittance T_{3M} and phase change ϕ_{3M} in $3M$-layer graphene. Using $T(I)$ and $\phi(I)$ for three-layer graphene, we can derive $T_{3M}(I)$ by

$$\begin{cases} T_{3M}\left(I_{\text{in}}\right) = T(I_{\text{in}})\prod_{k=1}^{M-1} T(I_{\text{out},k}) \\ I_{\text{out},k} = I_{\text{out},k-1} T\left(I_{\text{out},k-1}\right) \text{ and } I_{\text{out},0} = I_{\text{in}} \end{cases} \tag{22}$$

where I_{in} is the incident optical peak power density. The phase change $\phi_{3M}(I)$ from the phase at I_{in}=0 is given by

$$\phi_{3M}\left(I_{\text{in}}\right) = \sum_{k=0}^{M-1} \phi(I_{\text{out},k}) \tag{23}$$

Fig. 14 shows the transmittance and the phase change, where $N = 3M$, It is approximately estimated that the transmittance can be changed by $2K$ dB with $N = 30K$, where K = 1,..., 4. Phase change of K rad is accompanied by increasing the peak power density to $0.4 \sim 0.5\ \text{GW}/\text{cm}^2$. From these results, we can estimate the required layer number to satisfy the switching conditions when the graphene is vertically introduced in the waveguide. However, it is noted that the nonlinear phase shift discussed in this section is the phase shift of the incident intense light, that is, not for weak signal light whose wavelength or polarization is different from that of the control light. Therefore, nonlinear phase shift in the proposed switch structure has to be investigated in detail.

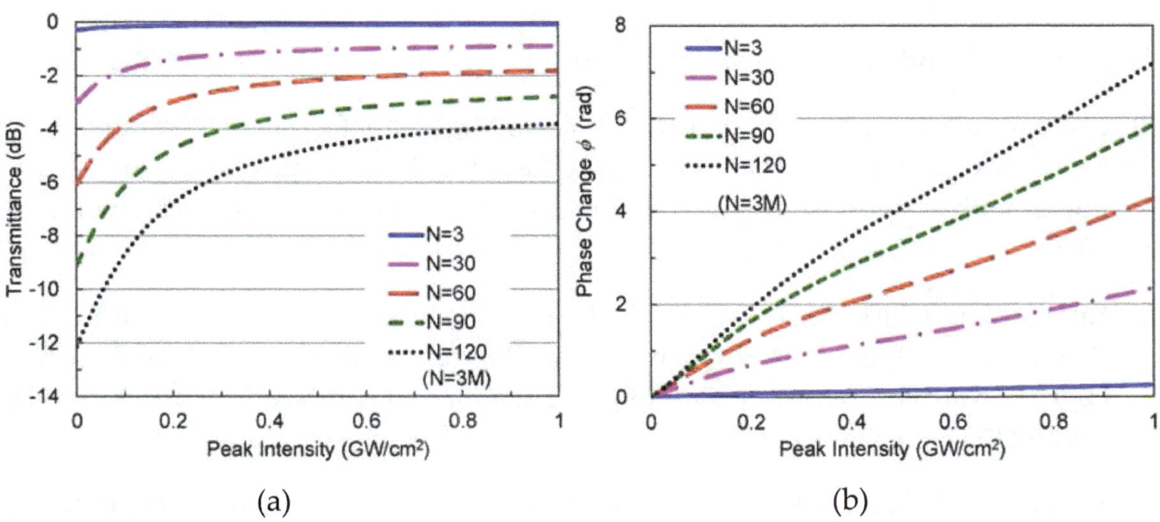

Figure 14. Transmittance and phase change through loosely multi-layered graphene.

8. Conclusions

Generalized switching conditions for the proposed switch were derived and confirmed by FD-BPM. It is found that the associated refractive-index change can reduce the required absorption decrease in saturable absorption. As a measured saturable absorption, nonlinear insertion loss along graphene-loaded waveguide was discussed. Also, evaluation of nonlinear transmittance and phase shift in multilayered graphene was described. Since conventional switches utilize only phase change to operate switching, the proposed switch, which uses both phase change and absorption change will give a new switching scheme.

Acknowledgements

This work was supported in part by JSPS KAKENHI (23656243 and 24360150).

Author details

Misaki Takahashi, Hiroki Kishikawa, Nobuo Goto * and Shin-ichiro Yanagiya

*Address all correspondence to: goto.nobuo@tokushima-u.ac.jp

Department of Optical Science and Technology, Tokushima University, Tokushima, Japan

References

[1] Doran N. J. and Wood D. Nonlinear-Optic Loop Mirror. *Optics Lett.* 1988, 13(1), 56-58.

[2] Nakamura S., Ueno Y., Tajima K., Sasaki J., Sugimoto T., Kato T., Shimoda T., Itoh M., Hatakeyama H., Tamanuki T., and Sasaki T. Demultiplexing of 168-Gb/s Data Pulses with a Hybrid-Integrated Symmetric Mach-Zehnder All-Optical Switch. *IEEE Photon Tech Lett.* 2000, 12(4), 425-427.

[3] Kitagawa Y., Ozaki N., Takata Y., Ikeda N., Watanabe Y., Sugimoto Y., and Asakawa K. Sequential Operations of Quantum Dot/Photonic Crystal All-Optical Switch With High Repetitive Frequency Pumping. IEEE/OSA *J Lightwave Technol.* 2009, 27(10), 1241-1247.

[4] Kishikawa H., Kimiya K., Goto N., and Yanagiya S. All-Optical Wavelength-Selective Switch Consisting of Asymmetric X-junction Couplers and Raman Amplifiers for Wide Wavelength Range. IEEE/OSA *J of Lightwave Technol.* 2010, 28(1), 172-180.

[5] Kishikawa H., Goto N., and Kimiya K. All-Optical Wavelength-Selective Switch by Intensity Control in Cascaded Interferometers. In *Frontiers in Guided Wave Optics and Optoelectronics*: Bishnu Pal (ed.), Intech; 2010. pp. 257-268.

[6] Oya M., Kishikawa H., Goto N., and Yanagiya S. All-Optical Switch Consisting of Two-Stage Interferometers Controlled by Using Saturable Absorption of Monolayer Graphene. *Opt Express*. 2012, 20(24), 27322-27330.

[7] Takahashi M., Ueda W., Goto N., and Yanagiya S. Saturable Absorption by Vertically Inserted or Overlaid Monolayer Graphene in Optical Waveguide for All-Optical Switching Circuit. *IEEE Photon J*. 2013, 5(5), 6602109.

[8] Takahashi M., Kishikawa H., Goto N., and Yanagiya S. All-Optical Switch with Cascaded Two-Stage Mach-Zehnder Interferometers Using Saturable Absorption Accompanied by Refractive-Index Change in Graphene. *IEEE/OSA J of Lightwave Technol*. 2014, 32(21), 3624-3630.

[9] Zhang H., Virally S., Bao Q., Ping L. K., Massar S., Godbout N., and Kockaert P. Z-scan Measurement of the Nonlinear Refractive Index of Graphene. *Opt Lett*. 2012, 37(11), 1856-1858.

[10] Geim A. K. and Novoselov K. S. The Rise of Graphene. *Natur Mater*. 2007, 6, 183-191.

[11] Iechika, Y. Application of Graphene to High-Speed Transistors: Expectations and Challenges. *Sci Technol Trends Quarter Rev* 2010, 37, 76-92.

[12] Sun Z., Hasan T., Torrisi F., Popa D., Privitera G., Wang F., Bonaccorso F., Basko D. M., and Ferrari A. C. Graphene mode-locked ultrafast laser. *ACS Nano* 2010, 4(2), 803-810.

[13] Bao Q., Zhang H., Ni Z., Wang Y., Polavarapu L., Shen Z., Xu Q.-H., Tang D. and Loh K. P. Monolayer Graphene as a Saturable Absorber in a Mode-Locked Laser. *Nano Res*. 2011, 4(3), 297-307.

[14] Yamashita S. A Tutorial on Nonlinear Photonics Applications of Carbon Nanotube and Graphene. *IEEE/OSA J Lightwave Technol*. 2012, 30(4), 427-447.

[15] Li W., Chen B., Meng C., Fang W., Xiao Y., Li X., Hu Z., Xu Y., Tong L., Wang H., Liu W., Bao J. and Shen Y. R. Ultrafast All-Optical Graphene Modulator. *Nano Lett*. 2014, 14(2), 955-959.

[16] Ferrari A. C., Meyer J. C., Scardaci V., Casiraghi C., Lazzeri M., Mauri F., Piscanec S., Jiang D., Novoselov K. S., Roth S., and Geim A. K. Raman Spectrum of Graphene and Graphene Layers. *Phys Rev* Lett. 2006, 97, 187401-1-4.

Permissions

All chapters in this book were first published in OMD, by InTech Open; hereby published with permission under the Creative Commons Attribution License or equivalent. Every chapter published in this book has been scrutinized by our experts. Their significance has been extensively debated. The topics covered herein carry significant findings which will fuel the growth of the discipline. They may even be implemented as practical applications or may be referred to as a beginning point for another development.

The contributors of this book come from diverse backgrounds, making this book a truly international effort. This book will bring forth new frontiers with its revolutionizing research information and detailed analysis of the nascent developments around the world.

We would like to thank all the contributing authors for lending their expertise to make the book truly unique. They have played a crucial role in the development of this book. Without their invaluable contributions this book wouldn't have been possible. They have made vital efforts to compile up to date information on the varied aspects of this subject to make this book a valuable addition to the collection of many professionals and students.

This book was conceptualized with the vision of imparting up-to-date information and advanced data in this field. To ensure the same, a matchless editorial board was set up. Every individual on the board went through rigorous rounds of assessment to prove their worth. After which they invested a large part of their time researching and compiling the most relevant data for our readers.

The editorial board has been involved in producing this book since its inception. They have spent rigorous hours researching and exploring the diverse topics which have resulted in the successful publishing of this book. They have passed on their knowledge of decades through this book. To expedite this challenging task, the publisher supported the team at every step. A small team of assistant editors was also appointed to further simplify the editing procedure and attain best results for the readers.

Apart from the editorial board, the designing team has also invested a significant amount of their time in understanding the subject and creating the most relevant covers. They scrutinized every image to scout for the most suitable representation of the subject and create an appropriate cover for the book.

The publishing team has been an ardent support to the editorial, designing and production team. Their endless efforts to recruit the best for this project, has resulted in the accomplishment of this book. They are a veteran in the field of academics and their pool of knowledge is as vast as their experience in printing. Their expertise and guidance has proved useful at every step. Their uncompromising quality standards have made this book an exceptional effort. Their encouragement from time to time has been an inspiration for everyone.

The publisher and the editorial board hope that this book will prove to be a valuable piece of knowledge for researchers, students, practitioners and scholars across the globe.

List of Contributors

Fernando de S. Campos, José Alfredo C. Ulson, José Eduardo C. Castanho, Bruno Albuquerque de Castro and Rudolf Riehl
Univ. Estadual Paulista "Júlio de Mesquita Filho" (UNESP) - Bauru campus, Brazil

Sergei L. Pyshkin
Academy of Sciences of Moldova, Republic of Moldova
Clemson University, USA

Oscal T.-C. Chen, Yi-Yang Lee and Robin R.-B. Sheen
National Chung Cheng University Chiayi, Taiwan

Jin-gang Liu, Hong-jiang Ni, Zhen-he Wang and Shi-yong Yang
Laboratory of Advanced Polymer Materials, Institute of Chemistry, Chinese Academy of Sciences, Beijing, People's Republic of China

Wei-feng Zhou
BOE Technology Group Co. Ltd., Beijing, People's Republic of China

Barylo Hryhoriy, Gotra Zenon, Ivakh Mariya, Kozhukhar Oleksandr and Virt Volodymyr
The ED Department, Lviv Polytechnic National University, Ukraine

Makara Ivanna
The IM Department, Lutsk National Technical University, Ukraine

Yi Gu and Yong-Gang Zhang
State Key Laboratory of Functional Materials for Informatics, Shanghai Institute of Microsystem and Information Technology, Chinese Academy of Sciences, China

Henrique M. Salgado and Pedro J. Batista
Faculdade de Engenharia da Universidade do Porto, Portugal
Instituto de Engenharia de Sistemas e Computadores - Tecnologia e Ciência - INESC TEC, Porto, Portugal

Rúben E. Neto and Luís M. Pessoa
Instituto de Engenharia de Sistemas e Computadores - Tecnologia e Ciência - INESC TEC, Porto, Portugal

Jian-Chiun Liou
National Kaohsiung University of Applied Sciences (KUAS), The Department of Electronic Engineering, Taiwan

Lung-Chien Chen
Department of Electro-optical Engineering, National Taipei University of Technology, Taipei, Taiwan

Misaki Takahashi, Hiroki Kishikawa, Nobuo Goto and Shin-ichiro Yanagiya
Department of Optical Science and Technology, Tokushima University, Tokushima, Japan

Index